Data Quality for Analytics Using SAS®

Gerhard Svolba

The correct bibliographic citation for this manual is as follows: Svolba, Gerhard. 2012. *Data Quality for Analytics Using SAS®*. Cary, NC: SAS Institute Inc.

Data Quality for Analytics Using SAS®

Copyright © 2012, SAS Institute Inc., Cary, NC, USA

ISBN 978-1-61290-227-2 (electronic book)
ISBN 978-1-60764-620-4

All rights reserved. Produced in the United States of America.

For a hard-copy book: No part of this publication may be reproduced, stored in a retrieval system, or transmitted, in any form or by any means, electronic, mechanical, photocopying, or otherwise, without the prior written permission of the publisher, SAS Institute Inc.

For a Web download or e-book: Your use of this publication shall be governed by the terms established by the vendor at the time you acquire this publication.

The scanning, uploading, and distribution of this book via the Internet or any other means without the permission of the publisher is illegal and punishable by law. Please purchase only authorized electronic editions and do not participate in or encourage electronic piracy of copyrighted materials. Your support of others' rights is appreciated.

U.S. Government Restricted Rights Notice: Use, duplication, or disclosure of this software and related documentation by the U.S. government is subject to the Agreement with SAS Institute and the restrictions set forth in FAR 52.227-19, Commercial Computer Software-Restricted Rights (June 1987).

SAS Institute Inc., SAS Campus Drive, Cary, North Carolina 27513-2414

1st printing, April 2012

SAS® Publishing provides a complete selection of books and electronic products to help customers use SAS software to its fullest potential. For more information about our e-books, e-learning products, CDs, and hard-copy books, visit the SAS Publishing Web site at **support.sas.com/publishing** or call 1-800-727-3228.

SAS® and all other SAS Institute Inc. product or service names are registered trademarks or trademarks of SAS Institute Inc. in the USA and other countries. ® indicates USA registration.

Other brand and product names are registered trademarks or trademarks of their respective companies.

*For my three teenage sons and their permanent effort
in letting me share the many wonderful moments of their life,*

*and without whose help
this book would have probably been completed a year earlier.*

You are the quality of my life.

Acknowledgments

Martina, for supporting me and even the crazy idea to write this book in that period of our family life, which is probably the busiest one.

My parents, for providing me with so many possibilities.

The following persons, who contributed to the book by spending time to discuss data quality topics. It is a pleasure to work and to discuss with you: Albert Tösch, Andreas Müllner, Bertram Wassermann, Bernadette Fabits, Christine Hallwirth, Claus Reisinger, Franz Helmreich, Franz König, Helmut Zehetmayr, Josef Pichler, Manuela Lenk, Mihai Paunescu, Matthias Svolba, Nicole Schwarz, Peter Bauer, Phil Hermes, Stefan Baumann, Thomas Schierer, and Walter Herrmann.

The reviewers, who took time to review my manuscript and provided constructive feedback and suggestions. I highly appreciate your effort: Anne Milley, David Barkaway, Jim Seabolt, Mihai Paunescu, Mike Gilliland, Sascha Schubert, and Udo Sglavo.

The nice and charming **SAS Press team** for their support throughout the whole process of the creation of this book: Julie Platt, Stacey Hamilton, Shelley Sessoms, Kathy Restivo, Shelly Goodin, Aimee Rodriguez, Mary Beth Steinbach, and Lucie Haskins.

The management of SAS Austria for supporting the idea to write this book: Dietmar Kotras and Robert Stindl.

August Ernest Müller, my great-grandfather, designed one of the very early construction plans of a helicopter in 1916 and was able to file a patent in the Austrian-Hungarian monarchy. However, he found no sponsor to realize his project during World War I. I accidentally found his construction plans and documents at the time I started writing this book. His work impressed and motivated me a lot.

Contents

Introduction ix

Part I Data Quality Defined 1

Chapter 1 Introductory Case Studies 3

Chapter 2 Definition and Scope of Data Quality for Analytics 21

Chapter 3 Data Availability 31

Chapter 4 Data Quantity 47

Chapter 5 Data Completeness 59

Chapter 6 Data Correctness 71

Chapter 7 Predictive Modeling 83

Chapter 8 Analytics for Data Quality 95

Chapter 9 Process Considerations for Data Quality 107

Part II Data Quality—Profiling and Improvement 121

Chapter 10 Profiling and Imputation of Missing Values 123

**Chapter 11 Profiling and Replacement of Missing Data in
a Time Series** 141

Chapter 12 Data Quality Control across Related Tables 159

Chapter 13 Data Quality with Analytics 167

**Chapter 14 Data Quality Profiling and Improvement with
SAS Analytic Tools** 181

Part III Consequences of Poor Data Quality—Simulation Studies . 199

Chapter 15 Introduction to Simulation Studies 201

**Chapter 16 Simulating the Consequences of Poor Data Quality for
Predictive Modeling** 207

**Chapter 17 Influence of Data Quality and Data Availability on
Model Quality in Predictive Modeling** 219

**Chapter 18 Influence of Data Completeness on Model Quality in
Predictive Modeling** 231

**Chapter 19 Influence of Data Correctness on Model Quality in
Predictive Modeling** 243

Chapter 20 Simulating the Consequences of Poor Data Quality in Time Series Forecasting	**255**
Chapter 21 Consequences of Data Quantity and Data Completeness in Time Series Forecasting	**265**
Chapter 22 Consequences of Random Disturbances in Time Series Data	**273**
Chapter 23 Consequences of Systematic Disturbances in Time Series Data	**281**
Appendix A: Macro Code	**289**
Appendix B: General SAS Content and Programs	**301**
Appendix C: Using SAS Enterprise Miner for Simulation Studies	**305**
Appendix D: Macro to Determine the Optimal Length of the Available Data History	**311**
Appendix E: A Short Overview on Data Structures and Analytic Data Preparation	**319**
References	**327**
Index	**329**

Introduction

Rationale and Trigger to Write This Book

The first impulse
In November 2005, shortly before I finished the full draft of my first book, *Data Preparation for Analytics Using SAS*, I was asked whether I wanted to contribute content and knowledge to the topic of data quality for analytics. At that time it was too late to include data quality into my first book. It also would not have been advisable to do so, as this important topic would have gone beyond the scope of the book on data preparation.

When *Data Preparation for Analytics Using SAS* was published in late 2006 I had already begun thinking about starting a new book on the topic of data quality. However, it wasn't until 2008 that I started collecting ideas and opinions on the book you are reading now. After I received the green light to start writing the book from SAS Publishing, I started work at the end of 2009.

Focus on analytics
My intention was not to write another book on data quality in general, but to write the first book that deals with data quality from the viewpoint of a statistician, data miner, engineer, operations researcher, or other analytically minded problem-solver.

Data quality is getting a lot of attention in the market. However, most of the initiatives, publications, and papers on data quality focus on classical data quality topics, such as elimination of duplicates, standardization of data, lists of values, value ranges, and plausibility checks. It will not be said here that these topics are not important for analytics; on the contrary, they build the foundation of data for analysis. However, there are many aspects of data that are specific to analytics. And these aspects are important to differentiate whether data are suitable for analysis or not.

For classical data quality, books, best practices, and knowledge material are widely available. For the implementation of data quality, SAS offers the DataFlux Data Management Platform, a market-leading solution for typical data quality problems, and many methods in the established SAS modules.

Symbiosis of analytic requirements and analytic capabilities
In many cases, analytics puts higher requirements on data quality but also offers many more capabilities and options to measure and to improve data quality, like the calculation of representative imputation values for missing values. Thus there is a symbiosis between the analytical requirements and the analytical capabilities in the data quality context.

Analytics is also uniquely able to close the loop on data quality since it reveals anomalies in the data that other applications often miss. SAS is also perfectly suited to analyze and improve data quality.

- In part II, this book shows software capabilities that are important to measure and improve data quality and, thus, close the loop in the data quality process and show how analytics can improve data quality.
- In part III, this book shows how SAS can be used as a simulation environment for the evaluation of data quality status and the consequences of inferior data quality. This part also shows new and unique simulations results on the consequences of data quality.

FOR analytics and WITH analytics

The book deals with data quality topics that are relevant FOR analytics. Data quality is discussed in conjunction with the requirements of analytical methods on the data. Analytics is, however, not only posing regulations on minimum data quality requirements. Analytical methods are also used to improve data quality. This book illustrates the demand of analytical methods but in return also shows what can be done WITH analytics in the data quality area.

Data quality for "non-analytics"

Much literature, publications, and discussions on general data quality exist from a broad perspective where the focus is not primarily on analytics. The chapters in this book, especially in the first part, include these typical data quality topics and methods as long as they are important for data quality for analytics.

The idea of this book is not to start from scratch with the data quality topic and to introduce all methods that exist for the simple profiling of data, like using validations lists and validation limits. These methods are introduced in the respective sections, but the focus of the book stays on analytic implications and capabilities.

Cody's Data Cleaning Techniques Using SAS, by Ron Cody [9], shows how to profile the quality of data in general. The book in your hand references some of these techniques in some chapters; however, it does not repeat all the data quality basics.

Data and Measurement

The term *data* not only appears in the title of this book but is also used throughout the text to discuss features and characteristics of data and data quality.

Measurement is very close in meaning to the word *data* in this book and could possibly be used as an alternative expression. Different from *data*, *measurement* also implies a process or an activity and, thus, better illustrates the process around data.

- Some data are **passively "measured"** like transaction data (calls to a hotline, sales in a retail shop) or web data (like social media sites). This also compares to an observational study of using measurements that are opportunistically collected.
- Other data are **actively "measured"** like vital signs in a medical study or survey data. This is usually the case in a designed experiment where measurements are prospectively collected.

In research analyses the "manufactured asset" for the analysis is usually called "measurement" instead of "data." The topics, methods, and findings that are discussed in this book thus apply not only to those who receive their data from databases, data warehouses, or externally acquired data but also to those who perform measurements in experiments.

Things are being "measured" actively and passively with many spillover benefits for uses not originally envisioned. Finally, the researcher or analyst has to decide whether his data fit for intended use.

Importance of Data Quality for Analytics

Consequences of bad data quality

Data quality for analytics is an important topic. Bad data quality or just the mere perception that data has bad quality causes the following:

- Increases project duration and efforts
- Reduces the available project time for analysis and intelligence

- Damages trust in the results
- Slows down innovation and research
- Decreases customer satisfaction
- Leads to wrong, biased, outdated, or delayed decisions
- Costs money and time
- Demotivates the analyst, increasing the risk of losing skilled people to other projects

Frequently used expression

Data quality is a frequently used expression. As a 21 September 2011 Google search reveals, *data quality* ranges with 10.8 Mio potential hits, behind terms like *data management* (30.9 Mio), *data mining* (28.1 Mio), or *data warehouse* (14.4 Mio). But it is still more prominent than terms like *relational database* (8.5 Mio), *regression analysis* (8.1 Mio), *data integration* (6.9 Mio), *ETL or extraction transformation loading* (6.2 Mio), *time series analysis* (3.6 Mio), *cluster analysis* (2.8 Mio), and *predictive analytics* (1.3 Mio).

The frequency of use of the term *data quality* reinforces the requirement for a clear definition of data quality for analytics. Chapter 2 of this book goes into more detail on this.

Trend in the market

Data quality is currently also an important trend in the market. David Barkaway [5] shows in his 2010 SAS Global Forum paper the 2009 results of Forrester Research. To the question, "Have you purchased any complimentary data management solution through your ETL vendor?" 38 percent replied that they had bought data quality management software.

Figure I.1: Complimentary data management solutions

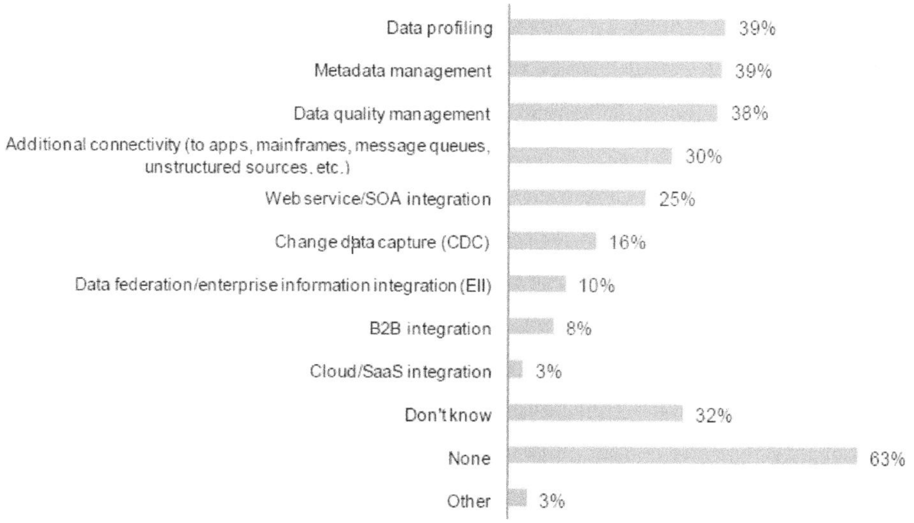

Solution	Percentage
Data profiling	39%
Metadata management	39%
Data quality management	38%
Additional connectivity (to apps, mainframes, message queues, unstructured sources, etc.)	30%
Web service/SOA integration	25%
Change data capture (CDC)	16%
Data federation/enterprise information integration (EII)	10%
B2B integration	8%
Cloud/SaaS integration	3%
Don't know	32%
None	63%
Other	3%

Source: Forrester survey November 2009, Global ETL Online Survey, "Trends in Enterprise and Adoption"

The Layout of the Book

Data quality process steps

There are different ways a process about data quality can be defined. Thus, different vendors of data quality software and different data quality methodologies present processes that differ to some extent.

The DataFlux Data Management Platform, for example, is built around five steps, which are grouped into three main buckets. The steps follow a logical flow that makes sense for the data quality and data cleaning process for which the tool is usually applied. These steps are shown in Figure I.2.

Figure I.2: Data quality process in the DataFlux Data Management Platform

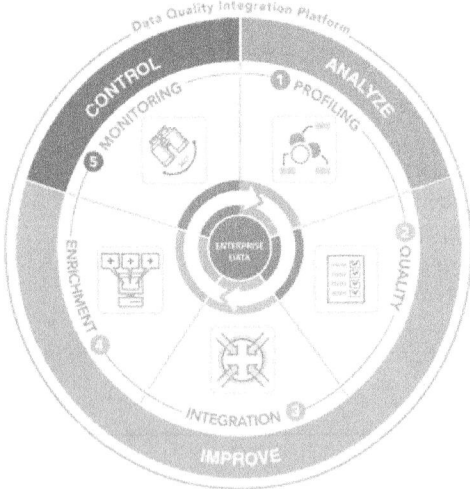

- Profiling
    - Get a picture about the quality status of data before beginning a project
    - Discover and check relations in the data
- Quality
    - Separate information into smaller units
    - Standardize, correct, and normalize data
- Integration
    - Discover related data
    - Remove duplicates
- Enrichment
    - Add data from other sources like address data, product data, or geocoding
- Monitoring
    - Detect trends in data quality
    - Track consequences of bad data quality

Main parts of this book

This book is divided into three main parts. The naming and ordering of these three parts and the respective chapters follow a process as well, but also consider a segmentation of the content of this book into well-defined parts and a good readable sequence of topics and chapters.

The three parts of this book are:

- Data Quality Defined
- Data Quality—Profiling and Improvement
- Consequences of Poor Data Quality—Simulation Studies

These three main parts can be represented as a data quality process as shown in Figure I.3 and that is described in the paragraphs that follow.

Figure I.3: Data quality process in this book

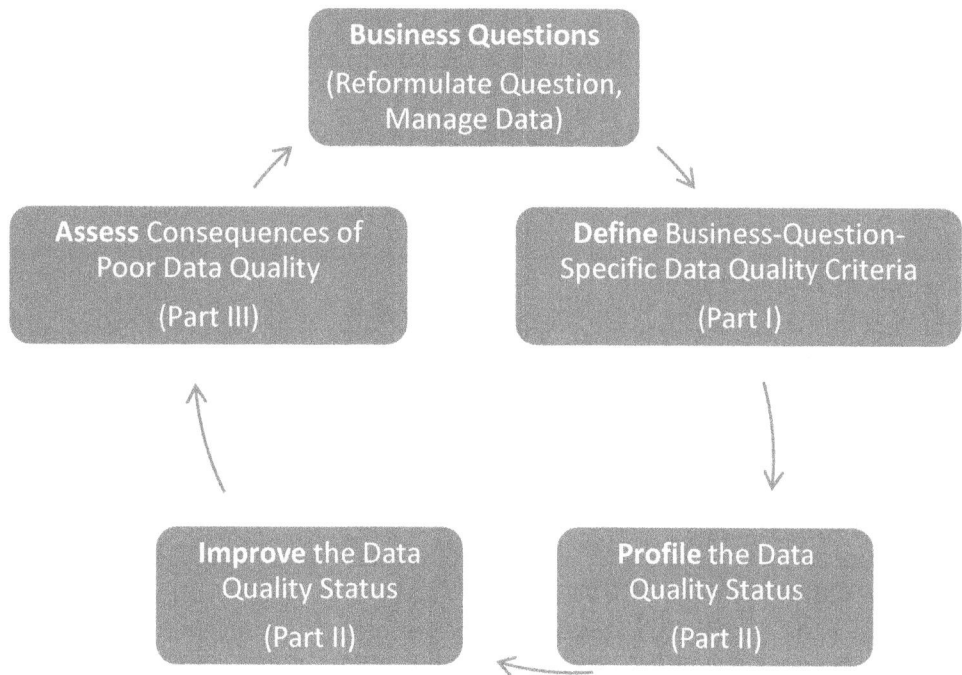

The logical order here is to **first define the requirements and criteria for data quality for analytics**. The first part of the book is therefore the conceptual part and contains text, definitions, explanations, and examples. This part is called "Data Quality Defined."

Based on these definitions the second part of the book focuses on **how the data quality status can be profiled** and how a picture of important criteria for advanced analytic methods and the data quality status of the data can be achieved. The second part also shows ways that data quality **can be improved with analytical methods**. The name of this part is "Data Quality—Profiling and Improvement."

As not all data quality problems can be corrected or solved (or the effort is not justifiable), the last part of the book deals with **consequences of poor data quality**. Based on simulation studies, general answers about the usability of certain analytical methods and the effect on the accuracy of models are given if data quality criteria are not fulfilled. The last part is named "Consequences of Poor Data Quality—Simulation Studies."

A cyclic approach

The process in this book, thus, also follows a cyclic approach, after the definition of criteria, the assessment and possible correction of the data quality status, and the consequences of the actual data quality status are analyzed. Based on the outcome the analysis is performed or measures are taken to fulfill the criteria or to relax the criteria by **reformulating the business questions**.

Selection of data quality criteria

The selection of the set of data quality criteria for this book has been made based on the practical experience of the author. Actually, there is no single definition that can be considered to be the golden standard for all applications. It can also be seen that many definitions highly overlap.

Gloskin [3], for example, defined the criteria Accuracy, Reliability, Timeliness, and Completeness. Orli [7] gives a longer list of criteria, which is Accuracy, Completeness, Consistency, Timeliness, Uniqueness, and Validity.

The data quality criteria that are defined in this book in chapters 3–9 are the following.

- Chapter 3, "Data Availability," starts with the question as to whether data are available in general.
- Chapter 4, "Data Quantity," examines whether the amount of data are sufficient for the analysis.
- Chapter 5, "Data Completeness," deals with the fact that available data fields may contain missing values.
- Chapter 6, "Data Correctness," checks whether the available data are correct with respect to its definition.
- Chapter 7, "Predictive Modeling," discusses special requirements of predictive modeling methods.
- Chapter 8, "Analytics for Data Quality," shows additional requirements of interdependences of analytical methods and the data.
- Chapter 9, "Process Considerations for Data Quality," finally shows the process aspect of data quality and also discusses aspects like data relevancy and possible alternatives.

These criteria are considered to be the most important ones in the context of this book and are shown in part I.

The Scope of This Book

Widespread expectations of this book

As already mentioned in a section above, data quality and data quality for analytics are very important topics that are discussed in many circumstances, projects, analysis domains, analytical disciplines, data warehouse communities, and across industries.

As a consequence, the expectations on the content of this book from people from these different areas are very diverse. Depending on the way people perform these analyses and acquire, prepare, and use data, the expectations may vary. A book titled *Data Quality for Analytics* thus bears the risk of not meeting the expectations of all people.

Consider the following roles, which have different perspectives on data quality and data quality for analytics and will likely have different expectations:

- An analyst who builds analytical models for customer behavior analysis for a retail bank
- An IT person who is in charge of maintaining the data warehouse and de-duplicating customer records in both the operational and the decision support system
- A researcher who conducts and analyzes clinical trials
- A statistician who works for the statistical office and creates reports based on register data

This section attempts to correctly set expectations. Chapter 2 also goes into more details on the scope of data quality for analytics.

Data cleaning techniques in general

In the book *Cody's Data Cleaning Techniques Using SAS*, Ron Cody shows a bunch of methods for profiling data quality status and correcting data quality errors. These methods include checking categorical, interval variables and date values as well as checking for missing values. Other topics include the check for duplicates in *n* observations per subject and work with multiple files.

The intention of this book is not to compete with other books but to complement other titles by SAS Publishing by offering a book that has a different point of view. The data quality checks presented in *Cody's Data Cleaning Techniques Using SAS* form an important basis for data quality control and improvement of analysis data in general. The emphasis of this book goes beyond typical basic methods and puts data quality into a more business-focused context and covers more closely the requirements of analytical methods.

The detailed presentation of typical methods to profile the data quality status is not a focus of this book. The methods shown in part II are more advanced to profile specific analytical data requirements.

The DataFlux Data Management Platform

SAS offers the DataFlux Data Management Platform for data quality management. This market-leading solution is well suited to profile data with respect to data quality requirements, to improve data quality by de-duplicating data and enriching data, and to monitor data quality over time.

The solution provides important features and strongly focuses on:

- profiling the distribution of variables of different types, the matching of predefined patterns, and the presentation of summary statistics on the data.
- methods to standardize data, for example, address data and product code and product name data, the controlled de-duplication of data.

The features are important in providing a quality data basis for analysis (see also David Barkaway [5]) and definitely focus on data quality. The aim of this book, however, is also to discuss data quality from a business- and analytical-methods-specific point of view in terms of necessary data histories and historic snapshots of the data and the reliability and relevancy of data.

Administrative records and data in statistical offices

In statistical analysis in statistical institutions and in the social sciences the use of administrative data sources has become an important topic over the last several years. On some parts of the population, administrative data provides more information than any survey data. Consequently, the data quality assessment of administrative data sources has become an important topic. Data like the Central Population Register or the Central Social Security Register are examples of such administrative records. For the assessment of this data, a quality framework has been defined for data quality. Compare [8] Berka et al.

While this is an important topic in data quality control, the details of this method go beyond the scope of this book, and the reader is referred to the respective literature.

Clinical trials

The analysis of clinical trials is also a field that has a strong relation to data quality. Compare also Case Study 2 in chapter 1 for an example. In this discipline there are also very specific topics like the **problem of missing values**.

While missing values are considered in many analysis domains, missing values in the clinical trial context, however, often mean that after a certain point in time no information about the patient and the patient status is available at all. In the diction of a predictive model, the target variable is affected by missing information here. If in an oncological trial the survival time between treatment groups is estimated it is important to have recent observations about the customer and his status.

The focus in clinical trial analysis is not on the ability to impute missing values of the input variables of a predictive model but rather on defining strategies of how to consider the respective observations in the analysis. Examples for such strategies include the pessimistic assumption that an event immediately after the last seen date or the "last observation carried forward" assumption (LOCF) is the last available state used for the analysis.

In the analysis of clinical trials, the differentiation between different types of missing values as "MISSING COMPLETELY AT RANDOM," "MISSING AT RANDOM," and "MISSING NOT AT RANDOM" is made and considered in different evaluation strategies. Compare Burzykowski [6].

In this book, missing values are considered as well, for example, in chapters 5, 10, 11, and 18; however, the missing value topic is not considered on such a specific level. For example, only the between two categories "random missing values" and "systematic missing" values are differentiated in the scope of this book.

Main focus of this book

The following points are the main focus of this book:

- Usability and availability of data for analysis
- Selection of the right data sources
- Explaining why analytical methods need historic data and also historic snapshots of the data
- Ensuring sufficient data quantity in terms of number of observations, number of event cases, and length of data history
- Typical analyses: predictive analytics and statistics, including time series analysis and time series forecasting
- Types of research are mainly observational studies, where existing data of the company or organization itself is used for the analysis (partly also controlled studies where the required data for the analysis is retrieved in a controlled way)
- Data completeness in terms of profiling of the number and the structure of missing values, finding special types of missing values, replacement of missing values, advanced imputation methods
- Consideration of the operational, data management, and analysis process point of view
- Data relevancy with respect to the definition of the data and the requirements of the analysis
- Data correction with focus on analytic profiling of outliers and complex data validation
- Methods in SAS on how to perform advanced profiling of the data quality status and what SAS can offer for the improvement of data quality
- Simulation studies and consequences of poor data quality for predictive modeling and time series forecasting

Areas with reduced focus

The following items are areas that are relevant for data quality in general but are not the primary focus of this book. Many of these points are, however, mentioned and discussed in a broader context:

- Data de-duplication, including data standardization and record matching
- Simple data quality profiling methods for the adherence of data to their definition and their defined data values
- Validity checks and format checks
- Data cleaning in terms of outliers based on simple rules
- Data quality in the process of market research and surveys, for instance, interviewer effects and the provision of accurate answers by the interviewed person
- Data entry validation and data quality checks directly in the operational system on data retrieval
- Data quality considerations in operational systems
- Technical interfaces for data cleaning in a real-life environment

About the Author

Gerhard Svolba was born in Vienna, Austria, in 1970. He studied business informatics and statistics at the University of Vienna and Technical University of Vienna and holds a master's degree. From 1995 till 1999, he was assistant professor in the department for medical statistics at the University of Vienna, where he completed his PhD on statistical quality control in clinical trials (the respective book is published in Facultas). In 1999 Gerhard joined SAS Institute Inc. and is currently responsible for the analytical projects in SAS Austria as well as the analytical products and solutions SAS offers.

In 2003, on his way to a customer site to consult with them on data mining and data preparation, he had the idea to summarize his experience in written form. In 2004 he began work on *Data Preparation for Analytics Using SAS*, which was released by SAS Publishing in 2006. Since then he has spoken at numerous conferences on data preparation and teaches his class "Building Analytics Data Marts" at many locations. He likes to be in touch with customers and exchange ideas about analytics, data preparation, and data quality.

Gerhard Svolba is the father of three teenaged sons and loves to spend time with them. He likes to be out in nature, in the woods, mountains, and especially on the water, as he is an enthusiastic sailor.

Gerhard Svolba's current website can be found at http://www.sascommunity.org/wiki/Gerhard_Svolba; he answers e-mails under sastools.by.gerhard@gmx.net. Blog entries can be found under http://www.sascommunity.org/wiki/Gerhard%27s_Blog.

Downloads and References

For downloads of SAS programs, sample data, and macros that are presented in this book as well as updates on findings on the topic of data quality for analytics, please visit: http://www.sascommunity.org/wiki/Data_Quality_for_Analytics.

This site also includes downloads of color versions of selected graphs and figures that are presented in this book. Graphs and figures that reveal their content much better in color are available.

The author will keep this site up to date and provide updates on the content presented in this book.

In addition, please also see the SAS author page for Gerhard Svolba at http://support.sas.com/publishing/authors/svolba.html.

In this book, references are keyed to the reference list using bracketed numbers. The number in brackets refers to the respective entry in the reference section.

Part I: Data Quality Defined

Introduction

General

The first part of this book focuses on the definition of data quality and the data quality characteristics that are important from an analytical point of view.

The first two chapters of this part extend the introduction by using example case studies and a definition of data quality for analytics.

- Chapter 1, "**Introductory Case Studies**," relates real-life examples to typical data quality problems, forming an example-oriented introduction to data quality for analytics.
- Chapter 2, "**Definition and Scope of Data Quality for Analytics**," defines data quality for analytics, discusses its importance, and provides examples of good data quality.

The next seven chapters discuss data quality characteristics that are at the heart of data quality for analytics:

- Chapter 3, "**Data Availability**," questions whether the data are available. Can the data needed for the analysis be obtained?
- Chapter 4, "**Data Quantity**," examines whether the amount of data are sufficient for the analysis.
- Chapter 5, "**Data Completeness**," deals with missing values for the available data fields from a data analysis perspective.
- Chapter 6, "**Data Correctness**," discusses whether the available data are correct with respect to their definition. Are the data what they claim to be and do they, in fact, measure what they are supposed to measure?
- Chapter 7, "**Predictive Modeling**," discusses special requirements of predictive modeling methods.
- Chapter 8, "**Analytics for Data Quality**," shows additional requirements of interdependences for analytical methods and the data.
- Chapter 9, "**Process Considerations for Data Qualitly**," shows the process aspect of data quality and also discusses considerations such as data relevancy and possible alternatives.

These chapters form the conceptual basis of the book (that is, the relevant features of data quality for analytics). The second part of the book uses this as a basis to show how the data quality status can be profiled and improved with SAS.

Chapter 1: Introductory Case Studies

1.1 Introduction 4

1.2 Case Study 1: Performance of Race Boats in Sailing Regattas 4
    Overview 4
    Functional problem description 5
    Practical questions of interest 6
    Technical and data background 6
    Data quality considerations 8
    Case 1 summary 10

1.3 Case Study 2: Data Management and Analysis in a Clinical Trial 10
    General 10
    Functional problem description 10
    Practical question of interest 11
    Technical and data background 12
    Data quality considerations 12
    Case 2 summary 14

1.4 Case Study 3: Building a Data Mart for Demand Forecasting 14
    Overview 14
    Functional problem description 14
    Functional business questions 15
    Technical and data background 15
    Data quality considerations 15
    Case 3 summary 17

1.5 Summary 17
    Data quality features 17
    Data availability 18
    Data completeness 18
    Inferring missing data from existing data 18
    Data correctness 18
    Data cleaning 19
    Data quantity 19

1.1 Introduction

This chapter introduces data quality for analytics from a practical point of view. It gives examples from real-world situations to illustrate features, dependencies, problems, and consequences of data quality for data analysis.

Not all case studies are taken from the business world. Data quality for analytics goes beyond typical business or research analyses and is important for a broad spectrum of analyses.

This chapter includes the following case studies:

- In the first case study, the performance of race boats in sailing regattas is analyzed. During a sailing regatta, many decisions need to be made, and crews that want to improve their performance must collect data to analyze hypotheses and make inferences. For example, can performance be improved by adjusting the sail trim? Which specific route on the course should they sail? On the basis of GPS track point and other data, perhaps these questions can be answered, and a basis for better in-race decisions can be created.
- The second case study is taken from the medical research area. In a clinical trial, the performance of two treatments for melanoma patients is compared. The case study describes data quality considerations for the trial, starting from the randomization of the patients into the trial groups through the data collection to the evaluation of the trial.
- The last case study is from the demand forecasting area. A retail company wants to forecast future product sales based on historic data. In this case study, data quality features for time series analysis, forecasting, and data mining as well as report generation are discussed.

These case studies illustrate data quality issues across different data analysis examples. If the respective analytical methods and the steps for data preparation are not needed for the data quality context, they are not discussed.

Each case study is presented in a structured way, using the following six subsections:

- Short overview
- Description of the functional question and the domain-specific environment
- Discussion of practical questions of interest
- Description of the technical and data background
- Discussion of the data quality considerations
- Conclusion

1.2 Case Study 1: Performance of Race Boats in Sailing Regattas

Overview

This case study explores a comprehensive data analysis example from the sailing sport area. Note that these characteristics of data quality not only apply to sailboat analysis, but they also refer to research- and business-related analysis questions. For a specific race boat, the GPS (global positioning system) track point data over different races and the base data (like the size of sails, crew members, and external factors) are collected for one sailing season. These data are then cleaned, combined, and analyzed. The purpose of the analysis is to improve the race performance of the boat by answering questions like the influence of wind and choice of sails or the effect of different tactical decisions.

Functional problem description

The name of the boat of interest is *Wanda*, and the team consists of a helmsman and two crew members. The boat participates in sailboat fleet races, where 10–30 boats compete against each other in 5–10 regattas per sailing season, and each regatta consists of 4–8 races. The race course is primarily a triangle or an "up-and-down" course, where the "up" and the "down" identify whether it is sailed against or with the wind.

The typical race begins with a common start of all participating boats at a predefined time. After passing the starting line, the boats sail upwind to the first buoy, then in most cases they go downwind to one or two other buoy(s), and then upwind again. This route is repeated two to three times until the finishing line is passed. Figure 1.1 illustrates an example race course.

Figure 1.1: Typical course in a sailboat regatta

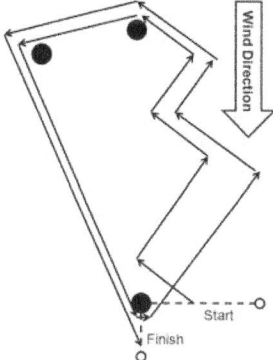

Sailing is a complex sport. In addition to the optimal sailing technique, the state of the sailing equipment, and the collaboration and physical fitness of the crew, many factors have to be considered to sail a good race. The most important factors are listed here:

- When going upwind, sailboats can sail at an angle of about 45 degrees with the true wind. To reach the upwind buoys, the boats must make one or more tacks (turns). The larger the angle to the wind, the faster the boats sail; however, the distance that has to be sailed increases.

- Depending on the frequency and the size of wind shifts, it might be better to do more tacking (changing the direction when going upwind) to sail the shortest possible course. However, tacking takes time and decreases speed.

- The specific upwind course of a boat is typically planned to utilize the wind shifts to sail upwind as directly as possible.

- The sailboat itself offers many different settings: different sail sizes and different ways to trim the boat. An average race boat has about 20 trim functions to set (for example, changing the angle and shape of the sails).

There is much literature available on sailboat race tactics and sailboat trimming. To successfully compete with other teams, these two areas deserve as much attention as the proper handling of the boat itself.

Based on this situation, many practical questions are of interest to get more knowledge on the boat handling, the impact of different tactical decisions, and the reaction of the boat to different trim techniques.

Practical questions of interest

Based on the factors described earlier, there are many practical questions of interest:

- How can sailors better understand the handling of their boats?
- How does the boat react to trim decisions?
- What are the effects of different tactical decisions?
- Can the specific route that is sailed for a given course be improved?

A comprehensive list would go far beyond the scope of this book.

For this case study, let us focus on questions that are of practical interest for learning more about boat speed, effects of trim techniques, and tactical decisions. These questions are sufficient to describe the case study from a data quality perspective:

- Tacking: how much time and distance are lost when tacking? During a tack, the boat must turn through the wind and, therefore, loses speed. Only when the boat reaches its new course and gets wind from the other side is speed regained. Depending on the time and distance required for a tack under various conditions, the tactical decision to make many or few tacks during a race can be optimized.
- How does the upwind speed of the boat depend on influential factors like wind speed, wind direction, and sail size? On various settings of the trim functions or on the crew itself? The boat, for example, gains speed if it is sailed with only 55 degrees to the wind. The question is whether this additional speed compensates for the longer distance that has to be sailed to get to the same effective distance upwind. What data are needed to optimize the angle for sailing to the wind?
- How does the maximum possible course angle to the true wind depend on influential factors like wind speed, sail size, and trim function settings? Different trim functions allow changing the shape of the foresail and the mainsail. The effective course angle and speed in setting these trim functions is of special interest. Given the crew members, their physical condition, the boat, its sailing characteristics, the weather conditions, and the sea conditions, what are the optimal trim settings over the route chosen for the given course?
- How do different tactical decisions perform during a race? When sailing upwind, for example, tactical decisions can include making only a few tacks and sailing to the left area of the course and then to the buoy, sailing to the right area of the course, or staying in the middle of the course and making many tacks.
- How does the actual sailing speed or the angle to the true wind deviate from other boats competing in the race? Comparing the effect of different course decisions between the participating boats is of special interest. We can then see which areas of the course have the best wind conditions or whether different boats perform in a different way under similar conditions.
- By comparing the performance across boats in a race, can the sailing abilities of the individual crews and boats be further analyzed and improved?

Technical and data background

The boat *Wanda* uses a Velocitek SC-1 device, which is a GPS device that collects the coordinates from different satellites in 2-second intervals. Based on these data, the device displays in real time the average and maximum speeds and the compass heading. This information is vital during a race to track boat performance. The GPS device also stores the data internally in an XML format. These data can then be transferred to a computer by using a USB cable.

The following data are available in the XML file with one row per 2-second interval: timestamp (date and time), latitude, longitude, heading, and speed. A short excerpt is shown in Figure 1.2.

Figure 1.2: Content of the XML file that is exported by the GPS device

```xml
<?xml version="1.0" encoding="utf-8"?>
<VelocitekControlCenter xmlns:xsi="http://www.w3.org/2001/XMLSchema-instance"
xmlns:xsd="http://www.w3.org/2001/XMLSchema" createdOn="2009-05-
25T18:29:02.65625+02:00"
xmlns="http://www.velocitekspeed.com/VelocitekControlCenter">
 <MetadataTags>
 <MetadataTag name="BoatName" value="Wanda" />
 <MetadataTag name="SailNo" value="0000" />
 <MetadataTag name="SailorName" value="xxxx" />
 </MetadataTags>
 <CapturedTrack name="090521_131637" downloadedOn="2009-05-
25T18:23:46.25+02:00" numberTrkpts="8680">
 <MinLatitude>47.773464202880859</MinLatitude>
 <MaxLatitude>47.804649353027344</MaxLatitude>
 <MinLongitude>16.698064804077148</MinLongitude>
 <MaxLongitude>16.74091911315918</MaxLongitude>
 <DeviceInfo ftdiSerialNumber="VTQURQX9" />
 <SailorInfo firstName="xxxx" lastName="yyyy" yachtClub="zzzz" />
 <BoatInfo boatName="wwww" sailNumber="0000" boatClass="Unknown" hullNumber="0"
/>
 <Trackpoints>
 <Trackpoint dateTime="2009-05-21T13:49:24+02:00" heading="68.43" speed="5.906"
latitude="47.792442321777344" longitude="16.727603912353516" />
 <Trackpoint dateTime="2009-05-21T13:49:26+02:00" heading="59.38" speed="5.795"
latitude="47.7924690246582" longitude="16.727682113647461" />
 <Trackpoint dateTime="2009-05-21T13:49:28+02:00" heading="65.41" speed="6.524"
latitude="47.792495727539062" longitude="16.727762222290039" />
 <Trackpoint dateTime="2009-05-21T13:49:30+02:00" heading="62.2" speed="6.631"
latitude="47.792518615722656" longitude="16.727849960327148" />
 <Trackpoint dateTime="2009-05-21T13:49:32+02:00" heading="56.24" speed="6.551"
latitude="47.792549133300781" longitude="16.727928161621094" />
 <Trackpoint dateTime="2009-05-21T13:49:34+02:00" heading="60.56" speed="5.978"
latitude="47.792579650878906" longitude="16.728004455566406" />
 <Trackpoint dateTime="2009-05-21T13:49:36+02:00" heading="61.57" speed="7.003"
latitude="47.792606353759766" longitude="16.728090286254883" />
 <Trackpoint dateTime="2009-05-21T13:49:38+02:00" heading="52.03" speed="7.126"
latitude="47.792636871337891" longitude="16.728176116943359" />
```

These data can be analyzed by using Velocitek software to visualize the course, speed, and heading of a boat and to perform simple analyses.

Other data processing systems can use these data to perform specific analyses. In this case study, the data have been imported into SAS by using a SAS DATA step to prepare the data for analysis. Different graphical and statistical analyses can be performed to answer the practical questions listed earlier.

Figure 1.3 is a line chart that has been produced using SAS/IML Studio. It shows the race course and the specific route that was sailed. The course is similar to the one shown in Figure 1.1. After the start, the boat goes upwind to the first buoy and then downwind to the second buoy. The circuit is repeated a second time, and then the finish line is reached. Note that some annotations are included to identify some features of the specific route that was sailed.

On the first upwind path, the boat obviously experienced a wind shift that slowed the progress to the upwind buoy. From an ex-post tactical viewpoint, the boat should have tacked again or it should have been farther to the right in the course area.

Figure 1.3: Line chart of one race

The **second data source**, in addition to the GPS track point data, is a logbook that contains crew-recorded data for each race. For example, it includes the names of crew members, the sailing area, the general wind direction, the general wind strength, and other meteorological values as well as the size and type of the sails.

Data quality considerations

Based on the practical and technical background, many aspects of the analysis can be discussed, but our focus is data quality:

- The GPS track point data are only available for two boats: the boat whose crew wants to perform analyses and for one additional boat. Most of the remaining boat teams either did not save the GPS track point data or they were unwilling to share the data with potential competitors. A few other teams did not use a GPS device. Thus, comparison between boats can only be performed in a limited way.

- The GPS device only collects data that are related to the position of the boat itself. Information about the wind direction and wind strength are not collected. In order to collect this information, a separate device is needed. Therefore, the questions that relate to the effect of wind strengths and wind direction shifts cannot be answered with the GPS track point data.

- Assuming a constant behavior of the boat itself and the way the helmsman pilots the boat, it is possible to infer the wind direction from the compass heading of the boat. However, if the wind shifts immediately before or during a tack, the analyst might not be able to identify if the tacking angle and the new heading after the tack are caused by a wind shift or by a different helmsman behavior.

- There is no timestamped protocol of the different settings of the trim functions of the boat. There is only a rough recording, in many cases based on personal memory, of the main trim function settings at the beginning of the race. It is therefore not possible to identify if a change in speed in the second upwind course is due to a different trim setting or to different wind or helmsman conditions.

- During the sailing season, only the GPS tracking data were recorded in a regular and structured way. Thus, at the end of the season, when the idea to perform this analysis arose, some of the other data, such as participating crew members, size of sails used, average wind speed, average wind direction, and other trim settings, were retrieved based on the memory of the crew members. So, clearly, some information was lost, and given the human factor some data were recorded with potential errors. (The probability of data accuracy and completeness is very high for data that are collected automatically through electronic systems. However, for data that are manually documented and entered into a system, the probability is lower—not because of systematic and malicious bias but due to environmental distractions and fatigue. In practice, the human source of error can be found in many other cases.)

These points reflect the situation of **data unavailability**. Some data that are desirable for analytical insights are simply not available, which means that some analyses cannot be done at all and other analyses can only be done in a reduced scope.

- To analyze the data using SAS software, the data must be exported to a PC from the GPS device as XML files. In this step, the data must have been correctly collected and stored in the GPS device itself and then exported correctly into an XML file.
- After the XML file is stored on the PC, it is read into SAS, which then validates that the file was imported correctly. Care has to be taken to correctly separate the individual fields and to correctly represent the date, time, and numeric values. Thus, before the data can be analyzed, there are multiple places where erroneous information can enter the data, but there are also multiple checks to ensure data quality.

The **correctness in data collection and the accuracy of their transfer** are vital to data preparation for analysis. Before data can be analyzed, data validation must be performed to ensure that the real-world facts, the data, are collected by the device correctly, stored correctly, and transferred correctly to the computer for use by the analysis software.

- GPS data for another boat are available, but before these data can be combined with the first boat, the time values must be realigned because the internal clock of the other boat was one hour behind. When the two data sets are aligned by the common factor of time, they can be merged, and the combined information can be used for analysis. Note that in many cases augmenting data can add important information for the analysis, but often the augmenting data must be prepared or revised in some way (for example, time, geography, ID values) so that they can be added to existing data.
- If three races are sailed during a day, the log file contains the data for the three races as well as the data for the time before, between, and after the races. To produce a chart as shown in Figure 1.3, the data need to be separated for each race and any unrelated data need to be deleted. Separating the races and clearing the non-race records is frequently quite complicated because the start and end of the race is often not separately recorded. To perform this task, the data need to be analyzed before as a whole and then post-processed with the start and end times.

These points show that prior to the analysis, **data synchronization and data cleaning** often need to be done.

- In a few cases, the GPS device cannot locate the position exactly (for example, due to a bad connection to the satellites). These cases can cause biases in the latitude and longitude values, but they can especially impact the calculated speeds. For example, if a data series contains a lost satellite connection, it can appear that the boat went 5.5 to 6 knots on average for over an hour and then suddenly went 11.5 knots for 2 seconds. These data must be cleaned and replaced by the most plausible value (for example, an average over time or the last available value).
- In another case, the device stopped recording for 4 minutes due to very low temperatures, heavy rain, and low batteries. For these 4 minutes, no detailed track point data were recorded. During this interval, the position graph shows a straight line, connecting the last available points. Because this happened when no tacking took place, the missing observations could be inserted by an interpolation algorithm.
- In another case, the GPS device was unintentionally turned off shortly before the start and turned on again 9 minutes later. Much tacking took place during this interval, but the missing observations cannot be replaced with any reasonable accuracy.
- Some of the above examples for "data unavailability" can also be considered as missing values similar to the case where information like sail types and settings and crew members were not recorded for each race.

These data collection examples show how some **values** that were intended to be available for the analysis can be **missing or incorrect**. Note that the wind direction and wind strength data are considered to be **not available** for the analysis because they were not intended to be collected by a device. The GPS track point data for the first 9 minutes of the race **are missing** because the intention was to collect them (compare also chapters 3 and 5).

- Practical questions to be answered by the analysis involve the consequences of tacking and the behavior of the boat when tacking. The data for all races contain only 97 tacks. If other variables like wind conditions, sail size, and trim function settings are to be considered in the analysis as influential variables, there are not enough observations available to produce stable results.

To answer practical questions with statistical methods, a representative sample of the data and a sufficient amount of data are required. The more quality data that are available, the greater the confidence we can have in the analysis results.

Case 1 summary

This example was taken from a non-business, non-research area. It shows that data quality problems are not limited to the business world, with its data warehouses and reporting systems. Many data quality aspects that are listed here are relevant to various practical questions across different analysis domains. These considerations can be easily transferred from sailboat races to business life.

Many analyses cannot be performed because the data were never collected, deleted from storage systems, or collected only in a different aggregation level. Sometimes the data cannot be timely aligned with other systems. Due to this incomplete data picture, it is often impossible to infer the reason for a specific outcome—either because the information is not available or because the effects cannot be separated from each other.

These aspects appear again in the following chapters, where they are discussed in more detail.

1.3 Case Study 2: Data Management and Analysis in a Clinical Trial

General

This case study focuses on data management and analysis in a long-term clinical trial. In general, the specifics of a clinical trial significantly impact data collection, data quality control, and data preparation for the final analysis. Clinical trials focus on data correctness and completeness because the results can critically impact patient health and can lead, for example, to the registration of a new medication or the admission of a new therapy method.

This case study only discusses the data quality related points of the trial. The complete results of the trial were published in the *Official Journal of the American Society of Clinical Oncology* 2005 [2].

Functional problem description

The clinical trial discussed in this case study is a long-term multicenter trial. More than 10 different centers (hospitals) recruited patients with melanoma disease in stages IIa and IIb into the trial that lasted over 6.5 years. Each patient received the defined surgery and over 2 years of medication therapy A or B. The trial was double-blind; neither the patient nor the investigator knew the actual assignment to the treatment groups. The assignment to the treatment group for each patient was done randomly using a sequential randomization approach.

During and after the 2 years of treatment, patients were required to participate in follow-up examinations, where the patient's status, laboratory parameters, vital signs, dermatological examinations, and other parameters that describe patient safety were measured. The two main evaluation criteria were the recurrence rate of the disease and the patient survival rate. Depending on their time of recruitment into the trials, patients were expected to participate in follow-up exams at least 3 years after the end of the therapy phase.

Patients were recruited into this trial in different centers (hospitals). All tasks in treatment, safety examinations, trial documentation into case-record forms (CRFs), and evaluation of laboratory values were performed locally in the trial centers. Tasks like random patient allocation into one of the two treatment groups (randomization), data entry, data analysis, and trial monitoring were performed centrally in the trial monitoring center.

The following tasks in the trial took place locally in the trial center:

- Recruitment of patients into the trial and screening of the inclusion and exclusion criteria.
- Medical surgery and dispensing of medication to the patients.
- Performance of the follow-up examinations and documentation in writing of the findings in pre-defined CRFs.
- Quality control of the accuracy and completeness of the data in the CRFs compared to patient data and patient diagnostic reports. This step was performed by a study monitor, who visited the trial centers in regular intervals.

The CRFs were then sent to the central data management and statistic center of the trial. This center was in charge of the following tasks:

- Performing the randomization of the patients into the treatment groups A and B with a software program that supports sequential randomization.
- Storing the randomization list, which contained the allocation patient number to treatment, in access-controlled databases.
- Maintaining the trial database that stored all trial data. This database was access-controlled and was logging any change to the trial records.
- Collecting the CRFs that were submitted from the trial centers and entering them into the trial database.
- Performing data quality reports on the completeness and correctness of the trial data.
- Performing all types of analyses for the trial: safety analyses, adverse event reports, interim analyses, and recruitment reports.

Practical question of interest

The practical question of interest here was the ability to make a well-founded and secure conclusion based on the trial data results.

- The main criterion of the trial in the per-protocol and in the intention-to-treat analysis was the comparison of the disease-free intervals between the treatment groups and the comparison of the survival between treatment groups.
- To achieve this, a sufficient number of patients, predefined by sample-size calculation methods, were needed for the trial. To check whether the recruitment of patients for the trial was on track, periodic recruitment reports were needed.
- Beside the main parameters, recurrence of disease and survival, parameters that describe the safety of the patients, was collected for the safety analysis. Here laboratory and vital sign parameters were analyzed as well as the occurrence of adverse events.

All these analyses demanded correct and complete data.

Technical and data background

The randomization requests for a patient to enter the trial were sent by fax to the monitoring center. The trial data were collected on paper in CRFs and entered into an Oracle database. This database did not only support the data entry, but it also supported the creation of data quality and completeness reports.

Data quality considerations

Based on the scope of a clinical trial presented here, the following aspects of data quality during data collection, data handling, and data analysis are of interest:

- To improve the correctness of the data provided through the CRFs, a clinical monitor reviewed and validated the records in each trial center before they were submitted to the monitoring center. In this case, very high data quality was established at the very beginning of the process as possible errors in data collection were detected and corrected before data entry for the records.

- Each information item was entered twice (that is, two different persons entered the data). Therefore, the data entry software had to support double data entry and verify the entered data against lists of predefined items, value ranges, and cross-validation conditions. It also had to compare the two entered versions of the data. This was achieved by online verification during data entry and by data quality reports that listed the exceptions that were found during the data checks.

- A crucial point of data quality in this clinical trial was the correctness of the values of the randomization lists in the clinical database. This randomization list translates the consecutive numeric patient codes into treatment A and treatment B groups. Obviously, any error in this list, even for a single patient number, would bias the trial results because the patient's behavior and outcome would be counted for the wrong trial group. Therefore, much effort was used in ensuring the correct transfer of the randomization list into the trial database.

- The randomization list was provided to the data monitoring center as hardcopy and as a text file in list form. Thus, the text file had to be manually preprocessed before it could be read into the database. Manual preprocessing is always a source of potential error and unintended data alteration. The final list that was stored in the database was manually checked with the originally provided hardcopy by two persons for correctness.

- As an additional check, two descriptive statistics were provided by the agency that assigned the double-blind treatments and prepared the randomization list, the mean and the standard deviation of the patient numbers. For each group A and B, these statistics were calculated by the agency from the source data and then compared with the corresponding statistics that were calculated from the data that were entered in the trial database. This additional check was easy to perform and provided additional confidence in the correctness of the imported data.

These practices indicate that in clinical trials there is an extremely strong emphasis on the correctness of the data that are stored in the clinical database. To achieve and maintain data correctness, the focus must be on validating and cross-checking the **data collection**, the **data transfer**, and the **data entry** of the input data.

- To trace changes to any field in the trial database, all data inserts, updates, or deletions of the trial database were logged. Based on this functionality, a trace protocol could be created for any field to track if, and how, values changed over time. An optional comment field enabled the insertion of comments for the respective changes. The commenting, logging, and tracing processes were very important in maintaining high data quality, especially for data fields that were critical for the study: the time until relapse, the survival time, and the patient status in general. The ability to perform an uncontrolled alteration of data does not comply with external regulations, and it is a potential source of intended or unintended biasing of the trial and the trial results.

- From a process point of view, it was defined that any change to the data, based on plausibility checks or corrections received at a later point, would only be made to the trial database itself. No alterations or updates were allowed at a later stage during data preparation for the analysis itself. This requirement was important to create and maintain a single source of truth in one place and to avoid the myriad coordination and validation problems of data preparation logic and data correction processes dispersed over many different analysis programs.

- Based on logging data inserts, updates, and deletions, it was also possible to rollback either the database or an individual table to any desired time point in the past. The historical database replication functionality is required by Good Clinical Practice (GCP) [10] requirements. It enables analysts to access the exact status of a database that was used for an analysis in the past.

For security and regulatory reasons, **tracing changes in the database** was very important. In addition to the support for double data entry, the trial database provided functionality for tracing changes to the data and for enabling the database rollback to any given date.

- Because there was no central laboratory for the trial, the determination of the laboratory parameters was done locally in each hospital. But the nonexistence of a central laboratory led to two problems.
  - Some laboratories did not determine all the parameters in the measurement units that were predefined in the CRF, but they did define them in different units. Thus, to obtain standardized and comparable measurements, the values had to be recalculated in the units specified in the CRF.
  - The normal laboratory values for the different laboratories differed. Frequently different laboratories have different normal laboratory values. To perform plausibility checks for the laboratory values based on normal laboratory values, a different lookup table for each trial center may have been needed.
- As it turned out, the usage of normal laboratory values was not suitable for plausibility checks because roughly 15% of the values fell outside of these limits. If the normal laboratory values had been used, the validation effort required would have been much too high and would result in the acceptance of the slightly out of limit value. The purpose of data validation was not to highlight those values that fell out of the normal clinical range but to detect those values that could have been falsely documented in the CRF or falsely entered into the database. Thus, it was decided to compute validation limits out of the empirical distribution of the respective values and to calibrate the values that way so that a reasonable amount of non-plausible values were identified.
- The primary evaluation criterion of the trial was the time until relapse. For each treatment group, a survival curve for this event was calculated and compared by a log rank test. To calculate this survival curve, a length of the period is needed, which is calculated from the patients' trial start until the date of their last status. In the survival analysis, the status on the patient's last date, relapse yes or no, was used to censor those observations with no relapse (yet). The important point here is the correct capture of the patient status at or close to the evaluation date. In a long-term trial, which continues over multiple years and contains a number of follow-up visits, the patients' adherence to the trial protocol decreases over time. Patients do not show up to the follow-up visits according to schedule. The reasons can be from both ends of the health status distribution. For some, their health status is good, and they see no importance in attending follow-up meetings; for others, their health status is bad, and they cannot attend the follow-up meetings. Therefore, without further investigation into the specific reason for not adhering to the trial protocol, identifying the patient's exact status at the evaluation snapshot date is complicated. Should the status at their last seen date be used? That is an optimistic approach, where if no relapse has been reported by those not adhering to the trial protocol, then no relapse has occurred. Or should it be based on the pessimistic assumption that a relapse event occurred immediately after their last seen date?
- Also, determining the population to be used for the per-protocol analysis is not always straightforward. The per-protocol analysis includes only those patients who adhered to all protocol regulations. A patient, for example, who did not show up at the follow-up visits for months 18 and 24 might be considered as failing to follow-up at an interim analysis, which is performed after 2.5 years. If, however, they showed up at all consecutive scheduled visits in months 30, 36, and 42, then they might be included in the final analysis after 4 years.

These points focus on the **correctness of the data** for the analysis. In the following, plausibility checks and rules on how to define a derived variable play an important role:

- In the respective study, a desired sample size of 400 patients was calculated using sample-size calculation methods. This number was needed to find a difference that is statistically significant at an alpha level of 0.05 and a power for 80%. Recruitment was planned to happen over 4 years (approximately 100 patients per year).
- After 9 months of recruitment, the clinical data management center notified the principal investigator that the actual recruitment numbers were far below the planned values and that the desired number of patients would only be achieved in 6.5 years. Continuing the study at this recruitment pace for the desired sample size would delay the trial completion substantially, about 2.5 years. But stopping recruitment and maintaining the 4-year schedule would result in too few patients in the trial. Based on this dilemma, additional hospitals were included in the trial to increase the recruitment rate.

In clinical research, much financial support, personal effort, and patient cooperation are needed. It is, therefore, important to ensure there is a reasonable chance to get a statistically significant result at the end of the trial, given that there is a true difference. For this task, sample-size planning methods were used to determine **the minimum number of patients (data quantity)** in the trial to prove a difference between treatments.

Case 2 summary

This case study shows the many data quality problems in a very strict discipline of research, clinical trials. There are two strong focuses: the correctness of the data and the sufficiency of the data. To obtain sufficient correct and complete data, substantial effort is needed in data collection, data storage in the database, and data validation. The financial funding and personal effort to achieve this result need to be justified compared to the results. Of course, in medical research, patient safety—and, therefore, the correctness of the data—is an important topic, which all clinical trials must consider. In other areas, the large investment of effort and funding might not be easily justified.

From this case study, it can be inferred that in all analysis areas, there is a domain-specific balancing of costs against the analysis results and the consequences of less than 100% correct and complete data.

1.4 Case Study 3: Building a Data Mart for Demand Forecasting

Overview

This last case study shows data quality features for an analysis from the business area. A global manufacturing and retail company wants to perform demand forecasting to better understand the expected demand in future periods. The case study shows which aspects of data quality are relevant in an analytical project in the business area. Data are retrieved from the operational system and made available in analysis data marts for time series forecasting, regression analysis, and data mining.

Functional problem description

Based on historic data, demand forecasting for future periods is performed. The forecasts can be sales forecasts that are used in sales planning and demand forecasts, which, in turn, are used to ensure that the demanded number of products is available at the point of sale where they are required. Forecast accuracy is important as over-forecasting results in costly inventory accumulation while under-forecasting results in missed sales opportunities.

Demand forecasts are often created on different hierarchical levels (for example, geographical hierarchies or product hierarchies). Based on monthly aggregated historic data, demand forecasts for the next 12 months can be developed. These forecasts are revised on a monthly basis. The forecasts are developed over all levels of the hierarchies; starting with the individual SKU (stock keeping unit) up to the product subgroup and product group level and to the total company view.

Some of the products have a short history because they were launched only during the last year. These products do not have a full year of seasonal data. For such products, the typical methods of time series forecasting cannot be applied. For these products, a data mining model is used to predict the expected demand for the next months on product base data like price or size. This is also called *new product forecasting*.

A data mining prediction model has been created that forecasts the demand for the future months based on article feature, historic demand pattern, and calendar month. For products that have a sufficient time history, time series forecasting methods like exponential smoothing or ARIMA models are employed. For many products, the times series models provide satisfactory forecasts. For some products, especially those that are relatively expensive, if they have variables that are known to influence the quantities sold, then regression models can be developed, or the influential variables can be added to ARIMA models to form transfer function models.

Functional business questions

The business questions that are of primary interest in this context are as follows:

- On a monthly basis, create a forecast for the next 12 months. This is done for items that have a long data history and for items that have a short data history.
- Identify the effect of events over time like sales promotions or price changes.
- Identify the correlation between item characteristics like price, size, or product group and the sales quantity in the respective calendar month.
- Identify seasonal patterns in the different product groups.
- Beyond the analytical task of time series forecasting, the system also needs to provide the basis for periodic demand reporting of historic data and forecast data and for planning the insertion of target figures for future periods into the system.

Technical and data background

In this case study, the company already had a reporting system in place that reports the data from the operational system. Data can be downloaded from this system as daily aggregates for a few dimensions like product hierarchy or regional hierarchy. These data have two different domains, the order and the billing data. Time series forecasting itself was only performed on the order data. For additional planning purposes, billing data also were provided.

Another important data source was the table that contains the static attributes (characteristics) for each item. This table contained a row for each item and had approximately 250 columns for the respective attribute. However, not all variables were valid for each item. Beside a few common attributes, the clusters of attributes were only relevant to items of the same item group.

Some additional features for each item were not yet stored in the central item table, but they were available in semi-structured spreadsheets. These spreadsheets did contain relevant information for some product groups that could be made available for the analysis.

Data quality considerations

The following features of the project had a direct relation to data quality:

- Historic order data and historic billing data for the last 4 years were transferred from the operational system to the SAS server. Given all the different dimensions over millions of rows, the data import was several gigabytes in size.
    - This amount of data cannot be checked manually or visually. To verify correctness of the data that were imported into the SAS system, a checksum over months, weeks, product hierarchies, and so forth was created. The checksum shows the number of rows (records) read in, the number of rows created, and so on. While in SAS virtually any checksum statistic can be calculated, only those

statistics that are also available in the original system (for example, a relational database) can be used for comparison. For some dimensions of the data, the checksums differed slightly.

- o Usually it is a best practice rule to investigate even small differences. In this case, however, most of the small deviations were due to a small number of last-minute bookings and retrospective updates that were shown in the life system on a different day than in the export files. This also made the comparison difficult between the exported data from the life system and the values in the life system itself. There is the possibility that immediately after the data was exported, the numbers had already changed because of new transactions. In a global company, it is not possible to export the data during the night when no bookings are made. From a global perspective, it is never "night."

These points reflect the **control of the data import process and correctness check** after transfer and storage in the source system.

- In the case described here, the order and billing data were complete. It is reasonable for the billing data to be complete in order to bill customers; otherwise, revenue would be lost.
- Static data like item features, other than product start date and price, were not as well-maintained because they are not critical for day-to-day business operations. However, from an analytical perspective, the characteristics of various items are of interest because they can be used to segment items and for product forecasting. The table containing item characteristics had a large number of missing values for many of the variables. Some of the missing values resulted when variables were simply not defined for a specific product group. The majority of the missing values, however, occurred because the values were not stored.
- The non-availability of data was especially severe with historic data. Orders from historic periods were in the system, but in many cases it was difficult to obtain characteristics from items that were not sold for 12 months.
- Because the company was not only the manufacturer of the goods but also the retailer, point-of-sale data were also available. Point-of-sale data typically provide valuable insight, especially for the business question on how to include short-term changes in customer behavior in the forecasting models. However, capturing these data for the analysis was complicated because only the last 12 months of point-of-sale data were available in the current operational system. For the preceding time period, data were stored in different systems that were no longer online. To capture the historic data from these older systems, additional effort in accessing historic backup files from these systems was required.
- Another source of potential problems in data completeness was that for some item characteristics, no responsibility for their maintenance and update was defined. This is especially true for data provided in in the form of spreadsheets. Therefore, in the analyses, because it was uncertain whether an updated version of these data would be available in a year, care had to be taken when using some characteristics.

These points refer to the **availability and completeness of the data**. While it is important to emphasize that existing data were transferred correctly into the system for analysis, it is also important to clearly identify the completeness status of the data. In this case, what was observed was typical for many data collection situations: Transactional data that are collected by an automatic process, like entering orders, billing customers, and forwarding stock levels, are more complete and reliable. Also, data that control a process are typically in a better completeness state. Data that need to be collected, entered, and maintained manually are, in many cases, not complete and well-maintained.

- For demand forecasting of products with a shorter history, classical time series forecasting methods could not be applied. Here, a repository of items was built with the respective historic data and item characteristics. As described earlier, predictive data mining models were built based on these data to forecast demand. The repository initially contained hundreds of items and increased over time as more and more items became available in the database. At first view, it might seem sufficient to have hundreds of items in the database. But in a more detailed view, it turned out that for some product categories only around 30 items were available, which was insufficient to build a stable prediction model.

- Individual product-group forecasting was necessary because products in different groups had different sets of descriptive variables. Also, products in different groups were assumed to have different demand patterns. Thus, separate models were needed. For the whole set of items, only six characteristics were commonly defined, and so the analyst had to balance data quantity against individual forecasting models. The analyst could decide to build a generic model on only the six characteristics, or they could build individual models with more input variables on fewer observations.
- In time series forecasting, for products with a long time history, each time series is considered and analyzed independently from the other series. Thus, increasing or decreasing the number of time series does not affect the analytical stability of the forecast model for a single series. However, the number of months of historic data available for each time series has an effect on the observed performance of the model.

These points refer to **data quantity**. For stable and reliable analytic results, it is important to have a sufficient number of observations (cases or rows) that can be used in the analysis.

- Deferring unavailable data. In some cases, data did not exist because it was not stored or it was not retained when a newer value became available or valid. Sometimes it is possible to recalculate the historic versions of the data itself. A short example demonstrates this:
    o To analyze and forecast the number of units expected to be sold each month, the number of shops selling the items is an influential variable, but it is often unavailable.
    o To overcome the absence of this important variable, an approximate value was calculated from the data: **the number of shops that actually sold the article**. This calculation can easily be performed on the sales data.

The content of the variable is, however, only an approximation; the resulting number has a deceptive correlation with the number of items sold because a shop where the item was offered but not sold is not counted. The inclusion of this variable in a model results in a good model for past months, but it might not forecast well for future months.

Case 3 summary

This case study shows features of data quality from a time series forecasting and data mining project. Extracting data from system A to system B and creating an analysis data mart involve data quality control steps. This process is different from a set of a few, well-defined variables per analysis subject because there are a large number of observations, hierarchies, and variables in the product base table. Given this large number of variables in the data, it is much more difficult to attain and maintain quality control at a detailed level. The case study also shows that for accuracy of results in analytical projects, the data quantity of the time history and the number of observations is critical.

1.5 Summary

Data quality features

This chapter discusses three case studies in the data quality context. These case studies were taken from different domains, but they share the fact that the results depend directly on the data and, thus, also on the quality of the data.

The quality of data is not just a single fact that is classified as good or bad. From the case studies, it is clear that there are many different features of data quality that are of interest. Some of these features are domain-specific, and some depend on the individual analysis question. The different features can be classified into different groups. For each case study, an initial grouping was presented.

These case studies are intended not only to whet your appetite for the data quality topic but also to highlight typical data quality specifics and examples of analyses.

A classification of data quality features that were discussed in the case studies follows. This classification is detailed in chapters 3 through 9.

Data availability

- GPS data were only available for two boats. No wind data were collected.
- No recording of the trim setting on the sailboat was done.
- Static information on the items to be forecasted was not entered into the system or maintained over time.
- Historic data or historic versions of the data (like the static information on the items from 12 months ago) were not available.
- Point-of-sale data from historic periods that were captured in the previous operational system could not be made available or could only be made available with tremendous effort.

Data completeness

- Some of the GPS data were missing because the device was turned on late or it did not record for 4 minutes.
- For a number of patients, no observations could be made for some follow-up visits because the patients did not return.
- Static information on the items to be forecasted was not completely entered into the system or maintained over time.

Inferring missing data from existing data

In some cases, an attempt was made to compensate for the unavailability of data by inferring the information from other data, for example:

- Estimating the wind direction from the compass heading on the upwind track.
- Approximating the unavailable number of shops that offered an item for sale from the number that actually sold them.

In both cases, a substitute for the unavailable data was found, which should be highly correlated with the missing data. This enables reasonable approximate decisions to be made.

Data correctness

- In a few cases, the value of the calculated boat speed from the GPS data appeared to be wrong.
- For the sailboat case study, when data for sail size and composition of crew members were captured post-hoc, the sail sizes could not be recaptured with 100% certainty. In this case, a most likely value was entered into the data.
- The source data on the CRFs were manually checked by an additional person.
- Data entry in the clinical trial was performed twice to ensure accuracy.
- The transfer of the randomization list for the trial followed several validation steps to ensure correctness.
- Transferring data from a GPS device to a PC text file and importing the file into the analysis software are potential sources of errors if data change.
- Any change in clinical trial data was recorded in the database to provide a trace log for every value.
- For each laboratory parameter, a plausibility range was defined to create an alert list of potential outliers.

- Transferring millions of rows from the operational system to the analysis system can cause errors that are hard to detect (for example, in the case of a read error when a single row is skipped in the data import).

Data cleaning

- After the GPS were imported, the values of the XML file needed to be decoded.
- The GPS data for the individual races needed to be separated.
- Implausible laboratory values were output in a report that the monitor used to compare with the original data.

Data quantity

- The database did not contain enough tacking data to analyze the tacking behavior of the boat in detail.
- In the clinical trial, a measurement in study control was taken to increase the number of participating clinical centers. Otherwise, not enough patients for the analysis would have been available.
- In the data mining models for the prediction of the future demand for items that have only a short history, only a few items had a full set of possible characteristics.

Chapter 2: Definition and Scope of Data Quality for Analytics

2.1 Introduction	**22**
Different expectations	22
Focus of this chapter	22
Chapter 1 case studies	22
2.2 Scoping the Topic Data Quality for Analytics	**22**
General	22
Differentiation of data objects	22
Operational or analytical data quality	23
General data warehouse or advanced analytical analyses	24
Focus on analytics	24
Data quality with analytics	25
2.3 Ten Percent Missing Values in Date of Birth Variable: An Example	**25**
General	25
Operational system	25
Systematic missing values	25
Data warehousing	26
Usability for analytics	26
Conclusion	27
2.4 Importance of Data Quality for Analytics	**27**
2.5 Definition of Data Quality for Analytics	**27**
General	27
Definition	28
2.6 Criteria for Good Data Quality: Examples	**28**
General	28
Data and measurement gathering	28
Plausibility check: Relevancy	28
Correctness	29
Missing values	29
Definitions and alignment	29
Timeliness	29
Adequacy for analytics	30
Legal considerations	30
2.7 Conclusion	**30**
General	30
Upcoming chapters	30

2.1 Introduction

Different expectations
Writing a book on data quality for analytics is an interesting but complex project. The interesting component obviously comes from the topic and its importance. The complexity involves the widespread points of view and the different expectations people, especially you as a reader, can have.

As already shown, the term *data quality* has many different meanings and interpretations in the data analysis area. The addition of the term *analytics* offers another dimension of possible meanings. Data quality is often just the existence or correctness of data for the analysis or for the operations process. Matching records and eliminating duplicates is also a prominent topic in data quality.

Focus of this chapter
This chapter starts by scoping out the topic *data quality for analytics*. Differences and similarities between analytical and non-analytical data quality are worked out and the importance of the topic is discussed.

Based on these points, a definition of data quality for analytics within the context of this book is provided. In the last section, example criteria for data quality are shown, which also complement the definition.

Chapter 1 case studies
The case studies in chapter 1 showed three examples of analyses and the related data quality criteria and problems. These case studies not only present data quality criteria that are used in the following chapters. They also offer some representative examples of the typical cases that are under closer consideration in this book.

2.2 Scoping the Topic Data Quality for Analytics

General
This section discusses the scope of data quality for analytics in the context of this book. Starting with a differentiation of data objects, it moves to the boundaries of *operational* and *analytical systems* and differentiating between general data warehouses and analytical and reporting analyses. This process reveals that the separation is not an either/or decision but more a bottom-up approach where one topic builds upon the other.

Differentiation of data objects
For the consideration of data in the data quality context, the differentiation of data objects is important. This subsection differentiates between fields, records, tables, and databases.

- A **data field** can be considered as a cell in a data matrix, where rows represent records and columns represent variables. For example, the age in years for a certain person is a data field. Considering data fields on their own, the possible data quality checks are checking whether:
    - the field is populated with a value
    - the values fall into predefined value ranges or lists of values
    - the values correspond to the predefined format for the field
- A **data record** is a row in the data matrix. Depending on the structure of the table, the record contains measured variables (features) for an analysis subject.
    - From a data quality point of view, rules can be applied here for plausibility checking. These rules go beyond simple value ranges or lists of values. They check the dependencies between different variables for one analysis subject. For example, a rule could check that values for the outcome of a pregnancy test must only be populated for female patients.

- The **data table** (or data matrix) is the collection of data records on the same domain. Examples are tables in a relational database like the CUSTOMER or PATIENTS table.
    - From a data quality point of view, it is possible to consider the number of observations in general and to compare values across analysis subjects, like checking the distribution of values and identifying outliers.
    - While the data quality checks in data records are based on fixed rules from a domain or business expertise, the data quality checks in a data table that go across observations are more analytic, like calculating the quantiles for a variable.
- In a **database** (also called a data mart, data warehouse, or registry), a collection of data tables is considered to have a relationship to each other (like in a relational database).
    - The existence of relationships allows you to check data quality features that go across tables, like the accordance of numbers of observations across tables or the data quality rules that use information from different tables.

Operational or analytical data quality

Data quality in an operational system has a different focus than data quality in an analytical or decision support environment.

- An **operational system**, like a hotel booking system or a call center system, focuses on the following data quality aspects:
    - Correctness of customers' names, addresses, and contact details
    - Response time of the system (in real time to be able to service the operations process)
    - Validity on the attributes of the customer, such as the rate plan, the balance on his account, and various transactions, to respond to his questions when contacting the call center and to give him correct answers and advice
    - Availability of a list of bookable hotel rooms, which reflects the current state and is not booked by a different system in parallel
    - Forwarding of the reservations details like reservation confirmation and booking order to the counterparts in the operations process chain
- An **analytical system** (or **decision support system**) for information retrieval typically uses data from the operational system. However, for reports and analytical analyses, often additional requirements on data are important:
    - Possibility to create a time series of the number of bookings per day
    - Ability to split the booking numbers across dimensions like room category, region, and customer segment
    - Availability of the historic booking data for each day to see how the booking behavior for a specific day evolves over time
    - Availability of the data not only as aggregates but also as detailed transactional data to create aggregations and derived variables from different viewpoints
    - Correctness of the data, such as the reason for the call to the call center. Is the call reason value the real reason or did the call center agent just choose the first value from a selection list?
    - Availability of a sufficient quantity of data to perform combined analyses and subgroup analyses, such as the analysis of the call length and the service success for each topic and call center agent

These points show that there are different requirements on data quality in the operational and analytical world. **This book focuses on the analytical and information retrieval aspect of data quality.** The quality of the data in the analytical system, however, heavily depends on the quality in the operational system. Thus, this is not an either/or situation, but more a focusing on the analytical and information retrieval area. Implicitly, the quality of the data in operational systems is a topic here because the analytical system consumes the operational data and, thus, also depends on its data quality.

General data warehouse or advanced analytical analyses

In the previous subsection, some examples of data quality considerations in the analytical area were shown. Here the analytical world is divided into general data warehousing and advanced analytics. This differentiation is not easy to perform because the distinction between analytical and advanced analytical areas is not straightforward. The following examples make this picture clearer:

- Data quality in data warehousing focuses, for example, on the following topics:
  - Standardizing data from different environments. Data that are imported from operational systems and made available for analysis need to be standardized so that values in the same column are based on the same definition and calculated to the same units.
  - Removing duplicate data. Data combined from different silos can cause the same analysis subjects to occur more than once in the database. Integrating these records into one record is not easy because usually different databases do not have the same identifier for the records. Here, methods of standardization and records matching on attributes of the data such as names, addresses, phone numbers, or bank account numbers are performed.
  - Validating data. Data values must fall into predefined ranges or lists of values.
  - Providing a data picture that is consistent and combined from different data sources.
  - Providing data in different aggregations over the time dimension.
  - Providing correctness and comparability. Results of reports that are performed on the basis of an analytical system must compare to values that can be retrieved from other systems (for example, directly from the operational system).
  - Each value must fall into a predefined list of categories.
- Advanced analytics goes beyond simple descriptive reports and data warehouse topics, as follows:
  - For analytics, a sufficient data quantity is needed to create representative results.
  - In predictive modeling, a set of significant predictor variables is needed to predict the target variable.
  - In time series modeling, data in the right granularity for each forecast dimension need to be available.
  - Implicit missing values like numeric codes that represent a missing value are expressed as explicit missing values (null values or missing values).
  - The distribution of the values in the analysis need to be transformed to conform to the requirements of some analytical methods.
  - Observations with missing values cannot be used for many analytical methods.

Similar to the situation of operational systems and analytical systems, the intention is not to separate analytical methods from data warehouse systems. Data warehouse systems are, moreover, often the basis for analytical analyses and, thus, rely on the data quality in these systems. **This book mainly focuses on data quality for advanced analytics.** However, there are many topics that are relevant to data warehousing, reporting, and analytics together.

Focus on analytics

This book focuses on using data for analytical methods like regression analysis, cluster analysis, survival analysis, and time series forecasting as well as sample extrapolations. The next chapter discusses data quality features based on analytical methods. These considerations are relevant to widespread applications, like clinical research, demographics and public statistics, market survey analysis and market research, data mining for risk analysis and customer behavior analysis, and demand forecasting.

The possible analytical data quality topics, however, are so broad that including and discussing them in a single book would extend the scope. This book, therefore, discusses data quality for analytics from a general perspective and focuses on the most prominent areas in business analytics: data mining, statistics, and time series forecasting. Many of the results, however, can be transferred and applied in other analytical domains.

As previously mentioned, analytical methods have different, often higher, requirements on data quality. However, analytical methods **do not only pose requirements** on data quality. In many cases, analytical methods help you to **solve data quality issues**. Examples include imputing missing values and detecting outliers.

Data quality with analytics

Analytic methods demand a higher degree of data quality and pose additional requirements on the data for the analysis. However, they also offer methods for data profiling that allow better insight into the data quality status and ways to improve it. Here are some examples:

- Outliers can be detected with statistical measures, and a most probable and most plausible value can be calculated. These methods not only include the definition of upper and lower limits. They also include the calculation of individual validation values based on the individually expected value for each record.
- Missing values can be imputed with methods that range from simple mean imputation to complex and individual imputations that are based on predictive models.
- Distributions can be transformed into a more appropriate shape using mathematical formulas and statistical measures.
- The methods for matching and de-duplication of records are based on analytical methods that describe the similarity and closeness of records, which are then used to decide which records in the databases can be combined.

2.3 Ten Percent Missing Values in Date of Birth Variable: An Example

General

To illustrate the different analysis areas and data origins that were discussed in the last section, this section offers an example based on the date of birth variable of customers in a database.

Operational system

The operational system (for example, the customer database for retail or banking customers) gets its date of birth data directly from the business process where new customers are entered into the system. They are entered into the system, for example, when they subscribe to a purchasing card or when they open an account. The date of birth variable may be missing for some customers because it has not been re-queried by the system, provided by the customer, or entered into the system yet.

The variable can be added to the system at a later stage (for example, in the call center) when the call center agent asks the customer on the phone, "May I ask you for your date of birth?" Or this information is added during voluntary data cleaning campaigns, where the customer can complete his or her data to get additional benefits or discounts.

In the operational system, the completeness of the date of birth data is needed to serve the customer better (for example, sending him birthday congratulations). If the operational process requires a valid value for the date of birth (for example, if certain products are only available at a certain age), it can be assumed that values are available for (almost) all customers.

Systematic missing values

If 10% of missing data of birth values only occur in a certain subgroup (for example, a certain region or product group), the values are systematically missing. For the operational process, there is no difference in the treatment of a random or systematic missing value. In the analytical area, however, the systematic missing values cannot just be replaced with a value that is calculated from the distribution of the existing values, as this would ignore the systematic behind the missing pattern.

Missing values can also occur when a sales agent does not find an existing customer record in the database and enters a new (duplicate) record. He intentionally leaves the date of birth field (and many other fields) blank because the data are available in the database. Imputing the missing records in this case causes an error in the data because the correct values are already available in the database.

Data warehousing

The data warehouse takes the data from the customer table in the operational system. If possible, the missing values are corrected in the data warehouse. Here are some examples:

- If the date of birth values for some customers have been unintentionally overwritten, older version of the data can be used to restore the value (for example, backups).
- Other data sources are available where the correct date of birth value can be retrieved. This table is included into the data warehouse and used to correct the data.
- In the previous subsection, duplicate records are inserted into the data, and the date of birth fields are not populated. The de-duplication process in a data warehouse can correct this data issue.

Usability for analytics

Deciding whether 10% of missing values are acceptable for analytical methods can only be answered in a domain-specific context.

- If a report for the age distribution of the customers will be created, the descriptive statistics in the report are based on the existing values. The report probably also includes information about the number or percentage of missing values. The fact that some values are missing, however, does not influence the fact that the report can be created.
    - Figure 2.1 shows a histogram with the frequency distribution of age as created in the variables explorer in SAS Enterprise Miner. Note that there is a separate bar for the missing values at the left side of the *x*-axis.

Figure 2.1: Frequency distribution of age

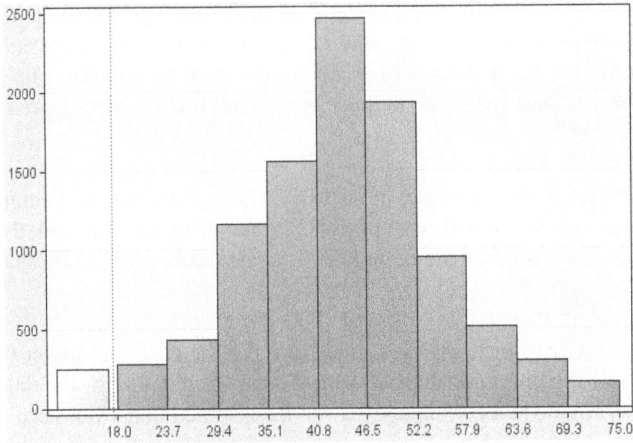

- In statistical methods like regression analysis or cluster analysis, however, observations that have a missing value for the analysis variable cannot be used. Those observations are either excluded from the analysis or their value is imputed by a missing value imputation technique. Whether missing values are imputed or not depends on the percentage of missing values and on domain specifics.
    - Even if replacing missing values is possible from a domain-specific point of view and it can be assumed that the values are missing at random, the percentage of missing values still determines whether you can expect reliable results. If only 10% of the cases will be replaced, you could assume that it is possible to achieve meaningful imputations from the other 90 % of existing

values. However, what happens if 30% or 50% of the values are missing? Chapter 18 shows a simulation study for predictive modeling in this context.

- o The 10% of missing age values in the clinical trial example can be a severe problem for the analysis. In clinical trials, you usually cannot impute missing values in order to use the records for the analysis. If you want to compare the trial outcome by the age class, 10% of the records cannot be used for the analysis. If a minimum sample size has been calculated, the inability to use 10% of the cases increases the risk that a difference cannot be verified statistically.

Conclusion

This example illustrates the consequences and features of different analysis areas and data origins for an example data quality feature—the fact that 10% of the date of birth values are missing.

2.4 Importance of Data Quality for Analytics

Data are becoming a more and more important asset in the information-driven world. Data are encountered in every field of daily life:

- Mileage information on the car display combined with an estimated drivable distance with our current tank filling schedule
- Number of available theater seats per category when booking tickets online
- Pulse, energy consumption, and blood pressure values on an electronic sports wristwatch
- Usage and last visit statistics on our favorite websites and portals
- Weather forecasts on television when the actual and forecasted temperatures and wind speeds are presented

This list can be easily extended to a long itemization of examples where data, and analyses on that data, influence our business and personal lives.

Data have not only become an important asset for us. They have become extremely influential in the life of an individual. Decisions are not only based on our experience and knowledge but also on facts about the past and forecasts about the future.

As data exert more influence on individuals, organizations, and businesses, a stronger dependence on the quality of the data is implied. Biases, deviations, and unavailability in the data affect our lives and our decisions. In general, the better the data, the better the decisions we can make. The adage "Garbage In, Garbage Out," which describes the fact that the outgoing results can only be as good as the incoming data, has been around for many years, but it has not lost its relevancy.

While the term *data quality* implies a technical topic perhaps best addressed by data engineers, data warehouse programmers, statisticians, and analysts, the importance of data quality nowadays extends well beyond this group. Businesspeople and information consumers understand that the validity of their results largely depends on the quality of their data.

Chapter 9, "Process Considerations for Data Quality," discusses this topic in more detail.

2.5 Definition of Data Quality for Analytics

General

This section defines data quality for analytics for the context of this book. Depending on the focus on different areas of data retrieval and analysis, there are several potential definitions of this term. The definition provided here aligns with the topics and data quality features that are discussed in this book.

Definition

Data quality in the context of this book is the degree of excellence of data to precisely and comprehensively describe the practical situation of interest in an unbiased and complete way. The data must be appropriate to answer the business or functional question of interest without reducing the scope of the question and the applicability of the results. The data should be available, complete, correct, timely, sufficient, and stable.

The data should be suitable for the analytical methods that will be used to answer the business or functional question. The data, therefore, must comply with the analytical requirements of the respective methods. For predictive modeling and time series forecasting, the data should have predictive power with respect to the values that will be predicted or forecasted. (Note that the Wikipedia definition for the general term *data quality* is consistent with the definition presented here.)

2.6 Criteria for Good Data Quality: Examples

General

This section lists many criteria that define good data. The items on this list reflect the author's experience in this area and the feedback and discussions with other researchers who are involved in analytical projects.

This list is not complete (it can never be 100% complete). But it does contain a good overview of criteria that are categorized into different groups. Not all criteria are of importance for every analysis question. For each analysis, a subset of these criteria is relevant. This list, however, serves as a good overview.

Data and measurement gathering

- Data are retrieved from a trusted source.
- Throughout the business process, there is no chance to amend the data without recording or tracing.
- The data's origin and correctness can be traced back in the source systems.
- The process of data collection is standardized over different data retrieval units.
- The data entry and retrieval system forces the entry of relevant values and checks for typical meaningless entries to fill a field (for example, a date of birth of 1.1.1900).
- The data have integrity across sources and analysis levels.
- Data retrieval takes part in a restrictive process. Unlike the analysis area, where creativity in the performance of different approaches is key, there should only be a single way to retrieve and enter data.
- When data are provided for the analysis, the way data are technically preprocessed by the data entry and retrieval systems (for example, the replacement of missing values and the treatment of outliers) is documented and known.
- Changes over time in the data collection process are known, for example:
  o Changes in the definition of fields.
  o Date when an attribute has first been recorded in the system.
- If only a sample of the data is available, the sample has been drawn randomly or the systematic method of data selection is known.

Plausibility check: Relevancy

- Data (for example, master data, transactional data) match the experience of a business expert in terms of distribution, number of missing values, outliers, and anomalies. This business expert should be a person who knows the data, the business process, and the data collection process over time.
- The business process of data retrieval, as well as potential psychological aspects in data collection that influence how data represent the truth, are known.

- Data, and the resulting patterns that can be derived from the data, are consistent with the subjective perception of business experts, comparisons with similar data sources, industry averages, and expert knowledge.
- The correlations, findings, and dependencies that are found in the modeling phase can be explained from a business point of view and withstand a plausibility check.
- Data are complete in terms of missing values and a track record of relevant business that may influence the data is available.
- Data are not only available on an aggregated level, but they are also available on a detailed level (for example, per customer, region, division, production group, or time interval).
- Transactional data and master data are available in sufficient quantities.
- The data reflect the actual state of the world they describe.
- The data contain the potential to produce the desired information.

Correctness

- The data are correct and precise.
- Deviation of the data does not exceed a certain threshold.
- There are no duplicate data.
- The data hold against plausibility checks on a record level (for example, if the number of children equals 2, the binary variable "ChildrenYesNo" cannot be "NO").
- Errors in the data are negligible or random and do not have a strong impact on the outcome.
- The data do not contain systematic errors.
- In parent/child relationships in the data, redundancies and missing records are eliminated and there are no contradictions between the values in the parent and the child record; for example, the open date of an account must not be older than the start date of the respective customer.

Missing values

- There are no missing values or the missing values do not exceed a certain threshold.
- There are no systematic missing values.
- Missing values can be explained.

Definitions and alignment

- There is a clear definition of the meaning of variables and the specification of units (values in $ or in 1,000€).
- Categories and groups are well-defined.
- Data to be compared are available in the same units (for example, currencies are converted to a single currency unit).
- Values that change over time have a valid from and until timestamp.
- The data cover the time period of interest.
- The data can be aligned with other data sources.

Timeliness

- The data are available in time for the analysis.
- The results can be created when they are needed.
- The data are available and the quality is checked in granular time units.

- The data picture or a certain interval or point in time is invariant against the date/time of data extraction. A consistent picture for a certain period can be retrieved, irrespectively when the data is extracted for that period.

Adequacy for analytics
- Patterns in the data can be detected and transferred to analytical models.
- There is sufficient data quantity: observations, history, and events.
- From an analytical point of view, the data are of good quality, if it is possible to build reliable models on them that predict and forecast future events and values. This can be tested by validating with past observations.

Legal considerations
- The data follow legal guidelines.
- The data are not sensitive and do not violate data privacy considerations.

2.7 Conclusion

General
This chapter scopes out the topic of data quality for analytics and defines the term within the context of this book. It has also been shown that data quality is a very heterogeneous and widespread topic, depending on the data retrieval area and the analytical domain.

The first chapter addressed the criteria for data quality on the basis of case studies. The previous section in this chapter presented example criteria for good data quality. The next chapters discuss various data quality criteria in detail.

Upcoming chapters
The criteria for data quality are discussed in the next seven chapters:

- Chapter 3 deals with **data availability**, where the general existence of suitable data is discussed.
- Chapter 4 focuses on **data quantity**, a topic of importance for any analysis but particularly for analytical methods.
- Chapter 5 discusses **data completeness** and gives examples of random and systematic missing values.
- Chapter 6 gives examples for **data correction**. Again, random and systematic errors are discussed.
- Chapter 7 focuses entirely on **predictive modeling**, a very important topic in analytics. Data requirements and the consequences on the results are shown.
- Chapter 8 shows a range of **specific analytical methods** that are relevant when you are deciding about the suitability of the data.
- Finally, chapter 9 discusses data quality for analytics from a **process point of view** and addresses the consequences of data quality features on the analytic process.

Chapter 3: Data Availability

3.1 Introduction **32**

3.2 General Considerations **32**
  Reasons for availability 32
  Definition of data availability 32
  Availability and usability 32
  Effort to make data available 33
  Dependence on the operational process 33
  Availability and alignment in the time dimension 34

3.3 Availability of Historic Data **34**
  Categorization and examples of historic data 34
  The length of the history 35
  Customer event histories 35
  Operational systems and analytical systems 35

3.4 Historic Snapshot of the Data **36**
  More than historic data 36
  Confusion in definitions 36
  Example of a historic snapshot in predictive modeling 37
  Comparing models from different time periods 37
  Effort to retrieve historic snapshots 38
  Example of historic snapshots in time series forecasting 38

3.5 Periodic Availability and Actuality **39**
  Periodic availability 39
  Actuality 40

3.6 Granularity of Data **40**
  General 40
  Definition of requirements 40

3.7 Format and Content of Variables **40**
  General 40
  Main groups of variable formats for the analysis 41
  Considerations for the usability of data 41
  Typical data cleaning steps 42

3.8 Available Data Format and Data Structure **42**
  General 42
  Non-electronic format of data 42
  Levels of complexity for electronically available data 42
  Availability in a complex logical structure 43

3.9 Available Data with Different Meanings **44**
  Problem definition 44
  Example 44
  Consequences 45

3.10 Conclusion **45**

3.1 Introduction

This chapter deals with the availability of data for the analysis. Obviously, analyses can only be performed if the required data exist. This chapter discusses different types of availability and the usability of data.

This includes:

- the availability of current and historic data
- the ability to provide the same structure of data in future time periods (periodic availability)
- the availability of data in the correct or appropriate format

3.2 General Considerations

Reasons for availability

There are many reasons why data are available. The most important include the following:

- Data have been **collected specifically for the analysis** that will be performed (for example, a controlled clinical trial).
- Data have been **observed** on the analysis subjects and collected for the respective analysis or for another analysis, like the number of large companies that went bankrupt over the last 30 years, detailed per geographic region and industry.
- Data have been collected to serve as a generic database for analyses from a specific domain. Examples are:
  - Demographic data that are provided by statistical offices per country.
  - Web data that are available throughout the Internet and on social media platforms.
- Data have been collected as part of the operational process to run the business, such as billing and stock keeping in a retail company.
- Data have been manually documented in paper files, such as the contract specifics of service agreements between medical doctors and insurance companies.

Definition of data availability

Data availability refers to the data needed to answer business or functional questions. This assumes that relevant data exist. Data are considered to be available if:

- They have been collected and stored in electronic systems, written down manually, or documented in log files and transactional histories.
- It is technically possible and legally and ethically permitted to access and process the data.

Availability and usability

This definition not only includes considerations that the data can be made available technically. It also includes the fact that these data are **relevant** to address specific business or functional questions.

Otherwise, any data collected from anywhere could be considered as available data for an analysis. This would mean that there is an infinite amount of data available for any analysis.

From a practical and domain-specific point of view, it is important to require that the data also be **usable for the analysis**. *Usable* means that the data can contribute relevant content to the business or functional questions.

The following list gives some examples of available but non-usable or irrelevant data:

- In a customer retention analysis project, regional meteorological data are available for the day when each customer joined the company. It is difficult to imagine that the weather on the day that a customer joined the company would affect the customer's cancellation behavior. These data could not be included in a meaningful analysis. Even if a correlation is found here, it may be assumed that this relationship is driven by other co-variables (for example, the calendar month when the customer joined and the fact that different seasonal campaigns attract different cohorts of customers with different behavior).
- When scoring applications for private loans, a bank agent knows the color and kind of shoes the applicant wears. Even if it could be proven that there is a statistical relationship between different types of shoes (for example, sport shoes and leather shoes), it is not practical to include this feature in the approval process.

In contrast, the following list shows cases where necessary data are not available for an analysis:

- To study sales demand patterns over time, the retail price and the time periods when marketing campaigns were run would be needed.
- In analyzing internal fraud at an electronic retailer, the events of internal theft by employees cannot be assigned to a certain day, but only to longer periods. Thus it is not possible to combine the theft data with the service schedule of the employees to create a list of suspicious persons.

Effort to make data available

In many cases, data availability is not just a yes or no issue. It is also a matter of effort to retrieve the data. For example, the data might be physically available, such as handwritten data in ring binders or data stored in a certain data processing system, but transferring the data into an electronic format in an analyzable structure might require much effort.

Consider the following examples:

- Historic demand data are technically available, but the system is no longer in use. In order to retrieve the data, the system must be re-launched and the data have to be exported as a full database dump. Even when these files are then transferred to the analysis platform, significant effort is needed to decode the logical structure of the data and transfer them into an analysis table.
- A similar example is the following: Point of sale data for the history months 13 to 24 are available weekly in a system that can be directly accessed. If there is a need for the transactional data itself, these data can also be made available; however, they first need to be imported from external backup storage. Both data sets are technically available; however, their availability has different timeframes.
- In retention analysis of customers in a telecommunications company, specific input variables about the call partners for each line must be made available. It is technically possible to query for each line the IDs of the top-10 call partners and use, for example, their mobile phone provider as a categorical input variable or to perform network analysis on these data. However, this results in considerable processing effort to calculate and refresh on a regular basis for thousands of customers.

Dependence on the operational process

Data availability also has a strong dependence on the operational process itself. If data are not stored in the operational system, these data are consequently not available for analytic purposes. In many cases, the demand of data availability for analysis purposes is not anticipated during the design of the operational system.

Consider the example of credit applications. The credit applicants' data are entered into the system to judge the applicants' credit worthiness using a scorecard rule. If the application is accepted, the data are consequently used for further processing of the credit contract. If the application is rejected, it is very likely that the data are not electronically stored.

Availability and alignment in the time dimension

Availability of data cannot only be seen as a static fact; it also has a time dimension. Therefore, availability has to be considered from different points in time.

- **Historic data** are data that are not only available for the current period but also for historic periods, like the number of transactions for a credit card account for the last 12 months. These data can, for example, be used in time series analysis or in comparing baseline data with data after the application of a certain intervention. Examples for such an intervention include marketing campaigns or medical treatments.
- **Historic snapshots of data** extend the concept of historic data. Not only are the data from previous periods available but the snapshot of the data at a point in time in the past is also available. Whereas historic data are mostly important for transactional data, the concept of historic snapshots is particularly important for base data like the status of the analysis subjects (tariff plan, number of products) or accumulated variables like the balance at the end of the previous year.
- **Periodic (future) availability of data**. If an analytical model is based on certain data or definitions of data, the data have to be made available also in future periods in order to be able to apply to the model.
- **Alignment of data**. The requirement of historic data can also be extended to the need to have data available that can be aligned to each other.
    - The series must cover the time period in question.
    - Historic data that are collected from different domains must have time stamps that allow aligning the available information together.

These terms are discussed in more detail in the following sections of this chapter. Chapter 7, "Predictive Modeling," also discusses an additional requirement of data availability. It outlines how training data for a model need to be available for the past, and data for the scoring, the application of the model, need to be available at the time of scoring (mostly in future periods).

The availability of data from different periods is also referred to as "timeliness" (compare [4] Gloskin 2009). Timeliness also includes the fact that for each observation a timestamp is available to help classify each observation and data value to a point in time.

3.3 Availability of Historic Data

Categorization and examples of historic data

Many analyses are not only based on actual data; they also require data from previous periods in order to answer business questions.

The requirement for historic data can be differentiated into two areas:

- **Transaction data**, where the value not only for the current period but also for historic periods is needed.
    - The most typical example of this is time series analysis. Here data values for each period are used to study the behavior over time. The data values can be quantities like the number of persons or the number of sold items. It can also be amounts like sales amounts or econometric data.
    - The demand for historic data, however, does not only come from time series analyses where data for many points in time are analyzed. It applies also to the case where data from the previous period is compared with data from the actual period in order to detect changes over time.
- **Base data** that have a valid-from and valid-to time stamp that indicates when the respective value has been the actual value.
    - An example is the previous address, bank connection, or Internet provider that a customer used.
    - Also, the customer's previous behavior segment, which may be the result of a data mining model, is a value of historic data.

- Analysis subjects that are no longer part of the analysis database; for example, customers that have already left the company. Depending on the operational system, the records for these customers are either deleted, transferred to a different table, or kept in the database and set to inactive.

The length of the history

The crucial point with historic data is the length of the history that can be provided. Many data are historically available, however, only for a limited time. Reasons for this fact include the following:

- **The storage amount**. Storing the full set of transactional point of sale data for a worldwide retail chain over several years results in an immense amount of data. In this case, the data are only kept for a limited period on the most detailed level. For older periods, data are accumulated to a sales-case, daily, or weekly aggregation.

- **Change of the IT system**. If over time the IT system is replaced by a new one, usually not all (historic) data from the old system are transferred. In most cases only data needed to proceed with the business operations are kept in the system. Thus it is often the case that historic data can only be found in the "old" IT system, and it is in a format that can be extracted and used only with a lot of effort.

- **Start of the electronic collection of the data**. In many cases, the limitation of the historic data is the start of data collection in the electronic system. Older data are very often only available in paper documentation.

- **Change of the underlying definition of data**. Sometimes no technical reason interrupts the availability of the data, but the definition of the underlying features changes, like the definition of the unemployment rate or the change from counting on-going sales to counting only closed sales. Consistent definitions of the data series are required for a meaningful analysis.

Note that the available data history must also cover the time period of interest. This includes the available length of historic data, but it also must not end too early and exclude more recent periods.

Customer event histories

The availability of historic data is also important if customer event histories will be analyzed. Analyses like survival mining, event history analysis, and time-to-event modeling are performed in many industries.

In addition to a proper business definition of events and steps in data preparation, these analyses also require the availability of historic data and stability of the data definitions.

Operational systems and analytical systems

Another point to discuss is the paradigm of an operational system in general. The demand for a long history of the data mostly comes from the analysis, not from the operational area. In order to perform business operations, a system does not need a history of a couple of years or more. In many cases, operational systems only need a very short history. They focus much more on the actuality and availability of the data of the current billing period.

Operational systems are built to perform and assess operational processes. Availability requirements on data for analyses (for example, the availability of history) are usually covered by analytical systems like data warehouses. In these systems, not only the actual version for the operational process is covered but also the historic data and historic versions of the data.

The following are important characteristics of data warehouse systems:

- Data warehouse systems handle time histories. They provide historic information so that the information consumer can retrieve the status of the data at a certain historic point in time. As previously mentioned, operational systems deal more with the actual version of the data.

- For example, customers who canceled their contracts with the company two years ago no longer appear in the operational system. The customers and their related data, however, are still available in a data warehouse (with a record-actuality indicator).
  - Data warehouses also track changes of subject attributes over time and allow the creation of a snapshot of the data for a given date in the past.
  - Data warehouse environments provide aggregations of several measures in predefined time intervals. Depending on the data warehouse design and the business requirement, the time granularity is monthly, weekly, or daily. Also note that due to the large space requirements, the underlying detail records are seldom maintained in the data warehouse for a long time in online mode. Older versions of the data might exist on external storage media.
- Data warehouse systems extend the time granularity. Billing systems create balances per customer and their individual bill cycle. The bill cycles, however, often do not coincide with calendar months. Therefore, sums from the bills cannot be used to exactly describe the customer's monthly usage. Data warehouses store additional information in the billing data, including monthly usage by calendar months, weeks, and days. The granularity itself depends on the data warehouse design.

For more details, see [1] *Data Preparation for Analytics*, chapter 3.

3.4 Historic Snapshot of the Data

More than historic data

For many analyses, however, it is not enough to only have the historic transactional data (for example, data aggregated to monthly buckets) available. Also, a historic snapshot of the data is needed for analysis:

- The historic snapshot of the master data for a certain period in the past. In order to relate the behavior of a customer in previous periods to his or her product portfolio or contract status, the respective data version at that time is needed.
- The balance of loyalty points for a retail customer is needed for each date of a visit to a branch. This value can be calculated backward from the transactional history.
- In case study 3 in chapter 1, the need for data on the number of shops that offered a certain product in each month of the last year was mentioned.
  - In many real-life situations, it may be possible to extract the number of shops that are offering a certain product currently. This information may be stored in the operational system because the definition of the product assortment is maintained in this system.
  - The operational system, however, usually does not store the number of shop assortments of the last months and earlier months because it is designed to support current business operations.
  - As long as there is no data management process that captures the assortment for the shops on a monthly basis and stores it in a separate table with a time dimension, the data will not be available in future periods.

Confusion in definitions

The two terms *historic data* and *historic snapshots* are often used interchangeably, which causes a lot of confusion in formulating data requests and explaining the requirements for analyses, especially for analytics.

There is a significant difference between historic data (like time series) and a historic snapshot of data (a single observation). It may cause a lot of confusion if analysts require data for analytics and want the historic version of the data, but they formulate their data requests for "historic data." Thus, you should clearly specify the data needs in the early stages of a project, especially with peers in the projects who have not yet provided data for analytical projects.

Example of a historic snapshot in predictive modeling

Examples of historic data are widespread. Especially in the most prominent data mining method, predictive modeling, the need for historic snapshots of the data is a central requirement. Consider the following example:

- An insurance company wants to analyze and predict the claim risk in car insurance for its customers.
- The analysis will show whether there is a correlation between the fact that a customer has a claim event and his personal data.
- The claim events were collected for a calendar year (year 2) and will now be analyzed together with the customer data. This is also called the target window.
    - From both a business and an intuitive point of view, however, it does not make sense to use the version of the customer data that will be used as explanatory data from the same year; the data might already be influenced by the claim event.
    - In order to logically follow the causality that certain customer attributes may lead to an event, the customer attributes from before the claim event are needed.
    - Also, in the application of the model, the input variables will be used to generate a score for future periods, not for the current periods. Thus, it is necessary to train the model with data from different time periods.
- Thus, the customer data can only be taken from a year earlier than the claim events. In our example this is year 1 (observation window).

Figure 3.1 illustrates this case. Here the model itself can be built in year 3 when the information about the target events is available.

Figure 3.1: Versions of customer data and claim events on the time axis without a separate period for model validation

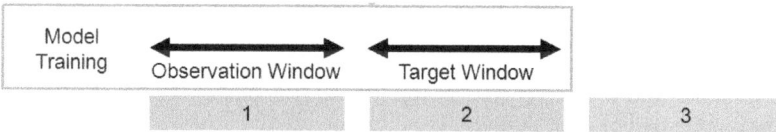

Comparing models from different time periods

Additionally, often a model will not only be trained and validated on one period, but it will be trained on at least two periods in order to be available to validate the robustness of the model logic on new data.

- In this case, the first analysis is performed with customer data from year 1 and predicts the claim events in year 2.
- Then the model and its logic are validated on customer data from year 2 to predict claim events in year 3.

Figure 3.2: Creation of data marts for different points in time

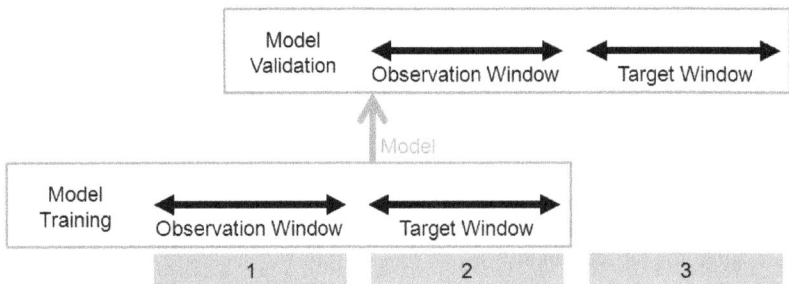

Here an analysis data mart for model training with input data from year 1 and target data from year 2 is built. The resulting model is then validated on data where the input data are taken from year 2 and the target data are taken from year 3.

Effort to retrieve historic snapshots

Providing a historic version of the data, however, is not always easy. From a business point of view, in the operational systems of the insurance company, it is important to have the actual contract status and bonus level in order to issue the correct premium notifications. Historic values of the variables are usually not stored if they are not needed for business purposes or processes.

Historic versions of the data are usually only available if the data already have been captured into a data warehouse, where for each attribute the time dimension is also considered. Historic versions of the data can then be retrieved from the warehouse. If no data warehouse exists, the only way to retrieve this older version is to access historic backup copies of the data. This process involves a lot of work and cooperation with IT departments, but depending on the importance of the analysis, the effort may be justifiable.

Even if historic versions of the relevant data are not available yet, capturing of these historic versions shall at least start now in order to have them in the future. Even if no data warehouse is in place it is still worthwhile to save a full copy of the customer data at the end of each period.

Example of historic snapshots in time series forecasting

For typical time series forecasting, it is sufficient to have historic data, like the sales quantities for each product for the last 36 months. This does not involve a historic snapshot of the data and can directly be calculated from transactional data.

However, consider the demand forecasting of the number of cars for a car rental service on a daily basis. For each day in the past, the number of rented cars can be calculated from the system. If the model, however, will also include how many bookings for a certain day have been made in the system one day before the date, two days before the date, and so on, then a historic snapshot is needed.

In this case, the model will include the already known demand for a certain day in the forecast. To create this input variable, it is not sufficient to aggregate historic data; a snapshot (or the ability to recalculate it) has to be available for each day, which includes the number of bookings for future calendar days. Table 3.1 illustrates this situation with example data.

Table 3.1: Example data with booking status for selected days

	January					
	22	23	24	25	26	27
Rented Cars	17730	17618	18959	16708	17899	16855
Bookings per Jan 21st	17729	17617	18912	16666	17619	16509
Bookings per Jan 22nd		17616	18853	16511	17675	16513
Bookings per Jan 23rd			18419	16512	17678	16707
Bookings per Jan 24th				16510	17727	16709
Bookings per Jan 25th					17728	16881
Bookings per Jan 26th						16843

The first row of the table, Rented Cars, can be created from historic data later. The booking status per each previous data for a target day, however, needs to be saved in the system for later use. Otherwise, the data will not be available.

3.5 Periodic Availability and Actuality

Periodic availability

Another aspect for availability of the time scale is periodic availability. Different from historic snapshots, periodic availability is forward-looking and addresses whether data that can be used for analysis today will also be available in the future.

For meaningful analysis, the data series must be consistent in meaning (what each variable represents), frequency of collection, granularity (at the appropriate level), and for other dimensions as needed.

If an analysis will only be run once, not much care has to be taken with the periodic availability of data. All data that are currently available can be used in this case. If, however, the analysis will be repeated over time and produce comparable results, you should consider whether the data can also be made available in later periods.

Examples are:

- Longitudinal analyses where certain statistics will be compared over time. In this case, make sure that the statistic is defined in the same way and that the data to calculate this statistic are available over time. In the case of demographic measures, for example, where ratios are calculated by dividing statistic A through statistic B, it is important to get periodic updates on both variables (for example, the number of available kindergarten spaces per married couple). It is not enough to provide the number of kindergarten spaces every year, but also an actual number of married couples.

- Statistical analyses that compare the product usage of customers with information that is gathered through market surveys. In this case, the product portfolio usage can be updated easily by accessing the respective data from the billing system. The market survey, however, might not be updated in regular intervals. In this case, actual customer behavior might be compared with two-year-old market survey data.

- Predictive models are usually built on training data and provide scoring logic. Based on the scoring logic, fresh data (for example, in future periods) will be scored and predictions for event probabilities, the time until events, or the size of values itself will be calculated.
    o In the case of a predictive model for the cancellation behavior of mobile phone customers, using a variable describing the newness of the actual handset is often a good predictor. This variable is also often called *handset grading*.
    o If, during analysis, a grading for each handset is performed that is available on the market, the respective variable usually has good predictive power in the analysis.

- If the handset grading, however, is not actualized in future periods, the predictive power of the variable decreases and the model provides inexact predictions. The variable needs to be actualized regularly because, over time, new handset types appear on the market that need to be graded and many handsets move down in the grading as newer, more trendy handsets appear in the market.

Actuality

Availability of data needs to be put into the context of time. The fact that data can be made available is, in some cases, not enough to be able to use them for analysis.

If data will be refreshed daily in order to calculate aggregated reports for the past or forecasts for the future, and these reports are needed on a daily basis, it is insufficient if the data retrieval and preparation takes longer than the length of the decision time window. If a decision on the data needs to be made the next day at 9 a.m., the preparation of the data, including updating the last day, must not exceed 9 hours.

3.6 Granularity of Data

General

The mere fact that data are available for a certain topic or domain may not be enough to perform the desired analysis on them. If the data contain information only on an aggregated level and the analytical method needs the data on an individual analysis subject level, the analysis may not be performed as desired.

For some descriptive analyses, pre-aggregated data are sufficient. But for most analytical methods, like statistics, data mining, or time series forecasting, expect the data on a detailed level.

Consider the following examples:

- In customer behavior analysis, the customers' usage of the product and service offering is studied. If the usage data are only available on an aggregated level over all customers, you cannot analyze the individual customer behavior. Even if the usage information is only available as aggregates for each customer segment, only the segments can be compared. Analysis on an individual customer level is hardly possible because all customers in the same segment will have the same value.
- If time series models will be built weekly on a product and regional level, the data must at least have the granularity week, product, and region. If one of these dimensions is aggregated on a higher level, the forecasts cannot be detailed as required.
- In a clinical trial for safety analysis, it is insufficient to know which side effects happened on an aggregated treatment group level. In the safety analysis itself, a side effect often must be interpreted in the context of the individual patient data, such as age, gender, laboratory values, and vital signs.

Definition of requirements

Similar to the definition of data requests for historic snapshots, the precise communication of data requirements with respect to granularity is very important in projects. These examples show that simply requesting whether certain data are available may result in much too highly aggregated data. Thus, data requests should always include the expected granularity to make sure that the desired analyses can be performed.

3.7 Format and Content of Variables

General

As previously mentioned, the format in which data can be provided is also very important for availability. This reflects the requirement that data can be processed directly from their origin in steps like data entry, data extraction from other systems, data restructuring, and data preparation. The format should not pose substantial additional effort on the data analysis process.

This section focuses on the technical representation of the data in the format and content of the variables. The next section deals with the technical representation and the structure of the table itself.

Main groups of variable formats for the analysis

Besides structural and technical format considerations (discussed in the next section), you should determine whether the data are available in the form of numeric values and well-defined categories or whether they are only in textual form (free-form text).

Analysis systems usually divide input data into three main categories:

- **Interval data**, such as age, usage duration, and claim amount. Based on this data, calculations like means, standard deviations, and medians and analytical methods like multivariate statistics or time series forecasting methods are applied.
- **Categorical data**, such as gender, race, treatment, diagnosis, price plan, and segments. Based on this data, only statistics like absolute and relative frequencies or distinct counts can be performed.
- **Text data**. This includes textual data that are not categorized and cannot be used in the same way as categorical data. These data are in many cases free-form text like comments or documents. Compare also to the SAS Text Analytics offering.

Considerations for the usability of data

You cannot start the analysis if the data is in free-form text fields. This requires additional effort to structure the data and make them available for data analysis.

Consider the following examples:

- Numeric values for interval variables like age or height shall be in numeric format with its numeric values "34" or "142," rather than being quoted in text fields with values like "34 years" or "~142 cm."
    - If the data are provided in different units, the unit will be provided in a separate column.
    - Interval variables will be made available as numeric columns in order to allow calculations on them.
- Categorical values will be made available as well-defined and standardized categories.
    - The marital status, for example, will not be identified in different ways, like "married," "marr.," "divorced," "div," "unknown," "none," or "m."
    - A standardized lookup table will refer each value to a single category.
- Additional comments (like "married, 5 years") will not be part of the category. If this additional information is relevant, it will be included in a separate variable or optional text field.
    - This text field needs to be pre-processed before the analysis. If information from this text field is relevant for the analysis, an additional variable can be created. You need to extract this information (manually) any time the data are refreshed (see also section 3.5, "Periodic Availability and Actuality").
    - Data in a text field (like "married—divorced—married") that provide additional information, for example, on the sequences of the categories, need to be decoded and put into special variables or standardized sequence chains.
- Information that is taken from multiple-choice replies should not be stored in a single column but in a column for each reply option. For each reply option, a binary indicator yes/no or 0/1 is provided. For example, the information in a single column "1 + 2 + 5" for the answer categories 1, 2, and 5 cannot be processed directly.

If these guidelines are not observed, the underlying data are classified as bad quality data, even if they contain the necessary data in general.

Typical data cleaning steps

Typical data cleaning and preparation steps in this context often include the following:

- If interval variables are stored in a text field, free the numeric value from potential textual additions and comments. Convert the character value into a numeric value. If values are quoted in different units, recalculate the value to a common unit.
- For categorical data, create a frequency list of all existing values. Based on this list, standardize the data to well-defined groups.
- If categorical data are already stored in the form of numeric codes, the list of available numeric codes must compare to the lookup table of the codes.
- If data are provided in text fields that contain information that cannot be processed by a single variable, the information of this variable has to be split into a set of variables.
- If textual fields contain free-from text, like lists of codes, lists of categories, and textual comments, **Text Mining** helps process this text and derive the content for analysis. SAS Text Miner, for example, offers a method that is based on singular values. See chapter 14 for more details.

3.8 Available Data Format and Data Structure

General

This section focuses on the format and structure of the available data. This includes the technical format of the data, whether the data are available electronically, and how the data are structured.

Non-electronic format of data

In the following example, an analysis cannot be performed because part of the data are not available electronically:

- In a social insurance project, medical doctor invoices will be analyzed in order to find out whether they are unexpectedly high for certain doctors.
- For the analysis, the invoices that were sent to the social insurance for the last 12 months are needed. These data are electronically available because the approval process and the settlement of the invoices are done electronically.
- In order to explain different invoice behavior and amounts between the doctors, it is necessary to have data on the doctors' contract. For example, some doctors may have the allowance to invoice special treatments, after they have completed special diploma courses.
- Some of this master data, however, is not yet stored electronically. It is only available in paper copies of their contracts or special negotiations.
- Thus, even if the data are available (the information can be picked up by reading the documents), the documents are not directly available for data analysis and need to be entered into a database first.

Levels of complexity for electronically available data

One extreme is data that are not available in electronic format and need to be entered online, which was shown above. But even the fact that the data are available in electronic format does not necessarily mean that they can be exported directly into the analysis system. The following needs to be considered in this case:

- What is the **technical interface** that can be used to transport the data from the originating system to the analysis system?
    - Is it possible to transfer data via the network?
    - Can external devices be plugged into the system?
    - Can data be burned onto CDs or DVDs?

- o Can other external drives be connected via USB connections?
- What is the **physical format** that determines how data in an electronic format can be exported from the originating system?
  - o Can data only be stored as text or delimited files?
  - o Is it possible to save data in semi-structured or structured formats like Microsoft Excel or Microsoft Access?
  - o Is it possible to connect to the system via standard methods like ODBC, OLEDB, or native database access?

Consider the following cases that illustrate different levels of complexity to access data:

- Import data from a Microsoft Access database that can be accessed via OLEDB or that is exported from Access to a tab-delimited file. The underlying data structure here, as well as the access of data in a relational database in the form of rectangular database tables, will already have a structure that can be converted into analysis tables by merging or aggregating tables.
- Data access to a hierarchical database on a mainframe system. Data need to be stored as text files and transferred from the mainframe to the analysis platform. The data in the hierarchical database need to be structured to fit into the logic of the relational tables.
- Data access to a proprietary IT system with a proprietary database. Historic data need to be captured from a system that is not modern but inherited from the early 1990s. Here the technical access as well as the decoding of the data structure is a complex task.

Availability in a complex logical structure

Not all data are available in the logical structure that is needed for the analysis. If, for example, data are stored in a hierarchical database, they need to be transferred from the source system to the analysis system, the data structure needs to be decoded, and the data need to be restructured for the analysis.

The output dump of a hierarchical database is characterized by the fact that different rows can belong to different data objects and, therefore, can have different data structures. For example, in a single table, lines with customer data can be followed by lines for account data for that customer.

An example of a hierarchical text file is shown here. There are rows for CUSTOMER data, and rows for USAGE data. The CUSTOMER data rows hold the following attributes: CUSTID, BIRTH DATE, GENDER, and TARIFF. The USAGE data rows hold the following attributes: DATE, DURATION, NUMBEROFCALLS, and AMOUNT. At the first character, a flag is available that indicates whether the data come from the CUSTOMER or USAGE data. Some example lines are shown here:

```
C;31001;160570;MALE;STANDARD
U;010505;8874.3;440;32.34
U;020505;1887.3;640;31.34
U;030505;0;0;0
U;040505;0;0;0
C;31002;300748;FEMALE;ADVANCED
U;010505;2345;221;15.99
U;020505;1235;221;27.99
U;030505;1000.3;520;64.21
C;31003;310850;FEMALE;STANDARD
U;010505;1100.3;530;68.21
U;020505;5123;50;77.21
U;030505;1512;60;87.21
```

How the data will be made available for analysis can be seen, for example, in the de-normalized data structure in Table 3.2.

Table 3.2: Data in a de-normalized data structure

	CustID	BirthDate	Gender	Tariff	Date	Duration	NumberOfCalls	Amount
1	31001	16MAY1970	MALE	STANDARD	01MAY2005	8874.30	440	32.34
2	31001	16MAY1970	MALE	STANDARD	02MAY2005	1887.30	640	31.34
3	31001	16MAY1970	MALE	STANDARD	03MAY2005	0.00	0	0.00
4	31001	16MAY1970	MALE	STANDARD	04MAY2005	0.00	0	0.00
5	31002	30JUL1948	FEMALE	ADVANCED	01MAY2005	234.50	221	15.99
6	31002	30JUL1948	FEMALE	ADVANCED	02MAY2005	123.50	221	27.99
7	31002	30JUL1948	FEMALE	ADVANCED	03MAY2005	1000.30	520	64.21
8	31003	31AUG1950	FEMALE	STANDARD	01MAY2005	1100.30	530	68.21
9	31003	31AUG1950	FEMALE	STANDARD	02MAY2005	512.30	50	77.21
10	31003	31AUG1950	FEMALE	STANDARD	03MAY2005	151.20	60	87.21

Even more complex examples exist, where the system from which data will be loaded holds the data in a complex structure that is suitable only for this particular system. Examples are:

- Key value structures where for each attribute and attribute subtype a separate row exists in the data.
- Data structures where the data for one analysis subject extends over more than one line in the text file.

3.9 Available Data with Different Meanings

Problem definition

Data from the same database field that is retrieved from the system may not necessarily have the same content. Depending on the adherence to operational processes and policies, this interpretation of the meaning of the field, the level of detail, or the effort to provide the correct information may differ across people, organizational units, regions, and time periods.

If data are collected and entered, for example, in different organizational units, data that are found in the database for the same value sometimes have a different meaning. The examples for this problem are very widespread: In data entry itself, human errors occur. In addition, data entry is different between different people, and the way certain facts are entered into the system may vary greatly.

- Consider customers who apply for a loyalty card in sales branches. In many cases, there is a difference between branches in how detailed data are requested and entered into the system—for example, geographic region, marital status, and birth date.
- Different customer care or call center agents may apply a different level of detail in categorizing customer complaints when entering them into the system.

Example

An international bank has subsidiaries in 12 European countries. The local banks in each country are heterogeneous because they were acquired over time. From a technical point of view, the countries have different technical systems and IT platforms. The data that are stored in the systems are heterogeneous for the following reasons:

- Because of historic and cultural reasons and legislative guidelines, different products are being offered in different countries. This means the naming of the products is not homogenous across countries.
- There is no standardization of the categorization of the core banding products yet because the subsidiaries were acquired at different points in time.
- The logical structure of the systems in terms of a customer view and a contract view is different between countries. Some of the countries are mature and able to define different hierarchical levels between their entities, like household, person, and contract. Others can only see and define their data on a contract level.

- Also, in data entry and data manipulation screens that are standardized across countries (for example, for credit approval), a different data entry behavior can be found per country. This is due to different organizational habits per country.
- Data need to be imported from different technical systems and IT platforms.

Building a central data warehouse in this context is a complex conceptual challenge. A lot of decisions need to be made about how the data between countries will be translated, not from a human language point of view but to create a central standardized version of the data that can be analyzed.

Consequences

The consequence of heterogeneity between countries is that even when data are centralized into a single system, not all available attributes can be used for comparison between countries. In addition, not all attributes can be used to form a global view over the entire banking group.

This example shows that even if data are available in different countries, it does not mean that the data can be directly used in the analysis.

For some analyses, using dummy variables might be sufficient because they would account for or mitigate the different intercepts and/or slopes for the different countries.

3.10 Conclusion

This chapter looked at the first data quality feature covered in this book: data availability. Data availability is a very important feature because often projects cannot start until the required data for the analysis is available. This fact is also referred to as bad data quality in many cases.

The chapter has also shown that data availability is not only a matter of getting data for the actual period. Availability also extends to historic and future periods. Data also have to have the appropriate granularity. This is especially true for analytical methods, where it is a prerequisite. Finally, the format of the data from a technical perspective was discussed.

Chapter 4: Data Quantity

4.1 Introduction **47**
    Quantity versus quality 47
    Overview 48
4.2 Too Little or Too Much Data **48**
    Having not enough data 48
    Having too much data 49
    Having too many observations 49
    Having too many variables 49
4.3 Dimension of Analytical Data **50**
    General 50
    Number of observations 50
    Number of events 51
    Distribution of categorical values (rare classes) 51
    The number of variables 52
    Length of the time history 52
    Level of detail in forecast hierarchies 53
    Panel data sets and repeated measurement data sets 53
4.4 Sample Size Planning **53**
    Application of sample size planning 53
    Sample size calculation for data mining? 54
4.5 Effect of Missing Values on Data Quantity **55**
    General 55
    Problem description 55
    Calculation 55
    Summary 56
4.6 Conclusion **56**

4.1 Introduction

Quantity versus quality

It seems to be cumbersome that in a book about data quality a chapter on data quantity exists. In colloquial speech as well as in technical language, the two terms *quality* and *quantity* are often used as antagonism. For example, it is said, "In your work you deliver a lot of quantity, however, the quality is missing," which shall reflect that mere quantity cannot replace the quality of a work, product, or service.

In the case of data analysis and analytics it is obviously also true that the quality of the analysis data cannot be increased just by adding attributes that are meaningless for the analysis of the business question. For instance, the predictive power of a data mart does not increase just by adding hundreds of variables that are randomly generated or taken from data sources that have a completely different business context. Only when the additional data contain additionally relevant aspects a gain in outcome is achieved.

However, especially for analytical methods like predictive modeling or time series forecasting, a strong link between data quality and data quantity exists. The more data that can be gathered about the analysis subjects and the richer in information content is, the more precise the predictions and forecast and the more usable the results will be. Therefore, there is a strong relationship between the amount of data and the quality of the analysis, which directly reflects to data quality.

Many analytical methods pose requirements on the available quantity of data. In order to perform estimation of parameters and to allow stable predictions, a certain minimum amount of data must exist. Depending on the respective method, these requirements, for example, include the length of the time history, the number of observations, the number of analysis subjects, the number of events, as well as the number of available attributes per analysis subject.

In many situations statisticians are known to be "greedy" in terms of trying to gather a lot of data quantity for their analysis.

- One reason is that the mathematical and statistical requirements of the respective methods need to be fulfilled as many methods demand a minimum quantity of data in order to perform well.
- The second reason is that the chance to build a good model increases if more data are available, where "good model" is often defined by correct predictions and stable estimates or forecasts that are close to the actual values or statistically significant results.

Overview

This chapter shows that data quantity is also a data quality characteristic.

The term **data quantity** is defined in more detail and the data quantity requirements of typical analytical methods are discussed. This is based on the dimension of analytical data, which are shown as well.

4.2 Too Little or Too Much Data

Having not enough data

In most cases in analysis, the data quantity requirement refers to situations in which there is insufficient data. In many practical situations, the analyst is confronted with the fact that not enough data for the analysis are available. For example, more events are needed for predictive modeling to train a stable model, more test persons are needed to evaluate an experiment, or the available time history in time series forecasting needs to be longer.

Not having enough data is often strongly linked to the topic of the previous chapter, data availability. Additional effort might need to be invested to gather a longer history or more variables on the analysis subject. In some cases, the retrieval of additional data is only possible when the analysis is postponed and additional months of data are collected. Or, if in a controlled experiment, additional analysis subjects might need to be included (for example, in a clinical trial).

In predictive modeling, for example, methods like bootstrapping can help to get more stable models on data with fewer observations.

Having too much data

Having too much data, on the other hand, can also be a problem. For example, in the case of an analysis data mart for predictive modeling, the problem has two dimensions. There can be the case of:

- having too many observations
- having too many variables

The two dimensions are explained here.

Note that as computing power increases and analysis algorithms become more efficient, having too much data is less problematic, and it continually takes more data to be classified as "too much data." The high performance software recently introduced by SAS enables users to analyze all their data, not just a subset. In addition, the need for sampling has been reduced.

Users probably will still want to have a holdout sample to test fitted models and to avoid overfitting. In data mining terminology, fit, test, and validate data sets are still useful to ensure the model generalizes well for new observations.

Having too many observations

Having too many observations slows the performance of the analysis because too many observations need to be processed. Also, the analysis data mart consumes too much disk space. The problem of having too many observations does not occur with hundreds or thousands of observations but rather with millions of observations.

The problem of too many observations is usually overcome with sampling. The most prominent methods include **random sampling** or **stratified sampling**. In stratified sampling, the number of observations for one segment can be reduced without losing much content from the data.

For example, in the prediction of events, the proportion of cases with an event is usually lower than the cases without an event. Here the number of non-event cases is usually reduced in sampling because no additional information is retrieved for the model from the excess of non-event cases. If the original data have a low event rate (for example, around 1-%), a common practice is to oversample the event cases in the data and reduce the non-event cases to achieve a proportion of 5% of events in the data.

Having too many variables

Many analysis data marts have more variables than the number of variables in the final model should include. Some of these variables, for example, cannot contribute additional content to the model, as in predictive modeling where variables do not have predictive power. Other groups of variables have a strong correlation among each other, so the final set of variables will not contain too many variables that are highly correlated.

Having too many variables can become a data quality problem, if the number of candidate variables is so large that it is hard to select and verify a subset for the final analysis. This is the case, for example, in genetic analyses, where a huge number of features are measured per gene. In this case, classical statistical methods of analyzing the influence of possible input variables and variable selection methods reach their limits. Principal components analysis and more elaborate analytical methods like variable screening techniques are needed to cope with this high number of variables. Compare, for example, JMP Genetics or JMP Genomics for this case, as well as the GLMSELECT procedure in SAS/STAT or the LARS node in SAS Enterprise Miner. LARS here stands for Least Angle Regression. See also chapter 14.

4.3 Dimension of Analytical Data

General

Analytic data marts may have different structures. In many statistical analyses and in predictive modeling, the one-row-per-subject data mart is the typical structure. For time series forecasting, a time series data mart structure is typically used. And for association analysis or market basket analysis, a multiple-row-per-subject data mart is used. Compare [1] chapters 7–10.

These data mart structures have different dimensions that are of importance for the discussion of data quantity considerations. The most important dimension includes the following:

- Number of observations
- Number of events
- Distribution of categorical values (number of classes)
- Number of variables
- Length of time history
- Level of detail in forecasting hierarchies

In the following subsections, each dimension is defined and its data quantity requirements are discussed. The subsections only discuss the dimension. It is not possible to provide hard facts about the minimum data quantity here because this depends on the individual properties of the methods. In part III of this book, simulation studies are shown for predictive modeling and time series forecasting, which include results on the minimum number of observations and events and the length of the time history.

Number of observations

The number of observations usually refers to the number of analysis subjects that are available for the analysis. Examples include:

- The 40 patients who are included in a pharmaceutical study
- A calling list of 6,500 customers who are contacted via the outbound call center
- The customer base of 4,500,000 customers of a retail bank

Thus, the number of observations in most cases refers to the number of unique analysis subjects. In the case of a one-row-per-subject data mart, this is the number of rows in the data mart or analysis table.

The required number of observations depends on many factors. The following list is not complete, but it highlights the most important areas:

- The statistical method that is used for the analysis.
  - Simple statistical methods like descriptive statistics do not have a particular requirement on the number of observations needed (beside the fact that the variance can only be calculated when at least two observations are available). The criterion here is a decision on the generalizability of the results rather than a statistical criterion.
  - More advanced analytical methods require a minimum amount of observations in order to be able to estimate parameters and perform variable selections.
  - In hypothesis testing, for example, when two means will be compared, the minimum number of observations depends on the difference of the means, the standard deviation, and the level of the type I and type II errors.
  - Section 4.4, "Sample Size Planning," later in this chapter, discusses planning the appropriate sample size for typical statistical analyses, like the testing of hypotheses.

- The fact of whether the available data are split into a training, validation, and test data partition for the analysis. If such a split sample validation is performed not all observations can used for the model training analysis, but are reserved for validation purposes.
- The number of variables that are analyzed. The necessary number of observations increases with the number of variables that are analyzed.
    - If 200 variables are available in a data set for predictive modeling with 500 observations, it is hard to build a stable model, even if variable selection is performed beforehand. Although the selection of the variables is performed in a univariate way, the ability to identify the correct set of predictor variables when 200 yes/no decisions have to be made is relatively low.
    - Strictly speaking, it is not just the number of variables that is important but the number of degrees of freedom. An interval input variable consumes one degree of freedom; a categorical input variable with eight classes consumes 7 degrees of freedom.

Number of events

The number of events is a dimension that is specific to predictive modeling with a binary target. However, because this case is very frequent in data mining and statistical analysis, it is addressed here as an extra topic. Also note that the term **event** is also used in time series forecasting to flag a certain behavior in the time series data, such as a level shift or a peak like a promotion or a strike.

In event prediction, the number of observations alone is not enough to judge the quantity of the data.

Consider a case with 67,451 observations. Of that number, the target event is observed in only 134. In this case, the limiting number is not 67,000, but 134. A predictive model will only slightly change its predictive power if the number of the 67,417 is doubled or divided by two. The analytical methods "learn" the relationships from the events and non-events. Thus, an excessive amount of non-events does not contribute as much as an increase of the events in this example.

As mentioned earlier in this chapter, in predictive modeling, it is a common practice to oversample the events. This is done in order to receive a distribution of events versus non-events of, for example, 5:95 or 10:90 in the analysis sample, even if the true proportion of events in the original data is only 1%. Thus, some non-events are excluded from the analysis because they do not contribute much additional information.

The question about the necessary number of events in typical data mining analyses often arises. The necessary number of events depends on the same factors as for the number of observations. An approximate answer to this question is very often that at least 100 to 250 events should be available. More details are shown in the simulation studies in part III of this book.

Note also that in predictive modeling data are often split into training data and validation data. Thus, the effective number of observations with events for model training is reduced. If, for example, only 72 occurrences of a certain event have been collected and the data is split into training and validation data with the ratio of 2:1, only 48 events are available in the training partition.

Distribution of categorical values (rare classes)

Using categorical variables like region, price plan, and product hierarchy in the analysis as input variables can cause data quantity problems even if the number of observations seems to be sufficiently high.

Consider the case with a categorical variable with a high number of classes like the geographical region on a low aggregated postal code level or the product hierarchy on an item subgroup level with hundreds of different classes (for example, 262).

Using a regression model on this data means that for 261 of the classes (262-1), a regression coefficient has to be calculated. For some of the classes, there is a sufficient amount of observations that fall into the class; for other classes, only two or three observations are available. Estimating the parameters for these classes then gets very unstable and could even lead to non-estimable parameters. A frequent way to tackle this problem is to run a univariate analysis on this variable first in order to group the levels (or at least the levels with small observation numbers).

Even with variables that have fewer classes than the postal code or the item subgroup, this problem can occur. This is the case when interactions are analyzed between the categorical values.

Consider a case with two variables: age class with 8 values and price plan with 12 values. Both variables on their own do not seem to be problematic based on their number of classes. If, however, the effect of how age class differs over the regions will be analyzed, an interaction term has to be included in the analysis. This results in the creation of 96 buckets in a two-dimensional table and the consumption of 77 (=(8-1)x(12-1)) degrees of freedom for the analysis.

As mentioned earlier, each additional class in a categorical value consumes 1 degree of freedom. This means that a categorical variable with 12 classes consumes the same number of degrees of freedom as 11 interval variables together. Thus, the maximum number of categories has to be taken into account for data quantity considerations.

The number of variables

The number of variables is an important data quantity dimension. The more variables in the data shall be analyzed in a univariate or multivariate way, the more observations are needed. Details have already been shown in the subsections "Number of observations" and "Distribution of categorical values (rare classes)" above.

Length of the time history

Classical time series forecasting methods use historic data to discover trends, seasonal cycles, or other patterns in order to produce forecasts for the values in future periods. The data quantity aspect here applies primarily to the length of the time history in the data.

In order to discuss minimum requirements for data history, it is necessary to differentiate between the technical requirements to calculate a forecast and the business requirements to make a meaningful forecast.

Consider time series forecasting on monthly data and assume a seasonal pattern in the forecast value:

- If only 1 month of time history is available, it is possible to follow the naïve approach and to use the current value as the forecast for the next period; in this case, however, it is not possible to model seasonality, and only a constant forecast can be produced.
- Even if 7 months of data history are available, no conclusion about the seasonality can be made.
- As soon as 12 months of data are available, a seasonal forecast can be produced just by replicating the historic values for each calendar month.
- Starting with 13 months of data, it is technically possible to train a seasonal trend model; however, in this case, the trend component of the model only depends on the difference between the value in month 1 and month 13.

SAS Forecast Server, for example, needs at least two full seasonal cycles in the historic data in order to model a seasonal trend model. For a pure trend model, at least two observations are needed. Some special types of forecasting methods, such as ARIMA models, need a time history of at least two seasonal cycles in order to produce results. From a business point of view, the technical minimum requirements are in many cases not enough to make a reasonable forecast:

- For a seasonal trend model on monthly data, for example, it can make sense to lower the minimum requirement of 24 months. As soon as 16 months of history are available, for example, the trend and the season can be estimated.
- However, in this case, only 4 data points (Number of Months – 12) can be used to estimate the trend. The values in months 1, 2, 3, and 4 and in months 13, 14, 15, and 16 are used to estimate the trend.

The length of the time history is, however, not only relevant to time series methods itself. Time history plays an important role in predictive modeling as well. In the case of transactional data that is aggregated to a one-row-per-subject data structure, derived trend variables are created to describe the analysis subjects' behavior over time. Simple derived variables like means or sums are calculated as well as complex derived variables such as trends over time.

Level of detail in forecast hierarchies

Time series forecasting is often performed on different hierarchical levels. Examples of these hierarchies are geographical regions, product hierarchies, or sales structures. In the case of product hierarchies, for example, forecasts are created on the product level, the product subgroup level, the product group level, and the product main group level.

For some products, perhaps those that are not sold often, some periods might have zero quantity. For these products, the lowest level in the hierarchy in the forecast model might not produce a meaningful forecast due to many zero intervals and small occasional quantities. In this case, it makes more sense to perform the forecasting on a higher level in the hierarchy in order to have a more stable and better populated time series.

Thus, the level of detail in the forecast hierarchy also needs to be considered in the area of data quantity.

Panel data sets and repeated measurement data sets

Larger data structures allow more detailed analyses and tests. Some of the classical data structures (like panel data sets) are enlarged to include more dimensions. As analysis methods and algorithms improve, repeated measures data sets are commonly used for finer analyses of correlated input data and fixed and random effects.

4.4 Sample Size Planning

Application of sample size planning

In many disciplines that employ statistical methods for analysis, planning the necessary sample size and the resulting power of the statistical test plays a central role. The most frequent applications are calculating the necessary sample size or the resulting power (= 1 – beta error) based on parameters like alpha-errors, means, standard deviations, proportions, and other features. A frequently used example calculates the necessary number of observations per group to find the statistically significant difference of the half standard deviation in the two-sample t-test. The result is 64 observations per group.

This is important because before executing the study, the researcher wants to know whether with the given parameters and assumptions, for example, a significant difference between two groups can be found. Findings from power and sample size planning are then used to adapt the settings from the trial specifications.

Thus, sample size planning is an important tool to ensure the necessary data quantity for statistical analysis. It helps researchers to decide early whether a trial should be executed with the current sample size or whether it should include additional analysis subjects. Note that in some disciplines, such as medical research, it can take months to include 20 additional analysis subjects (for example, if they have a rare disease).

In large pharmaceutical and clinical trials, for example, sample size planning is a key preparation component. Long-term trials usually involve a large amount of financial and personal effort. Therefore, all involved parties are interested in acquiring a sufficient number of patients to achieve statistically significant results if a true difference between therapy groups exists.

Expected input parameters like means, standard deviations, or odd-ratios that are necessary for the sample size calculation are typically taken form pilot studies, statistical publications, or expert knowledge.

Figure 4.1 shows a screenshot of the SAS Power and Sample Size Application.

Figure 4.1: SAS Power and Sample Size Application for logistic regression

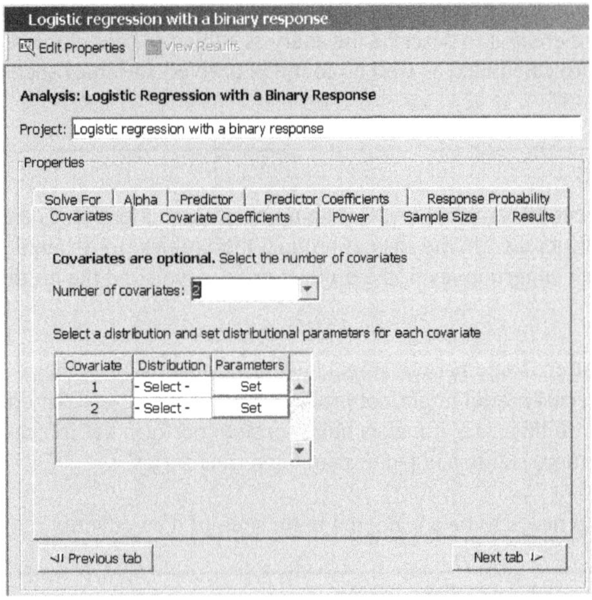

Sample size calculation for data mining?

In classical statistical analysis, the (limited) list of potential explanatory variables is defined a priori, their influence is statistically measured in the form of p-values, and the analysis usually follows a pre-defined analysis scheme. Therefore, methods for power and sample size calculation are regularly applied.

Data mining analyses, however, follow a different analysis paradigm. Compare [1] chapter 2:

- The focus is not on hypothesis testing and the in-depth analysis of single variables but on model quality measures like response rates, lift values, or average squared errors.
- During data preparation, the focus is often to build an analysis data mart with hundreds of variables. In the analysis, a subset of variables is selected that has the highest predictive power.

Thus, data mining and time series forecasting analyses usually do not apply methods of power and sample size calculation when judging the minimum number of observations.

Consider a logistic regression analysis:

- In the case of statistical analysis in a clinical study, there is usually a short list of explanatory variables usually up to 8.
- These variables are included for well-defined business reasons.
- For each variable, prior knowledge about the distribution and the expected relationship to the target variable needs to be provided to calculate the necessary sample size.
- Providing this information involves some work and analysis on pre-study data for each variable. As the inclusion of each variable in such analyses follows prior variable selections from expert knowledge anyhow, this effort is usually justifiable.

In predictive modeling, however:

- There is usually an initial set of hundreds or even thousands of variables.
- These variables are often generated in a semi-automated way in the data preparation process.
- Following the paradigm of data mining, the analysis is designed to find out from many candidates the most important and predictive input variables.
- Thus, there are usually more variables in the data that should be technically included into sample size calculations.
- And, more importantly, often it is not possible to provide the necessary knowledge about distribution and expected relationship for each variable.
- In data mining analyses, usually the researcher is not interested in statistical hypotheses testing by defining type I and type II errors, but rather in finding a model with a high business impact and a correct response level or lift value.

Thus, methods of sample size application are usually not found in typical data mining analyses, and rules of thumb based on experiences and expert knowledge are used for the minimum sample size considerations.

An overview on the SAS Power and Sample Size Application can be seen in appendix B.

4.5 Effect of Missing Values on Data Quantity

General
The last section in this chapter deals with missing values and thus leads into the next chapter, "Data Completeness." This section discusses how missing values in multivariate data can affect the number of observations.

Problem description
Most analytical methods cannot deal with missing values in the analysis data. This is the case for many prominent statistical methods, like regression analysis, neural networks, cluster analysis, and survival analysis. The only exception from this list of important data mining methods is decision trees, which can handle missing values by creating a separate branch for missing values. For the other methods, missing values reduce the number of observations.

As soon as one variable is missing for an analysis subject in an analytical data mart, the entire analysis subject is excluded from the analysis. Technically speaking, the entire row is excluded from the table if at least one variable contains a missing value. Note that for stepwise regression, whether an observation is included in the analysis is checked initially based on all the variables that are candidates for variable selection. Thus, an observation is excluded even if the variable for which it has a missing value will probably not be used in the final model.

In multivariate analysis, where many variables are analyzed together, no values can be missing; otherwise, the record will not be included in the analysis. Thus, the risk for excluding one record increases if more variables are used for the analysis.

The following subsection shows a calculation of the expected remaining proportion of records when k variables are in the analysis table and the probability for a missing value is $p\%$.

Calculation
Assuming independence, the binomial distribution can be used to calculate the probability that an observation can be used for analysis, given that there are n variables with a missing value probability of p. An observation can only be used for analysis if none ($k=0$) of the variables are missing. The binomial distribution with $k=0$ is used to model this analysis. Table 4.1 shows some results.

Table 4.1: Expected proportion of observations that are effectively available for analysis if a certain percentage of missing values occurs for a certain number of variables

Proportion Missing Values	Number of Variables											
	1	2	3	4	5	10	15	20	25	30	40	50
1%	99.0%	98.0%	97.0%	96.1%	95.1%	90.4%	86.0%	81.8%	77.8%	74.0%	66.9%	60.5%
2%	98.0%	96.0%	94.1%	92.2%	90.4%	81.7%	73.9%	66.8%	60.3%	54.5%	44.6%	36.4%
3%	97.0%	94.1%	91.3%	88.5%	85.9%	73.7%	63.3%	54.4%	46.7%	40.1%	29.6%	21.8%
4%	96.0%	92.2%	88.5%	84.9%	81.5%	66.5%	54.2%	44.2%	36.0%	29.4%	19.5%	13.0%
5%	95.0%	90.3%	85.7%	81.5%	77.4%	59.9%	46.3%	35.8%	27.7%	21.5%	12.9%	7.7%
6%	94.0%	88.4%	83.1%	78.1%	73.4%	53.9%	39.5%	29.0%	21.3%	15.6%	8.4%	4.5%
7%	93.0%	86.5%	80.4%	74.8%	69.6%	48.4%	33.7%	23.4%	16.3%	11.3%	5.5%	2.7%
8%	92.0%	84.6%	77.9%	71.6%	65.9%	43.4%	28.6%	18.9%	12.4%	8.2%	3.6%	1.5%
9%	91.0%	82.8%	75.4%	68.6%	62.4%	38.9%	24.3%	15.2%	9.5%	5.9%	2.3%	0.9%
10%	90.0%	81.0%	72.9%	65.6%	59.0%	34.9%	20.6%	12.2%	7.2%	4.2%	1.5%	0.5%
15%	85.0%	72.3%	61.4%	52.2%	44.4%	19.7%	8.7%	3.9%	1.7%	0.8%	0.2%	0.0%
20%	80.0%	64.0%	51.2%	41.0%	32.8%	10.7%	3.5%	1.2%	0.4%	0.1%	0.0%	0.0%
25%	75.0%	56.2%	42.2%	31.6%	23.7%	5.6%	1.3%	0.3%	0.1%	0.0%	0.0%	0.0%
30%	70.0%	49.0%	34.3%	24.0%	16.8%	2.8%	0.5%	0.1%	0.0%	0.0%	0.0%	0.0%

From the values in the table, you can see how quickly the number of observations reduces, as non-missing values for any of the observations is allowed. For example:

- Even if there are no more than 1% of missing values, the expected number of observations in the analysis reduces to 74% for 30 variables and to 60% for 50 variables.
- An average missing value rate of 5% and 10 variables does not leave more than 60% of the observations in the analysis data mart.
- Having four variables with 10% missing values means that one-third of the observations cannot be used for the analysis.

Summary

This section shows the importance of providing a complete data table for the analysis. The next chapter on data completeness deals with this situation in more detail. In chapters 10 and 11, methods in SAS for imputing missing values are described. Chapter 18 shows the results of simulation studies in the presence of missing values for predictive modeling. Appendix B contains the program that has been used to calculate the values for this table.

4.6 Conclusion

This chapter has shown that sufficient data quantity is a prerequisite that identifies whether the desired analysis can be performed. Therefore, data quantity is also a characteristic of data quality, even if these terms imply opposite, or even mutually exclusive, characteristics.

Data quantity has been discussed for both too much and too little data. And data quantity features have been detailed for various dimensions, like the number of observations and events and the length of the data history, to name a few. It has been shown that sample size calculation methods can help in some analyses.

The last section has shown the effect of missing values on data quantity in multivariate analyses.

Chapter 5: Data Completeness

5.1 Introduction...59
5.2 Difference between Availability and Completeness.................................60
　Availability..60
　Completeness...60
　Categories of missing data...60
　Effort to get complete data..61
　Incomplete data are not necessarily missing data.....................................61
　Random or systematic missing values...62
5.3 Random Missing Values..62
　Definition..62
　Handling...62
　Consequences..62
　Imputing random missing values...63
5.4 Customer Age Is Systematically Missing for Long-Term Customers........63
　Problem definition..63
　Systematic missing values..63
　Example...64
　Consequences..64
5.5 Completeness across Tables..64
　Problem description...64
　Completeness in parent-child relationships..65
　Completeness in time series data..66
5.6 Duplicate Records: Overcompleteness of Data..67
　Definition of duplicate records...67
　Consequences..68
　Reasons for duplicate records..68
　Detecting and treating duplicates..69
5.7 Conclusion...70

5.1 Introduction

This chapter deals with data completeness, which is defined here as the fact that the available variables have observations for all records (or example, for all analysis subjects or for all points in time). A fully complete data set consists of a full matrix: all cells of the rows and columns are populated with a value.

This chapter discusses the difference between data availability and data completeness, and it also describes the difference between random and systematically missing values. Completeness is discussed not only for a single table but also across multiple tables. Finally, duplicates are discussed as a special topic on overcompleteness.

5.2 Difference between Availability and Completeness

Availability

Availability and completeness are two related concepts. As explained earlier, availability deals with the question of whether data can be made available in general. This includes:

- The existence of additional features (variables) describing the customer demographics
- Transactional data from historic periods
- Survey data for particular market research questions

Completeness

Completeness here is defined as whether information that is intended to exist in the database exists for all observations. Completeness is, therefore, a concept that applies on the row level and judges whether information exists per observation. Completeness can also be understood as availability of the desired data on the row level.

Examples are:

- The variable postal code is missing for 18% of the customers.
- Historic data are available in general, but for some product groups, some months are missing.
- A market survey has been performed on customer loyalty; however, many of the questions have only been filled out in 40% of the cases.

In a purely technical sense, availability deals with the fact that a variable (column) exists in a table because then information can be made available. Completeness deals with the fact that values are available for each observation (row) in the table.

Categories of missing data

Missing data can be categorized into the following groups:

- The feature is not applicable for some subjects. From a business or practical point of view this means that for certain analysis subject(s), no value for a certain feature can be set. The respective fact is not applicable for this record.
    - Consider the example where the number of births per patient is recorded. This value only makes sense for female patients. For male patients, no value can be inserted. This also means that no value will be imputed for the observations.
- The feature is applicable, but it has not been retrieved. The data would be available; however, they have not been provided in the data collection or business process. The reasons here can be that:
    - The information was not provided. For example, a person did not provide the number of people in his household.
    - Certain data have been provided; however, they have not been processed further, for example, in data collection or data entry. If the existence of the true value has high priority, additional efforts can be invested to receive this value (for example, new data entry, interviewing the customer again in a market survey, or inspecting the paper forms).

- o Data have been entered into the system, but the value or the record has been deleted during data transfer or data management. If the original data still exist, data transfer can be executed again, and backup copies of the data can be retrieved and processed with revised programs.
- For various reasons, some data values have been entered incorrectly and can be immediately recognized as nonsensical or obviously wrong. This is not only a correctness problem, but it is also a completeness problem because these values often need to be set to missing.

Effort to get complete data

Similar to data availability, data completeness is often not simply a yes/no decision. The fact that data are not complete does not mean that they might not be in the near future.

The completeness of data is also often a question of the effort that is put into data acquisition. If, for example, age values are missing for 10% of the analysis subjects in the database, you must decide how much effort to put into the acquisition of the data:

- For example, a company could contact every customer with a missing age value via its call center and ask for the date of birth.
- An incentive campaign could be run to offer the customers a special discount or a certain free usage of a service if the date of birth is provided.

After completing these initiatives, the age values may only be missing for 4% of the customers. It then can be again a question of priorities of how much additional effort to put into the data acquisition (for example, by conducting personal interviews or by buying additional data).

Incomplete data are not necessarily missing data

In many cases, completeness of the data can be derived from the fact that values exist for a certain column in a table. However, to judge completeness only from the fact that values exist or not may lead in the wrong direction:

- If a non-missing value exists in a table, it does not necessarily mean that a valid value exists for a certain observation. In many cases, missing values are entered as values like 0 or 999, which from a technical point of view is non-missing but not from a business point of view.
- Also, for categorical variables, a missing value may be coded with a separate category.

In contrast, the fact that a variable has missing values does not mean that no information is available here:

- The data representation of answers to multiple choice questions in a survey is often only entered as 1 value for those categories that were checked. The non-checked items are left as a missing value in the data. However they truly mean "not checked" or "does not apply." In this case, no value other than replacement value zero (0) makes sense, and there is no need for a complicated imputation method to correct the error.
- For a customer who has no calls to the call center, a variable holding the number of calls may show a missing value, indicating that no call has been made. This value should then be replaced by zero.
- In the context of transactional or time series data, frequently the count for intervals or categories with no observations is represented as a missing value, which, however, should be interpreted as a zero value. See chapter 11 for a method in SAS to implement this.
- Also a variable may be missing for some observations, but its values can be calculated from other variables of the same analysis subject.

Thus, you need to check whether values can be treated as usable values or how missing values can be filled by deriving the values from other variables.

Random or systematic missing values

Missing values can be categorized into two categories: *systematic* and *random*. The differentiation between the two groups is important in order to decide the impact of that value on the analysis results and on the options in the treatment of missing values.

5.3 Random Missing Values

Definition

Random missing values are defined by the fact that each observation in the data has the same probability of having a missing value for a certain variable. The fact that a value is missing does not depend on other variables in the data mart or on other causalities. If all data for observations were available initially, random missing values could be generated by deciding for each observation whether the value should be set to missing.

Handling

Handling applies to numeric and categorical data equally. In descriptive statistics, this usually leads to a separate specification of the number of missing values and the calculation of the respective statistics from the non-missing values.

The decision about randomness in the occurrence of a missing value is, however, usually not performed on very strict criteria. Any missing value can be considered to have a causal background. It thus can be a philosophic discussion to decide whether there is any case where a missing value occurs purely at random, or based on a possibly hidden systematic. For example:

- 12% of the customers did not specify additional information on their household situation (for example, number of persons, number of children). There are definitely causal reasons why this variable is not answered by these people. In data analysis, however, it is often decided to treat the missing values as random missing values.
- In data entry, some data entry personnel did not fill in the required fields as they should be based on the process. If it can be assumed that the cases are not systematically assigned to certain data entry persons, the missing fields can be considered to be random.
- Because data collection policies were not followed in one regional division, the financial income values for customers are missing in this division. Even if the missing values occur systematically only in this division, it can be assumed that the income values in this region do not differ in their distribution from the other regions. Therefore, the missing values can be considered to be randomly missing values.

Consequences

Random missing values decrease the amount of information in the analysis database. Valid observations are available for fewer observations. This means that even if effort has been made to include many observations in the analysis or analysis database, a subset of them cannot be used.

Missing values that are truly random do not damage the data, their distribution, their relationship to other variables, and the inferences that can be draw from analyses because such missing values do not introduce bias (the picture that is seen is incomplete, but it is not wrong). Still, truly random missing values affect the precision of the analysis, but in large sample sizes, the effect is quite small.

In analytical methods like regression analysis, for example, observations with missing values cannot be used in the analysis because no value can be inserted into the regression equation. In multiple regression methods, where more than one variable is used, a missing value for one variable, however, means all the observations cannot be used in the analysis, and the existing information for other variables is ignored as well. In these cases, it is better to decide how missing values can be replaced with a most likely value and to use the observation in the analysis. Decision trees are not as vulnerable to missing values because they treat a missing value as a separate category.

In the case of random missing values, the missing values are either considered as a separate category with the assumption that the statistics shown for the non-missing values are representative or an imputation logic can be found on how to derive a most representative replacement value.

The second example happens in data mining analyses or selected types of statistical analysis. Note that in some disciplines like clinical research, the imputation of missing values is not performed at all because the analysis results must be based only on real data.

Imputing random missing values

A good point with random missing values is that not only does the existing data represent a true picture for the analysis, but also that imputation methods can be used to impute the observations with missing values.

SAS offers a range of methods to impute missing values, including methods for time series and methods for one-row-per-subject data marts. Some of these methods impute a static value for all observations; other methods impute an individual most likely value. These methods are discussed in more detail in chapter 14.

The challenge in this case is to define the percentage of missing values that is still acceptable for imputation and allows you to get meaningful results or representative replacement values.

In practice, this decision for imputation on a yes/no basis is based on intuition. Based on experience and gut feeling, analysts decide whether the percentage of missing values still allows missing value replacement.

- In data mining analysis, a variable with 10% or 20% of missing values can usually be replaced by an algorithm and used in the analysis.
- In the case of 60% or 70% missing values, analysts usually do not use replacement algorithms, but treat the variable in a different way.
- Chapter 18 deals with the consequences of poor data in predictive modeling and shows a simulation study on the effects of random missing values and the replacement of missing values.

5.4 Customer Age Is Systematically Missing for Long-Term Customers

Problem definition

Customer age is a desirable variable in many disciplines of customer analytics. It can be assumed that at different ages customers show a different behavior in product usage, loyalty, customer care expectations, churn risk, or ability to pay back a loan. Therefore, the age variable is usually requested in many analyses such as customer segmentation, customer profiling, predictive modeling, and simple reporting.

In practice, it is not unusual to find customer databases where 20% to 70% of the customers do not have an age entry. To be more precise, in many cases the birth date variable is not available, which is the basis for age calculation.

If it can be assumed that the birth date for a customer is missing at random, the applicability of imputation methods for missing values can be discussed or the decision can be made to use analytical methods that can deal with missing values (for example, decision trees). The applicability of imputation methods, if a certain percentage of missing values is exceeded, is not discussed here, but it is the topic of the simulation studies in chapter 18.

Systematic missing values

In this example missing values do not occur randomly, but follow a systematic structure. Consider the case where the process of customer interaction changed three years ago (for example, product sale or supply of certain services) and since then requires the sales or customer care agent to collect the customer's birth date during the contract phase. Before that period customer age was only collected at a voluntary basis.

This shift in policies will cause the effect that there will be (almost) no missing values for age with fresh customers, which are most likely younger customers. Customers, however, who opened their contract more than three years ago and did not change it since then will have many missing values. An imputation logic that ignores this systematic and treats all missing values as random missing values will most likely not provide informative replacement values and distort the true distribution of the variable.

Systematic missing values are more difficult to deal with because the non-missing values cannot be considered as a representative sample of the population. Thus, more elaborate replacement methods and strategies need to be applied.

Crucial input to these methods is knowing about the origin of the data, the business process, the data collection process, and the reason that they are missing, to name a few. This knowledge is needed to formulate an imputation logic for the missing data.

Example

An analysis for a fixed-line telecommunications company shows that no birth date is available for 36% of their customers. The subscription process for new services only changed 12 years ago, when entering the birth date was required for every new and changed contract.

More detailed analysis reveals that the missing values are not evenly distributed over the customer database. Some 93% of the customers without an entry for birth date started their contract with the company sometime between the 1950s and the 1980s. Most of these customers have not changed their product or services since then. It can, therefore, be assumed that customers with missing values for birth date are predominantly those in higher age groups (they started their contract 40 years ago) who did not upgrade or change services over the last 12 years.

Consequences

From this example, it can be seen that treating missing values as random cases and replacing them with average values will bias the data and the resulting analysis. Thus, more thought needs to be put into handling these data in order to overcome this data quality issue of missing information.

It may, for example, be possible to compare the distribution of the existing dates of birth in the customer database with the assumed distribution. This assumed distribution can be based on business considerations or on market research. A system of rules for the replacement of the missing values can then be built so that replacing the missing values results in the assumed distribution.

Another consideration here is the following: Is this variable actually needed for the analysis? Or can another variable serve as a proxy variable. For example, if the date of birth is unavailable, is the date of high school graduation available? If such an alternative variable is available, then either it can be used in the analysis or a very good proxy variable can easily be constructed. See chapter 17 for the results of a simulation study on the availability of variables.

5.5 Completeness across Tables

Problem description

Completeness of data is not only relevant to a single table where the number of non-missing values in a column is counted; it is also relevant for comparisons across tables. Data can even be incomplete if in the respective table(s) no missing values occur.

There are two typical situations where data completeness has to be considered in a context that goes beyond a single table:

- data in a parent-child relationship, where a missing value is derived when a related record exists in another table
- data in a time series, where observations for each time interval need to be available

These two situations are explained in more detail in the following subsections.

Different from data profiling in a single table where only the number of missing values needs to be counted, the situations that are described here demand more complex profiling methods. These methods are shown in chapters 10, 11, and 12 in part II.

Completeness in parent-child relationships

In a relational database, the tables are connected by so-called relationships to each other. These relationships are represented by primary and foreign keys that identify the related records. This relationship is also referred to as a parent-child relationship. Here are some examples:

- For each account record, the customer ID is stored in order to assign the account to the correct customer.
- For each transaction, the account ID is stored in order to correctly link the transaction to the right account.

These keys are used to join tables together or to aggregate data to a higher level (for example, the balance of all transactions for a given customer). Compare Figure 5.1 for a physical data model.

Figure 5.1: Physical data model with a CUSTOMER and an ACCOUNT table

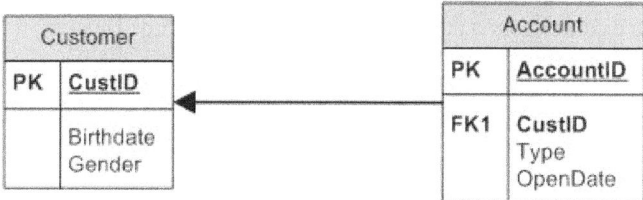

Assume that data are delivered in two tables as shown above: CUSTOMER and ACCOUNT.

- One customer can have one or more accounts.
- Each account must belong to one customer.

Depending on the requested analysis, these data can now be

- joined together to a table that holds a row for each account and the referring customer information is copied for each row
- aggregated on the customer level and joined with the customer table

See chapter 6 for more details on relational data structures.

Completeness in this respect refers to the fact that for each ACCOUNT record, a CUSTOMER record must exist. Sometimes the fact that the record is not found at a higher hierarchical in the data is referred to as lost children.

In Tables 5.1 and 5.2, data for CUSTOMER and ACCOUNT are shown. It can be seen that:

- In the ACCOUNT table a CUSTD=4 exists that has no parent record in the CUSTOMER table. This record is a LOST CHILD record.
- A CUSTOMER with CUSTID=3 exists that has no child records in the ACCOUNT table. This can be the case if the customer has not yet opened an account. From an analysis point of view, the decision needs to be made whether this record will be considered for analysis.

Table 5.1: Content of CUSTOMER table

CustID	Birthdate	Gender
1	16.05.1970	Male
2	19.04.1964	Female
3	31.03.1943	Male

Table 5.2: Content of ACCOUNT table

AccountID	CustID	Type	OpenDate
1	1	Checking	05.12.1999
2	1	Savings	12.02.2001
3	2	Savings	01.01.2002
4	2	Checking	20.10.2003
5	2	Savings	30.09.2004
6	4	Checking	21.01.2007

Usually relational database systems require and verify integrity by so-called integrity constraints, which, for example, make sure that no account record can exist without a corresponding customer record. Thus, in operational systems, child records should not exist without parent records.

Lost child records in analysis data marts can occur when the data in the parent table and the child table have been captured at a different point in time. Lost child records are a sign of bad data quality and the researcher needs to decide whether the child records without parent records should be used in the analysis or whether additional effort is needed to retrieve the corresponding parent records. Chapter 12 shows some ideas on how lost records can be detected in the data.

Completeness in time series data

In order to explain the necessity for checking and correcting the completeness of time series data, it is important to differentiate between *transactional* and *time series* data.

- Transactional data are time stamped data that are collected with no particular frequency over time. Examples of transactional data include Internet data like website visits or point of sale data in retail.
- Time series data, in contrast, is time stamped data that is collected at a particular frequency over time like web hits per hour or sales amounts per day.

Tables 5.3 and 5.4 show examples of transactional data and time series data. The data have been aggregated on an hourly basis for the time series data.

Table 5.3: Transactional data from a web server

	Session Identifier	requested_file
1	43d0a4da826149b5 2002-02-17 08:38:12	/Home.jsp
2	43d0a4da826149b5 2002-02-17 08:38:12	/Cookie_Check.jsp
3	43d0a4da826149b5 2002-02-17 08:38:12	/Home.jsp
4	43d0a4da826149b5 2002-02-17 08:38:12	/Corporate_Relations.jsp
5	43d0a4da826149b5 2002-02-17 08:38:12	/Retail_Store.jsp
6	43d0a4da826149b5 2002-02-17 08:38:12	/Store/Store_Locations.jsp
7	43d639ebce6c73d8 2002-02-17 23:43:16	/Home.jsp
8	43d639ebce6c73d8 2002-02-17 23:43:16	/Cookie_Check.jsp
9	43d639ebce6c73d8 2002-02-17 23:43:16	/Home.jsp
10	43d639ebce6c73d8 2002-02-17 23:43:16	/Department.jsp
11	43d639ebce6c73d8 2002-02-17 23:43:16	/Department.jsp
12	43bb8704bb370e09 2002-02-17 13:44:04	/Home.jsp
13	43bb8704bb370e09 2002-02-17 13:44:04	/Home.jsp
14	43bb8704bb370e09 2002-02-17 13:44:04	/Subcategory.jsp
15	43bb8704bb370e09 2002-02-17 13:44:04	/Product.jsp
16	43bb8704bb370e09 2002-02-17 13:44:04	/Department.jsp
17	43bb8704bb370e09 2002-02-17 13:44:04	/Product.jsp
18	43bb8704bb370e09 2002-02-17 13:44:04	/Department.jsp

Table 5.4: Aggregated web server data per hour (time series data)

	Time	NumberOfReqestedFiles
1	1:00:00	116
2	2:00:00	93
3	3:00:00	17
4	4:00:00	158
5	6:00:00	30
6	7:00:00	66
7	8:00:00	210
8	9:00:00	130
9	10:00:00	143
10	11:00:00	298
11	12:00:00	239
12	13:00:00	145

Note that in the time series data, there is no record for the time interval 5:00:00 because no web server traffic occurred during that time. For time series forecasting, however, the records need to be equally spaced at the time axis. The property of the existence of all consecutive periods is called CONTIGUITY. Time series data that are not contiguous cause problems in analysis because time series methods will not work properly.

The missing observations in time series data described here are similar to the parent-child relationships described earlier. In the case of time series data, a virtual table with a contiguous series of date intervals exists. For each observation, a record must exist in the time series data in set intervals. In the case of cross-sectional by groups in the time series data, a respective record must exist in each group. Section 11.3 in part II shows a method to profile the contiguity of time series data. (Compare Svolba [1], chapter 10, for more details on time series data formats and cross-section by groups.)

5.6 Duplicate Records: Overcompleteness of Data

Definition of duplicate records

Duplicate records occur when information that should be contained by a single record in the database is dispersed or duplicated over a number of records.

Duplicate records do not refer to the fact that in a multiple-row-per-subject data mart, a number of records for a given analysis subject can exist by definition. In that case, data are not duplicated over records but arranged according to a defined data structure (for example, time series data where data are stored for each analysis subject at different points in time).

Duplicate records means that for an analysis subject where only a single record is expected, more than one record exists. This is typically the case in situations of base data like the customer table or the patient table.

Duplicate, in this context, does not necessarily mean that a record is completely identical to the other. It means that records exist that belong to the same analysis subject and hold the same or similar information, but those records may have different variables or the variables are populated in a different quality.

Often, duplicate records in the data represent the typical example of bad data quality. In the strict sense of data quality as it is defined here, duplicate records are special cases of non-quality data because they do not fall into the categories of completeness or correctness.

- There is no problem with completeness in general because all information that is needed is available. However, maybe the information is dispersed over different records.
- There is also no problem with the correctness of the data as long as there are no contradictory entries for an ID in different records. The fact that a street name in an address field is written or abbreviated in a different way does not necessarily mean that it is incorrect.

Before consolidating or modifying the data, ensure that you have encountered actual duplicate data, not valid multiple observations treated as duplicates and falsely combined into a single record. This decision involves data analysis and business knowledge.

Consequences

The consequences of duplicate entries affect data preparation and data analysis as well as the usage of analysis results and the data in the operational process. Here are some examples:

- The number of unique records is counted falsely because there are more records in the database than unique analysis subjects. Associations might appear stronger than they actually are.
- The business process, which is based on that data, will usually not work properly. If analysis subjects occur twice, marketing offers are made twice to customers and patients are invited for follow-up visits twice.
- If transactional and behavioral data for one analysis subject are dispersed on two different records, you might not be able to derive a full picture of the analysis subject from the data.
  - Associations in the data may disappear. This may, for example, affect the analysis of customer behavior as well as the analysis of medical data and adverse events if patients occur twice in the database.
  - The true total amount of purchases or service usage by a customer may be underestimated if the usage for product A is linked with customer record 101 and the usage for product B is linked with customer record 102, where 101 and 102 truly refer to the same customer.
- Table joins of the respective tables may lead to wrong or unexpected results.
- Data from transactional data that are aggregated by the analysis subject are probably joined to both of the occurrences in the base table. Thus, the picture of the aggregated data on the analysis subject level is biased.
- Many analytical methods are not designed for data where analysis subjects occur twice. The assumption of independence is violated. However, here the statistical problem is less severe compared to the base problem of having wrong data.

Reasons for duplicate records

Duplicate records can be caused by problems in the operative process, in data entry, or in data processing.

Duplicate records can, for example, occur because of the following reasons:

- In the operational system, the same subject is entered more than once. Often customer databases hold duplicate records for the same subject.
  - A customer who is shopping in a branch has forgotten his loyalty card. Instead of searching his record in the database, the sales assistant issues a new loyalty card under the same name.

- On a website that is open to registered users only, people who do not know their login name and password or who do not have the time or will to search for it just repeat the registration procedure.
- Each time an existing customer applies for an additional insurance contract, a new customer is created in the database instead of adding a new contract for the existing customer entry.
- Databases exist that do not have common subjects IDs, for example:
  - A company stores its customer data in two separate product-specific databases. There is no common customer ID for these two tables. Joining these two tables requires the analyst to define matching criteria such as name, address, and bank account number. In many cases, however, due to different versions of personal names and street names, not all physically identical persons can be matched to the same record.

Detecting and treating duplicates

Detecting duplicate records in the data is straight-forward if there are unique IDs for each analysis subject. In this case, a query can be performed to select those subjects for which a duplicate ID exists.

If no unique ID for each record is available, it is not directly possible to define duplicate records. In this case, matching algorithms can be applied to assess the similarity or closeness of the records, as follows:

- These methods can be used to find out which records potentially belong to the same ID in a table. Based on this information similarity, a suggested cluster of records is created. These clusters are used as surrogates for the non-existing analysis subject ID.
- These methods can also be used to define which records in two separate tables can be merged because they belong to the same ID. Here so-called surrogate match codes are created that assign a surrogate ID to each subject in different tables. Consequently, this information can be used to merge customer data together.

Usually this procedure is performed in two steps:

- First, the data (for example, name, title, street name, postal code, city, and country) are **standardized** using formats, lookup tables, and other standardization schemes.
- Then the standardized values are used to **analyze similarities**.

SAS offers a powerful solution, the SAS Data Quality Server, which is perfectly suited to standardize and de-duplicate data. Chapter 14 describes more details on that solution.

If real or surrogate IDs are available, then the respective records with the same ID can be summarized into one record. Domain-specific knowledge is necessary here to judge how contradicting information or different meanings in fields will be treated or whether data in different records will be aggregated. For example, such rules can include:

- One data source is defined as the data source with the most trust. In the case of conflicting values, the value of this data source overrides the values of the others.
- The most recent version for the field is used, providing that the age of the information is available for each information source.
- If more than two data sources are combined, the version that occurs most often can be used.
- In an example from Europe, Austria moved from a full census to a register-based approach in 2011. For the execution of a census based on data registers, seven main data sources were combined. For each of the attributes, different values were found in the respective sources, even for the gender variable. A set of business rules was set up to define which value should be used for the respective attributes (compare Burka, Humer, and Lenk et al. [8]).

Whichever of these alternatives is used, the decision to use one involves the application of business and domain expertise and know how.

5.7 Conclusion

This chapter has shown that completeness of data differs from availability of data. Completeness refers to data that will be available for all analysis subjects.

Many factors can cause missing data. From an analysis point of view, these factors have to be analyzed, whether they are random or systematic. Random missing values allow the analyst to use different methods for missing value imputation than systematic values, where an imputation based on domain-specific knowledge has to be built.

This chapter has also described how data completeness must not only be considered from a single table point of view but also within the context of connected tables, like parent-child relationships.

Finally, the issue of duplicates has been discussed. Strictly speaking, duplicates are not a matter of completeness of data. They are discussed here in the context of overcompleteness.

Chapter 6: Data Correctness

6.1 Introduction 72
6.2 Correctness of Data Transfer and Retrieval 72
    General 72
    Data entry 72
    Data transfer 72
    Minimize the number of data transfer steps 73
    Comparing data between systems 73
6.3 Plausibility Checks 73
    General 73
    Categorical values 74
    Interval values 74
    Business Rules 75
    Plausibility checks in relationships between tables 75
    Process step where plausibility is checked 75
6.4 Multivariate Plausibility Checks 75
    General 75
    Multivariate definition of outliers 75
    Outlier definition in the case of trends 76
6.5 Systematic and Random Errors 76
    General 76
    Random errors 77
    Systematic errors 77
6.6 Selecting the Same Value in Data Entry 78
    Problem definition 78
    Using a default value in data entry 78
    Example 79
    Consequences 79
6.7 Psychological and Business Effects on Data Entry 79
    Problem definition 79
    Example 79
    Consequences 80
6.8 Interviewer Effects 80
    Problem description 80
    Geographic or interviewer effect 80
    A similar example 81
    Consequences 81
6.9 Domain-Specific and Time-Dependent Correctness of Data 81
    General 81
    Correctness from a domain-specific point of view 81
    Correctness from a time-dependent point of view 81
6.10 Conclusion 82

6.1 Introduction

Data correctness refers to the fact that the available data reflect the situation or real-world status that they are supposed to describe. Data correctness also has a conceptual component: the intended design of the available data. The same data can correctly describe a given real-world feature but inaccurately describe another such feature. Chapter 9 goes into more detail on this topic.

This chapter discusses the correctness of the existing data:

- under the assumption that the data are describing the intended situation
- in data transfers
- as determined by different data evaluation methods and plausibility checks
- in the light of systematic and random errors, including examples of systematic errors

6.2 Correctness of Data Transfer and Retrieval

General

Data entry and data transfer are frequent sources where errors are introduced. Here, either the manual data entry process or the technical data transfer process changes originally correct data into incorrect data.

Data entry

Many errors in the data occur during data entry itself. If data need to be entered manually into an electronic system, the potential for bias is usually higher than in electronic data retrieval, where data are read from bar codes or responses of survey data are electronically scanned.

For sensitive data, the entry is performed twice. In clinical trials, for example, data must be entered twice into the system. The two versions in the system are then compared against each other and error lists are created to be manually checked. Another method is manually re-checking the entered data with the original data.

In some cases, full double data entry or full manual re-check is not possible or justifiable because of the great effort required. The data check is then performed on a representative sample of the data. This does not allow a correction of validation of the whole data, but it gives at least an indication about the correctness status. A decision about additional data validation steps can be based on these results.

- Consider the example of entering data manually to fill out an online form and then pressing the Enter key to submit the data only to receive a message about correcting the items marked in red. This forces you to correct and augment the entries before the data are accepted. This initial check of the data greatly reduces entry errors and simultaneously ensures better quality data. But, of course, the entered data can meet the entry criteria and still be incorrect. Users haven been known to enter data they deem flattering or to use adjusted data values to be secretive or elusive.
- Any firm that depends on entered data for sales and revenues will take steps to ensure that at least plausibly correct data are entered, so that they can avoid lost sales. Financial data must be highly accurate, and clinical medical data should be highly accurate because health and lives depend on these data. Other data should also be highly accurate, but there might be less incentive to ensure it is, if funds, sales, revenues, and lives are not directly related to the recorded values.

Data transfer

Checking two versions of the data against each other is not only relevant for data that is manually entered, but it is also relevant for electronic versions of the data that reside in different systems. Especially during the transfer of data between systems, care has to be taken to ensure that a correct copy of the data is stored in the target system.

Data import and export are sources of data errors. Such errors can arise from, for example, the following:

- Using wrong or different formats for date values in the two systems.
- Shifts of the decimal point or using different numeric representations.
- Using wrong or different lookup tables for categorical data that is coded into groups.
- Observations as a whole get lost during transfer. These observations (for example, individual orders) are then not counted in aggregations.
- ID values like account or customer IDs can be lost or receive invalid values. The respective data then cannot be assigned to an analysis subject or they are assigned to the wrong analysis subject.

Minimize the number of data transfer steps

Usually, the risk increases if data are passed over many systems along the information supply chain. The more interfaces that need to be passed between systems before the final analysis can be performed, the higher the risk of a bias in the data. In this case, the correctness of data may be initially good, but it will get worse because the data are converted repeatedly.

Integrated business intelligence systems like SAS, which enable data integration, preparation, analysis, and reporting in one central system, minimize the number of interfaces and reduce the risk of a reduction in data quality along the information supply chain.

Comparing data between systems

Data transferred between two systems are also often checked for correctness with checksums. Aggregation statistics like counts and sums or frequency distributions for categorical data are created and compared between the two systems to determine the correctness.

The mere use of sums and means, however, does not necessarily show that two versions are identical. For a numeric variable, for example, changing the data for +1 units for 100 observations and for -1 units for 100 different observations results in the same mean or sum. In this case, calculating higher order moments like variances can give a better view on whether the data are correct. However, this requires that the variance (or standard deviation) can be calculated in the respective systems.

Comparing data between two systems is, however, not always easy. If data are retrieved from an operational system and stored in an analysis database, the value of the validation statistic can be time dependent. As operational system transactions occur constantly over time, the comparison of data versions with a time delay of even only seconds will give different results.

These time delays can also cause slight differences in the checksums between systems. If such slight differences arise, use domain to decide whether the data are correct or correct enough for the analysis.

6.3 Plausibility Checks

General

Performing plausibility checks requires a set of possible and plausible values against which the values in the data will be checked. For categorical variables, these are usually lists of possible categories; for interval variables, the sets are usually value ranges in which the value should fall.

The following subsections show examples of validating data. This is not a complete list on data validation methods. Chapter 13 shows some selected methods.

Categorical values

The determination of validation lists for categorical variables is usually retrieved from the definition of the variable content, for example:

- list of possible contract types or price plans, taken from the operational system
- list of postal codes, regions, or county IDs for address data
- set of possible diagnostic codes in clinical trials or classifications of side effects
- list of sales regions and branches from the base data
- product hierarchies, product groups, and product IDs

In many cases, however, these lookup or validation lists cannot be used to validate the entire data. Usually these lists need to be subgrouped because different sets of possible values apply for different categories. For example:

- The list of possible values for product subgroups depends on the selected product main group. In a do-it-yourself market, the product list for items in the gardening group will look different from the electronic group.
- Different contract types and price plans are valid depending on whether the customer is a private, small business, or large business customer.
- For a worldwide operating manufacturer, different packing units and cargo sizes apply for different sales regions around the world.

Interval values

In the case of interval variables, the definition of validation limits is getting more interesting from a statistical point of view. Here different methods of defining or calculating the appropriate value ranges come into play. The following list contains the most commonly found methods to define values ranges:

- Using moment-based location and variance parameters like **mean and standard deviations**. Each value that falls outside a certain number of standard deviations added or subtracted to the mean is considered an outlier. The disadvantage of this method is that if the mean and the standard deviation are biased by outliers, the respective limits are biased as well.
- **Using quantiles**. Values that fall outside a certain lower or higher quantile are defined as outliers. The disadvantage of this method is that for any distribution, even for data with no outliers from a business point of view, a certain proportion of the observations are classified as outliers.
- **Interquartile ranges**. The method that is used to define outliers in many statistical packages for box-and-whisker plots can also be used to define validation limits. Observations falling outside the 1.5-fold interquartile distance added to the 3rd quartile and subtracted from the 1st quartile are considered outliers. Here more robust statistics, like quartiles and interquartile distance, are used for the calculation than in the example of means and standard deviations.
- Other methods include the usage of trimmed means or special robust estimates of location.

These validation limits, however, only reflect properties from the data that can be statistically measured. A statistical validation limit for the variable AGE can cause observations to fall outside the range that are still relevant from a domain-specific point of view. Or the reverse can be true if, for example, the business process defines that people who are eligible for blood donations must be between 19 and 65 years, a calculation of the limits from a statistical point of view does not make sense. For the AGE variable, for example, the statistically calculated lower validation limit may even be negative, which can never be true in this case.

However, care has to be taken because the truly interesting cases at the extremes of the data series might be mistakenly identified as incorrect data when, in fact, they are correct, and they are the most interesting and important observations. For example, we want to avoid labeling the biggest orders as error-laden and incorrect or the data from the most ill patients who recovered as data entry errors.

Business Rules

Another important group for plausibility checks includes business rules or validation rules in general. Here, the correctness of a certain value does not depend only on the value itself, like comparing it against a list of values or of possible ranges for that variable, but also on other variables. Rules check the validity of the data independent of other values of the same analysis subject. Examples follow:

- The claim date cannot be after the policy or customer start date.
- The field for the number of births is missing or zero for male patients.
- The list and digit patterns of valid postal codes match with the country-specific format.
- The number of calls to the call center cannot be zero if a complaint in the call center has been recorded.

Although statistically simple, business rules provide a powerful tool to check and secure the data quality status of the data. If requirements on data correctness can be formulated into rules, then these rules can provide a direct decision (correct or not correct).

Plausibility checks in relationships between tables

Plausibility checks can also extend to different tables in a database. For example, in order to check whether the claim date is later than the policy start data, it may be necessary to check data in both the claim database and the policy database. In databases, integrity constraints or triggers can be defined that check across tables whether a certain value is plausible in the context of other values in other tables.

In the analysis itself, it is often necessary to combine these tables to check validation rules that are based on a relational data structure. Chapter 12 shows an example of how to implement this in SAS in an efficient way.

Process step where plausibility is checked

Plausibility checks may be performed before the analysis to check the correctness and validity of the data. In this case, however, only the analysis benefits of the corrected data and the corrected data are not loaded back to the originating system.

It is better to check and correct the data as soon as possible in the data generation and data integration process. This can, for example, be performed by integrity rules in the data entry screens or by the operational system itself.

The target should be to detect and eliminate errors in the data as soon as possible in order to avoid requiring that each department or analysis solve this problem separately.

6.4 Multivariate Plausibility Checks

General

The methods that were presented earlier define the validation limits only for the values of the variable itself. There are many situations where it makes sense to calculate more specific validation limits that consider individual specifics of different groups or segments of observations in the data. Two examples are presented here, one for the multivariate definition of outliers in different seasons and one for the definition of outliers when considering a time series.

Multivariate definition of outliers

Intuitively whether a value is an outlier or not depends not only on the variable itself in a univariate way but also on the value of other variables. In many cases, this other variable is a variable that defines a group or segment.

Consider the following examples: If customer demand follows a seasonal pattern, the value of 39,920 units is

- a very large outlier for January
- a moderate outlier for November and December
- a value close to the lower validation limit for July and August
- a value in the center of the validation limits for the other months

In this case, the definition of constant validation limits would not make sense or it would fail the business need of validation. Here, individual validation limits per month need to be used. Figure 6.1 provides an illustration:

Figure 6.1: Seasonal patterns shown by monthly averages in gray bars and a solid line for the value 39,920

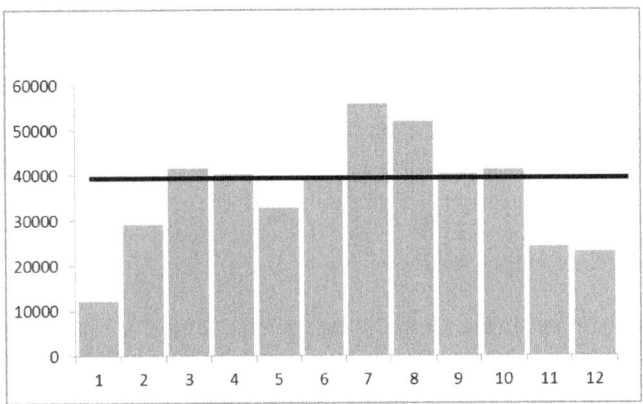

The need for individual values is not only the case for seasonal patterns but also for subgroups like:

- specific limits for laboratory values for each clinical center
- validation limits for the number of orders per region
- domain-specific limits for the number of page hits in websites

Outlier definition in the case of trends

Seasonal patterns are not the only way to identify the need for multivariate validation limits. If a time series shows a trend over time, validation limits must adapt to the mean location of the expected value for a certain period. To achieve these limits, a trend is modeled in addition to the seasonal pattern. Thus, the validation of data in a time series month should be based not only on the monthly means but also on a potential trend over time.

Chapter 13 includes examples of how to calculate individual reference values and expected values in time series analyses (see sections 13.4 and 13.5).

6.5 Systematic and Random Errors

General

Similar to missing values, as shown in chapter 5, errors in the data can occur randomly or systematically. This section explains general points for random and systematic errors. The following three sections show examples of systematic errors.

Random errors

Random errors or random biases in the data are defined as those observations where the value deviates from the true value. Both the selection of the subset of these observations as well as the size and direction of the deviation are random. They do not depend on any other variable in the data or on any other fact that is related to the analysis subject.

Similar to the situation with random missing values, hidden and untraceable factors may exist that explain why a certain random error occurs for a given analysis subject. Examples for random errors are

- 0.041% of the booking codes are wrong due to typos. The typos can be assumed to occur randomly.
- In a customer care center, which performs retention calls, the reason customers cancel is sometimes not entered correctly and a different value from the selection list is used by the call center agent.

Compared to systematic errors, random errors do not introduce an additional factor into the data that changes the direction relationships between variables and affects the underlying logic in the data. For example, the mean of a distribution for a variable is usually not affected by the existence of random errors in the data.

Random errors in the data reflect an incorrect picture for individual values, but the mean of numeric values or the frequency distribution of categories can be assumed to remain unchanged. Thus, estimations that are based on randomly biased data might have inaccurate confidence intervals because the standard deviation is different. Also, predictions for individual cases may be wrong. But the effect of the error can be assumed to be symmetric and the mean predictions are correct.

If the random errors increase, the size of the relationships between the variables decreases. Thus, the signals that data send out get smaller. In effect, it becomes harder to detect relationships, to train analytical models, and to draw conclusions from the data.

Systematic errors

Systematic errors can have one or both of the following properties:

- They occur only for a systematically selected subset of observations.
- The deviation from the true value follows a systematic pattern.

Consider the following examples:

- A customer segment gains more attention by the sales staff, and the data are, therefore, retrieved and maintained with more accuracy.
- For patients in a higher age group, the side effects are recorded with more detail.
- For patients in a certain diagnosis group, certain adverse events are expected and graded in a different way than for other patients.
- In a particular sales region, the forecast value for the next week is rounded to the closest 100.

Systematic errors cause an asymmetry in the data, which biases not only the individual value but also the general shape of the data. In order to detect and qualify systematic errors, domain and process knowledge are needed.

Consider some real-life examples for systematic errors in the data:

- Default dates: For persons in the database with an unknown date of birth, the value 1/1/1900 is inserted automatically into the system instead of a missing value. In the worst case, this value may not be visible to the data entry personnel. The database resets a missing date value to the smallest possible value.
- Customer tenure: Every time a customer changes his accounts (adds an account or modifies account conditions) but uses the same bank, the length of the relationship is reset to zero. Depending on the

available data during the analysis, this error may be detected because the customer tenure provided by the system and the calculated tenure from the account history differ.

The following three sections show more detailed examples of systematic data errors.

6.6 Selecting the Same Value in Data Entry

Problem definition

To get more insight into this issue, it is often desirable to subgroup an analysis for a certain category, for example:

- Not only is the overall cancellation rate shown, but also a frequency table for the cancellation reason that was retrieved from the customer during the cancellation entry appears.
- The reason for hospital admission for patients is tabulated per month.
- The originating country from which the booking request for a car rental in an online application was made shall be used to subgroup the analysis.

Performing such subgroup analyses often reveals unexpected results. In many cases, a dominant category exists that has significantly higher frequencies than the other categories. The high frequency of this category often cannot be explained from a business or practical point of view. In many cases, the surprising high frequency does not reflect the real-world situation, but it is an artifact in data entry. How to handle these results ranges from deciding not to use the analysis because the data are not trustworthy on one side to the enthusiastic presentation of new findings, which do not reflect the real world, on the other.

In general, the subgroup categories should be clearly defined and mutually exclusive; there should be no doubt about the proper subgroup for a specific case. The subject is either male or female; the subject is either 65 and over or under 65. Mutually exclusive subgroups and consistency of subgroup assignment are crucial for meaningful analysis. If the subgroups are not clearly defined and mutually exclusive, then the design must be adjusted so that the overlapping subgroups can be assigned to the proper subgroup.

Further, those classifying the data or entering the subgroup values must know the possible values and their meaning to correctly enter the data. These individuals must be trained and their assignments to the subgroups validated to ensure that the data are correct and ready for analysis.

Using a default value in data entry

Various factors explain why, during data entry, an accumulation of cases in the default category can occur.

- The default value is simply accepted, perhaps as a time-saving measure.
- Those performing the data entry need further training and more awareness about the importance of correct data.
- The data are unavailable to make the proper subgroup classification. If this is a consistent problem, then an adjustment to the research design should be considered.

If, for example, people enter data for their user profile in a web application where

- the number of fields is large
- the data entry process is annoying
- the time needed to enter this information in the data is extensive

It often happens that any value is entered or selected from the list box. In many cases, the first available value is chosen. From a process point of view, it may be inappropriate to query too many attributes and facts from a new customer who applies for a service.

This data quality problem is not limited to the selection of the easiest navigable item in a selection list. It also occurs frequently with date values where for the day of the month the first of the month is primarily used. This occurs in cases like the following:

- The day of the month is either unknown or can only be identified with some effort.
- The day of the month is known, but it is more convenient to accept a possible default value of 1 or to select the first value from the selection list for the day of the month.

Example

In a post-hoc analysis of a large consulting project, the billable and non-billable hours were analyzed per project member and project phase. Additionally, an analysis of the activity was performed. Possible values for activity included, among others, Customer Care, Development, Meeting, and Prototyping. Surprisingly, the analysis showed that 78% of the billable hours were found in the Customer Care activity. This result was surprising because this was a software development project and only a small percentage of the time was assumed to be non-billable customer care time performed by the project managers.

An inspection of the data entry form of the time registration system showed that the first value in the Activity selection list was Customer Care. The majority of the consultants rushed through the entry screen when recording their project times and selected the first available value in the list.

Consequences

Based on data that are biased by data entry, as described above, these subgroup analyses cannot be performed because their results do not make sense. It may be possible to derive replacement values for such dominant categories using a survey or applying business knowledge. However, the reliability of such results is usually very limited.

6.7 Psychological and Business Effects on Data Entry

Problem definition

The example in the previous section can also be extended to the case where date values need to be entered. In this case, it occurs frequently that for the day of the month primarily the first of the month is used in data entry. This occurs in cases like the following:

- The day of the month is unknown.
- The day of the month can only be identified with some effort.
- The day of the month is known, but it is more convenient to accept a possible default value of 1 or to select the first value from the selection list.

On the other hand, date values can accumulate at the end of a period (for example, at the end of a month or quarter). This often happens for reasons that are intrinsic to the business process (for example, if all the data need to be entered by the end of a month or if all the orders need to booked into the system by the end of the quarter).

In this case, the time scale of the data entry (for example, the creation date of an order in the system) does not reflect the original order sequence but a random or biased sequence.

Example

Customer demand analysis on a daily level at a worldwide manufacturing company showed a demand peak at the end of some months. From a business point of view, no reason could be found why customers should increase their demand during the last four days of these particular months.

After some analysis of the data-gathering process, the company found that the accumulation at the end of the month reflected the fact that salespeople wanted to achieve their monthly targets. If the monthly order total was below their target, they put in additional work at the end of the month to increase orders. This produced the outliers for some months in the data.

Consequences

The analysis consequences of this type of situation only occur if the data need to be analyzed or forecast on a daily basis. In these cases, the daily values do not necessarily reflect the uninfluenced customer behavior. Instead, it is a mixture of customer behavior and other influences, like sales efforts. Deciding whether to skip the analysis on a daily basis and move to a higher aggregation level or whether to include the data as they are in the analysis depends on the specific situation and the business requirements.

If data are aggregated and analyzed on a monthly basis, for example, this issue is usually not a problem. Problems only occur when these data are only aggregated to weeks. Weeks at the end of the month will also reflect this problem.

6.8 Interviewer Effects

Problem description

In surveys, the interviewer can have a substantial impact on the quality, completeness, and correctness of the data collected. The interviewer can bias the data in different ways, for example, through:

- his/her personality
- the way he/she poses the questions and interviews the responder
- the ways questions are posed
- the answer options that are offered or promoted
- the way responses are documented (for example, differences in the level of detail)

Analogous to the training needed for those entering data, interviewers also need to be trained on how to ask the questions to solicit the information so that their personality does not affect the data received and recorded. Interviewers must know that the correct response is the response that accurately reflects the condition of the case and not the interviewer's pre-conceived notion of what the response should be. Also, the correct subgroup for the case is the subgroup that corresponds to the actual condition of the case as defined in the research design. For collecting and maintaining accurate data, these steps are crucial.

Geographic or interviewer effect

In one study, the attitude and expectations of modern communications technology in small and medium enterprises were surveyed. A representative sample of companies was selected for the study. The survey took place in the form of personal interviews.

In the analysis, the survey results showed that there was a remarkable difference between different geographic regions. Because the interviews were assigned to the interviewers based on regional aspects in order to minimize travel costs and efforts, it was finally determined that it was not possible to evaluate whether this result was due to the differences in regions or to the different answer behavior of interviewers.

The data quality itself was accurate because the information that was gathered during the interviews was correctly entered into the analysis database. The problem, however, arose from the fact that the interviewers should have been assigned across regions in order to be able to evaluate this effect.

A similar example

A similar problem can arise if in tax inspection different tax supervisors are responsible for different industries. Beside geographic factors, the rationale for this segmentation can be that supervisors need special training to understand industry-specific facts and to read important details in the balance sheets or financial documentation.

If, in data analysis, a difference between industries in the frequency of back duties or penalty payments in general is found, care must be taken in interpreting the results. The difference may be due to the different behavior per industry, but it may also be due to how exactly and stringently the individual tax supervisor executes the reviewing guidelines.

Again, the quality of the data in terms of mapping the collected facts into an analysis table can be correct; however, the collected facts do not necessary equally represent the real world because they were collected by different people.

Consequences

From an analysis point of view, although the data have been entered correctly and there are no missing values, certain desired analyses cannot be performed if the data collection did not take place in a highly standardized way.

6.9 Domain-Specific and Time-Dependent Correctness of Data

General

This section briefly discusses two special examples of data correctness: one from a domain-specific point of view and one from the time-dependent point of view. Both show that there are cases where a general decision about the correctness of the data cannot be made.

Correctness from a domain-specific point of view

Deciding whether an age value in the database is correct or not cannot easily be determined. The following considerations illustrate the dependence on the origin of the data or, more precisely, the properties of the population from which the data were taken. An age value of 81:

- lies within the limits of the common understanding of age values in the sense that this is a plausible age value in general
- may lie within the statistically calculated validation limits from the data for a customer database of a retail company
- will lie outside the statistically calculated validation limits and the business knowledge-based validation limits of a database of secondary school attendees
- will, in contrast, be too small for a meeting of the alumni association's 70-year high school celebration

This simple example shows that, in many cases, a decision on the plausibility and validity of a value depends heavily on the business context.

This is also an area where business rules become more and more important. Financial institutions, for example, that must comply with regulations by financial authorities (like the Basel regulations in banking and the Solvency regulations in insurance) build business rule repositories that check data quality along different criteria.

Correctness from a time-dependent point of view

Data correctness itself can also be judged differently over time. A value or an aggregation of values (for example, sales amounts for a particular month) can change the data's correctness status over time, even if the value stays the same. The reasons for this are, in many cases, retrospective updates to the data.

Retrospective updates means that the known truth for data for a particular time interval changes over time as more information becomes available. Consider the following example:

- The monthly billable amount for each mobile phone customer in a customer base is extracted on the second day of the following month.
- The values in the database for each customer are assumed to be correct because they have been extracted, transferred, and aggregated correctly.
- Calls that were made in foreign countries in the network of other mobile phone providers (also called roaming) are sometimes only billed a few days after the call.
- As a result, such calls, which were made at the end of the month, are not yet in the aggregation on the first or second day of the following month.
- If the same aggregation is repeated a few days later, different values will show up for some customers, but these values are correct as well.

Whether a value can be assumed to be correct can be based on different knowledge bases at different time points. In this case, the different values at different time points are not incorrect. It is more a question of the stability of the knowledge against which correctness can be measured.

Especially when loading historic data, this fact needs to be taken into account. Data (for example, order data from historic periods that are loaded for analytic modeling at a later point in time) can be more complete because all orders have already been booked into the system with the correct month than in the application of the model on the data of the actual month, where not all orders have been entered into the system yet.

This is especially important for data preparation for predictive models and is discussed in more detail in the next chapter.

6.10 Conclusion

This chapter has discussed specific points for data correctness. The chapter has shown methods to profile the plausibility of data in both a univariate and multivariate way. The difference between random and systematic errors has been highlighted, and data entry and data transfer have been discussed as potential sources of errors in the data. For systematic errors, three examples have been described in more detail.

In chapter 14, some of these points are picked up and put in context based on methods that SAS offers to check and verify data correctness.

This chapter completes the quartet of chapters that discuss general data quality considerations: data availability, data quantity, data completeness, and data correctness. The next two chapters continue to discuss data quality features but focus more on the analytical part.

Chapter 7: Predictive Modeling

7.1 Introduction..83
    A widely used method..83
    Data quality considerations..84
7.2 Definition and Specifics of Predictive Models 84
    The process of predictive modeling...84
7.3 Data Availability and Predictive Modeling 86
    General 86
    Historic snapshot of the data...86
    Illustrative example for need to separate data over time......................................87
    Multiple target windows 87
    Data availability over time 88
7.4 Stable Data Definition for Future Periods..88
    General 88
    Requirements for regular scoring 89
    Categories and missing values...89
    Change in distributions...89
    Checking distribution changes 90
    Output data quality...90
7.5 Effective Number of Observations in Predictive Modeling.............................91
    General 91
    The vast reduction of observations...91
    Pre-selected subsets of analysis subjects..92
7.6 Conclusion 93

7.1 Introduction

A widely used method

Predictive models are a very well-known and widely used method in data mining and in classical statistical analyses. Many practical questions are answered by the analysis of the relationship of input variables to a target variable. The resulting logic (mathematical model) is used to derive predictions for other analysis subjects or for the same analysis subject at a later point in time. Consider the following examples:

- The probability for a fraud event can be analyzed and predicted based on the customer base and the transactional data.
- The claim risk and the expected claim amount of insurance customers can be predicted based on the past behavior of customers with similar properties.
- The propensity to upgrade to a more advanced service can be calculated for mobile phone customers.
- The probability that a website visitor clicks on a certain banner or completes the registration dialogue can be predicted.
- The expected survival time for high-risk patients can be calculated based on their risk factors.

Data quality considerations

Data quality for predictive modeling is a very important topic as well. The data quality requirements of predictive models exceed, in some dimensions, the requirements of other analyses. Of course, all the other aspects of data quality that have been presented in previous chapters are important. However, the next sections show that, especially for aspects of data availability, data quantity and the predictive power of data are crucial points for predictive models.

On the other hand, predictive models enable, in some respects, a relaxation of some data quality requirements, by allowing missing values to be imputed and by compensating the missing effect or influence of a variable by another variable.

The next sections discuss in more detail the following points:

- the specifics of predictive models, the application, and the results
- the data availability requirements for historic, actual, and future periods in a stable definition
- the possibly vast reduction of the number of observations when selecting a wide range of variables
- the consideration of the outgoing data quality when applying the logic of predictive models. Here, data quality is not considered in the input data but in the data (scores, predictions, and so on) that are output by the model.

Chapters 16 through 19 use a simulation study to show how various features in the data, like the number of observations, the number of missing values, and the biases in the data, affect the quality of the resulting models.

7.2 Definition and Specifics of Predictive Models

The process of predictive modeling

An overview of the typical predictive modeling process is shown here. Note that only the most important building blocks are shown. From a best practice point of view, this can be detailed into finer steps, including exploratory data analysis and model fine tuning. For the purposes of this chapter, this level of granularity is sufficient.

- Building the analysis data mart. This step is also called the analytic base table (ABT), training mart, training data, and modeling data. Here the data are provided that will be used for the analysis.
- Performing the analysis. This step is also called analysis, modeling, model training, and definition of the model logic. Here the available data are analyzed to derive results.
    - These results can be findings like the direction of the correlation between input and target variables. These **findings are used as knowledge** in future situations (for example, knowing the drivers that are related to a product purchase or knowing the top risk factors that lead to a certain disease).
    - Additionally, these **findings can be formulated in a so-called score logic**. This is also called score rules or score code. This score logic can be used to apply the logic of the predictive model to new observations or the same observations at a later point in time. Based on the logic predictions like probabilities, values or time intervals are calculated.
- In order to perform the scoring, a scoring data mart must be available. This scoring data mart needs to have the same structure as the analysis data mart. This data mart can be a table with a certain number of observations in it. The scoring data mart can be a single observation as well, for which a prediction needs to be retrieved. Consider the example where a single credit loan application has to be processed or where a survival assessment has to be made for one patient.
- The score logic is applied to this scoring data mart. This process is called scoring, rating, or calculating predictions.

- The outcome of this process is a table with the score values, also called predicted values, ratings, or predictions, for each analysis subject in the analysis data mart.

Figure 7.1 illustrates this process. More details on analysis and scoring can also be found in [1], Svolba, chapters 2.6 and 23.2.

Figure 7.1: The analysis and scoring process

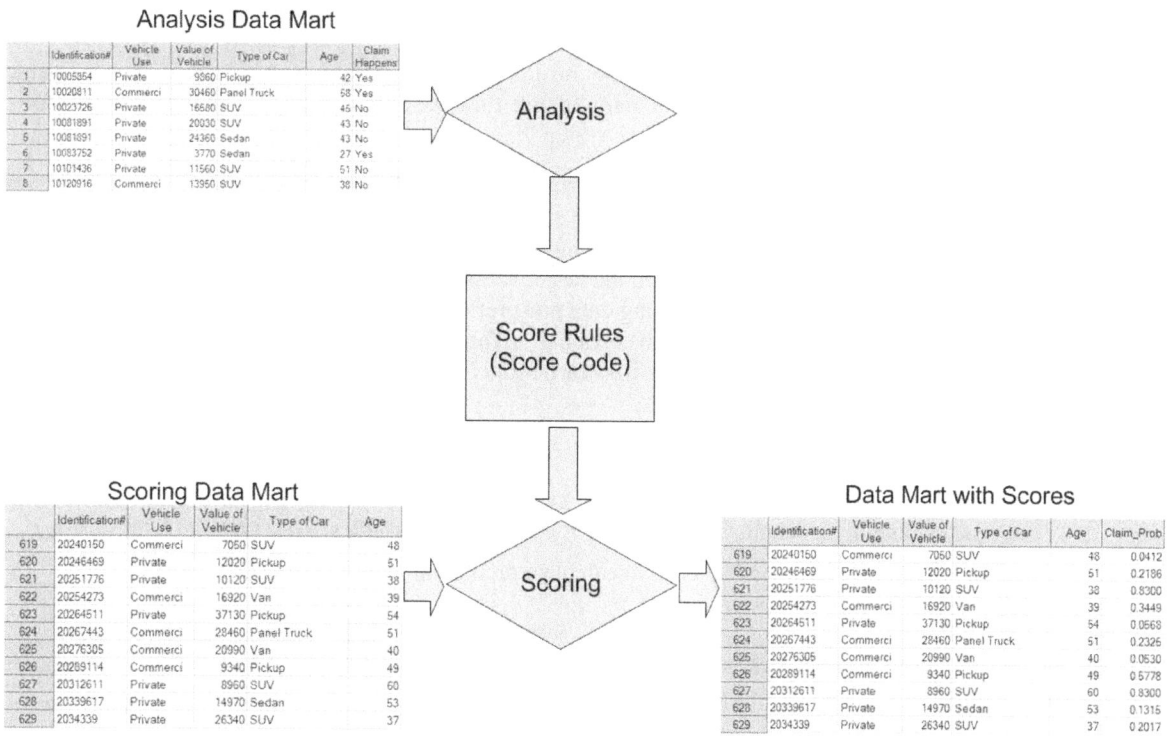

Referring to the process shown here, the following data quality requirements are of importance:

- The analysis data mart needs to have data in sufficient data quantity, **with respect to the number of observations**. If event prediction or classification is applied also a sufficient number of events is needed.

- In order to be able to analyze and train a model that allows predicting future values based on historic data, the training data (input data and target data) must be assembled from different time periods.

- If the resulting model will be applied for future periods, the data structure and the distribution of the scoring data need to correspond with the training data.

- If the scores that are output from the scoring process are used in consecutive steps in the analysis results (for example, in operational systems), the quality of the outgoing data (for example, the range of the score values) needs to be considered as well.

- Many analytical models do not allow missing values in the data. If data are available in general, but are incomplete for some records, the full set of analysis subjects and/or the full set of variables (characteristics) cannot be used.

These points are discussed in more detail in the next sections of this chapter.

7.3 Data Availability and Predictive Modeling

General

Data availability goes beyond the mere existence of data for the analysis. For predictive modeling, data need to be available from different periods. For model training, the input and the target data **need to be separated in time** from each other. This means that input data will not be taken from the same period, where the target event is already known, because the occurrence of the target event potentially influences the values of the input variables.

If the model will be used not only to get the analysis results and findings but to receive a scoring logic, which then is applied on the basis of the scoring data mart, **ensure that the data can also be made available for future periods**.

Historic snapshot of the data

In order to be able to analyze a relationship between input values and target values, the input variables must not be taken from the same or an earlier period than the target variable(s). Predictive modeling tries to predict future values of the analysis subjects. Thus, the training data must reflect this by training the logic in the same way. This requirement has already been discussed in section 3.4. The word "predictive" also contains the element "pre," indicating that the input data will come from earlier intervals.

In mathematical notation this means that

- the model f is trained on the known outcome Y(t) on basis of the inputs $X_i(t-1)$, ..., $X_i(t-n)$, where t denotes the time, n the length of the history, and i the input variables: $Y(t) = f(X_i(t-1), ..., X_i(t-n))$
- the resulting model f can then be used to calculate prediction in later periods:
  $Y(t+1) = f(X_i(t), ..., X_i(t-n+1))$

Note that depending on the data and on the domain-specific analysis question the required degree of accordance to the above principle can vary. Consider the following examples:

- If the input data only contain base data like age or gender, which are not subject to major changes over time (beside the fact that age increases year by year), the results from the predictive model will not change much, if it is trained with the current version of the input variables or with a version of one month earlier.
- In medical research, analyzing the influence of factors like age, gender, race, and current stage of melanoma on the expected survival interval is performed. In this case, the grading of the melanoma stage, which may get worse over time, usually has an impact on the predicted survival. If the model training is only performed on data from patients who are in their final stage, however, who were in a more moderate melanoma stages some months ago, the results will reflect the influence of the more recent values (= more severe melanoma stages) and will not predict well.
- If, in the telecommunications industry, the prediction of a cancellation event will be performed, it is essential to separate the data into the correct intervals. If, for example, usage data, which are taken from the interval where the cancellation already took place, are used, the model will use information that is only available at the time of model training and will overfit the model. Additionally, this data will not be available in the future in that actual version during model scoring.

These three examples show different levels of demand for a clear separation of data from different intervals in order to train reliable and stable analytical models. Therefore, historic snapshots need to be made available that distinguish the version of the data from the time interval where it is needed for the model.

Illustrative example for need to separate data over time

Training a model on data that are only known in the target period can be expressed in the following way in mathematical notation: $Y(t) = f(X_i(t), X_i(t-1), ..., X_i(t-n))$. This, however, means that predictions for future periods (t+1) can only be made on input data from these periods. $Y(t+1) = f(X_i(t+1), X_i(t), ..., X_i(t-n+1))$. These data are, however, not yet available.

For illustration, consider the following example:

- A student studies for a difficult math exam and uses test examples where parts of the approach and the final result are already shown (like examples in a textbook).
- He, thus, trains himself not only to use the exam task, represented by the $X_i(t-1) .. X_i(t-n)$, but he also relies on the fact that he has parts of the approach and the results, represented by $X(t)$, already available.
- He will most likely fail to create the correct results in his exam ($Y(t+1)$), because he is expecting to have the approach (now $X_i(t+1)$) available as well.

The situation is similar to the clear separation of training and target data. Training data, which are influenced by the target data, correspond to the existence of parts of the solution approach in the example above. The model is trained on data that make it too easy to predict the future and will probably fail in its application in future periods, when the data are not "confounded" by target data.

Like the student, models learn their logic only on data that are not influenced by values from the target interval (by the "exam solution"). The time period from which the input data are taken is also the observation window; the time period from which the target variables are taken is also called the target window. Figure 7.2 repeats the diagram shown in Figure 3.1.

Figure 7.2: Input and target data for the preparation of the data mart

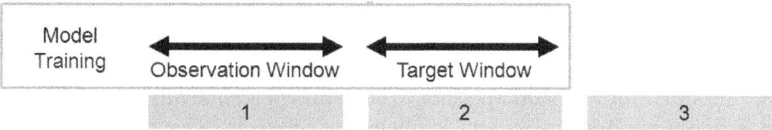

Multiple target windows

In order to build a predictive model, at least one historic snapshot of the data is needed. From this snapshot, input data are analyzed together with data about the target variables that are taken from a later time window.

In some cases, it is desirable to **have data snapshots at more than one point in time**. This can **improve model stability**, allow model validation, or increase the number of observations and events in the data. Here are some examples:

- When modeling monthly contract cancellations of customers, seasonal effects can occur that bias the model if the target events (= cancellation) are only taken from a single calendar month.
- Therefore, it is advisable to use the concept of multiple target windows ([1], Svolba, chapter 12.3) in order to include cancellation from different calendar months into the analysis.
- The structure of the analysis data mart may, for example, have the shape shown in Table 7.1. Snapshots are taken at different calendar months in order to analyze the customer behavior at different times of the year. The respective input data are taken relative to the target month.

Table 7.1: Analysis table with multiple target windows

Observation Month -4	Observation Month -3	Observation Month -2	Observation Month -1	Target Windows
July 2011	August 2011	September 2011	October 2011	November 2011
October 2011	November 2011	December 2011	January 2012	February 2012
March 2012	April 2012	May 2012	June 2012	July 2012

- Another example where more than one snapshot of the data is taken involves improving model quality. Assume you have a model that predicts insurance claims. In this case, the model is trained on input data from 2010 that predict claims that happened in 2011. Another snapshot is taken at the end of 2011, in order to analyze the performance of the model on claims happening in 2012. Thus, a model is built and model quality is objectively assessed with new future data.
- In some cases, data are combined from different periods in order to increase data quantity. Being able to collect more observations or events over time increases stability and improves validation. This is particularly true for analysis questions with only a small number of events in each period.

Data availability over time

The following situations with stable data definitions over time can occur in predictive modeling. These examples reflect cases where the availability of data in the same quality over time cannot be achieved or can only be achieved with great effort.

- Data can be made available for historic time periods, but in future certain variables cannot be made available any more. Here the model perfectly reflects the historic behavior, but the model cannot be applied technically in future periods.
- Data cannot be made available at the same data quality level in the future. The effort made for data cleaning of the historic data cannot be made in all consecutive periods. Similarly, the model can hardly be used for predictions in future periods.
- The model perfectly reflects the situation in historic periods. The data will technically stay the same, even in future periods. But in future periods, the underlying behavior of the analysis subjects will change so that the behavior and the relationships that have been trained on historic data will change over time.
- The data have lots of missing and incorrect values for the past periods. The training data that can be created based on these data will be inferior in data quality. For future periods, however, the data will have very good quality and they will be collected in a more structured and well-defined way.
    - This is, for example, sometimes the case in the creation and application of credit scoring models: Historic data can only be made available with some missing values and biases in the data. The application of the model (scoring), however, can always be based on complete data because a full record needs to be provided for application scoring.

7.4 Stable Data Definition for Future Periods

General

As already shown in a subsection above, predictive models can create a so-called **scoring logic**, which is applied in future periods. Not all analyses demand the application of the logic in later periods.

For example, in the manufacturing industry, the analysis of the relation between production factors and production quality is important. This is often done in a one-shot or ad-hoc analysis in long-term intervals, where the discussion of the findings about the influence between the factors are of interest rather than the creation of a regular score logic.

Another example is in the marketing industry, where an analyst might run screening models to determine which of the thousands of available input variables are influential and important in predicting target variables, such as which catalog recipients will place an order.

Requirements for regular scoring

If, however, the model will be applied in regular intervals, data quality also needs to be considered for the scoring data mart. This means that the data have to be available also in future periods, and they need to be prepared in the same structure and with the same definitions as in the modeling phase. This is important because the model logic derives the relationships based on the structure from the training data mart. It also needs to have this structure available in future periods. The following characteristics are important in this context:

- availability of data from the same domains and data sources as during training
- availability of actual versions of the data, which reflect the situation at or close to the scoring time point (see also section 3.5)
- stability in terms of the definition of the distribution variables, especially in terms of the units, the list of possible values, the number of missing values, and the distribution of interval variables

Categories and missing values

The **stability of the list of possible values** for categorical variables is an especially important data quality feature for scoring data marts. Consider, for example, a predictive model that has been trained on data that include categorical variables with the categories A, B, C, and D. If, in the training data, additional categories E and F are present, the model will not be able to produce a prediction for the observations with these categories because these values could not be trained in the training data. If the model logic does not consider an "other" group or a missing value imputation for these cases, those observations cannot be scored.

A similar situation occurs with missing values. If the score logic does not consider missing values, the scoring observations in the score data mart with missing values will fail.

In scoring, these two cases either receive an average score (for example, in SAS Enterprise Miner) or a missing value for the score. Chapter 10 shows how missing values can be imputed in SAS.

Change in distributions

Technically, new categories or missing values that occur in the scoring data can be easily detected. Then the data can be checked for missing values, lists of valid values, or valid ranges to identify the extent and magnitude of the problem. If, however, the data only change with regard to their content, a simple technical check is not enough. In this case, the distribution of the variables in the scoring data mart needs to be checked and compared with the distribution in the training data mart.

This is important because the model and the resulting score logic have been built assuming that the relationships found in the training data reflect the relationships in future periods as well. If the inherent behavior of the analysis subjects changes over time, a model might still be usable from a technical point of view, but it will not be usable from a functional point of view. Predicted values for the target are farther away from the actual target values, which is a signal to update the model.

Checking distribution changes

Usual methods to check the applicability of the scoring data for model scoring include:

- checking the number of missing values
- checking the distribution of categorical variables, especially the existence of new categories
- checking the distribution of values, both interval and categorical, from a content point of view

For implementation examples of these checks, compare [1], Svolba, chapter 23. SAS Model Manager provides some pre-checks for the scoring data from a technical point of view.

Output data quality

If scoring is performed on training data, the output of an analytical model is not only the results, like tables, graphs, and findings about the relationships, but also data. In this case, in the context of a predictive model, not only the input data quality is of importance, but also the output data quality of the score values is important.

Here, output data quality is understood as the ability of the predictive model, or its application, to deliver the output data (in most cases, the scores) in the definition and the format that it is intended to. Note that output data quality does not mean the same thing as accuracy of predictions but deals with the formal characteristics of the data that are output by a scoring model.

The scores that are created by a predictive model are, in most cases, used by other systems, like call center systems, campaign management systems, credit risk analysis environments, and demand planning systems.

Note that, for example, in the case of output data quality for demand planning systems, the context of predictive models can be extended to time series forecasting models. These models also output data that need to follow certain data quality requirements.

The following list shows criteria that are frequently used to check the outgoing quality of predictive models:

- **Score range check**: Are the output scores within the predefined range?
    - Probabilities for an extent, for example, should be in the range of 0 and 1.
    - Score points should be in a predefined range.
    - In demand forecasting and prediction, the predicted quantities should not be negative.
- **Distribution check**: Do the output scores follow an expected distribution?
    - the proportion of predicted YES events
    - the mean score or mean predicted probability
    - the shape of the distribution of the score or the predicted probability
    - the sum of the predicted quantities per subgroup
    - the distribution of the number of customers in each customer segment
- **Check over time**: How do the score values change over time?
    - The trend of the score values over time. Is there a clear explanation for the observed increase or decrease in the values or in the variation of the values?
    - Do the scores show large unexplained changes compared to the previous month?
    - Are there frequent and large migrations between the predicted segments of the customer data?

These deviations can be due to data quality problems in the input data (for example, if the data for a certain period are wrong or some variables are provided with different definitions).

In addition, the data may be correct, but the behavior of the analysis subjects (expressed by the data) has fundamentally changed over time. In this case, you need to decide from a context point of view whether to retrain the predictive model or whether its base assumptions are still correct.

7.5 Effective Number of Observations in Predictive Modeling

General

Chapter 4 shows the need for a sufficient number of observations and events. In chapter 17, simulation studies for the effect of the variation of these numbers of the predictive power are shown. Section 4.5 also shows that missing values reduce the number of usable observations for many methods.

Deciding on whether business questions can be answered on the basis of the relevant data and whether enough observations are available cannot, in many cases, be made just by looking at the number of rows in a table. The effective number of observations can decrease very quickly due to missing values for input and target variables.

The following real-life example shows how quickly a reasonable number of observations can reduce to a small set of analyzable subjects when some basic requirements for predictive modeling are considered.

The vast reduction of observations

- For the prediction of potential fraud behavior, data over four years have been provided for 89,342 companies where 264 variables and derived variables are available. The proportion of the fraud cases (= events) from historic control and monitoring activities is quoted to be around 4%. At first glance, this sounds like a good data mining exercise because there are a sufficient number of observations, a reasonable proportion of events, and a large number of variables to analyze the business questions.
- After some basic data analysis, however, it turns out that 89,342 is the number of observations of the full company database. Because not all companies have been controlled over the last four years, the information on whether a fraud event has occurred is only available for 43,591 companies. Control activities also took place before four years ago. However, the result of the control visit is not documented in the computer system.
    - Thus, the number of observations has been reduced by approximately 50% due to missing information about the value of the target variable due to non-conducted control visits or non-available data. Still, there is a sufficient amount of observations and events.
- Looking at the missing values of the input variables, it turns out that only for the last two years have specific variables about the companies' tax payment behavior, shortfalls in fee payments, and delays in submission of financial statements been collected. These variables will be included in the model because, from a business point of view, this makes sense. Preliminary data analysis has shown a high correlation of these variables with the fraud event. These data will also regularly be collected over the next months and will, therefore, be available for fraud scoring in the future.
- Excluding the analysis subjects that have missing values for these variables, however, means that only 5,842 observations reside in the analysis data set!
- The number of observations has fallen from a comfortable number of observations to a four-digit number of observations. The alternative would be to omit these variables from the analysis. In this case, however, input variables with good predictive power would be lost, and only a model with basic demographic data could be trained.
- In the first analysis, it turns out that 1,420 of the remaining companies have their local headquarters in a different district. Thus, different legislation applies for the headquarters of those companies. They are not under the control of the authority that monitors potential fraudulent behavior. Thus, they need to be excluded from the analysis because the subsidiaries may show different behavior than headquarters.

This example shows how the number of observations decreased from almost 90,000 observations to a small number of 4,500 observations. Figure 7.3 shows the reduction of observations graphically.

Figure 7.3: Reduction of the number of observations

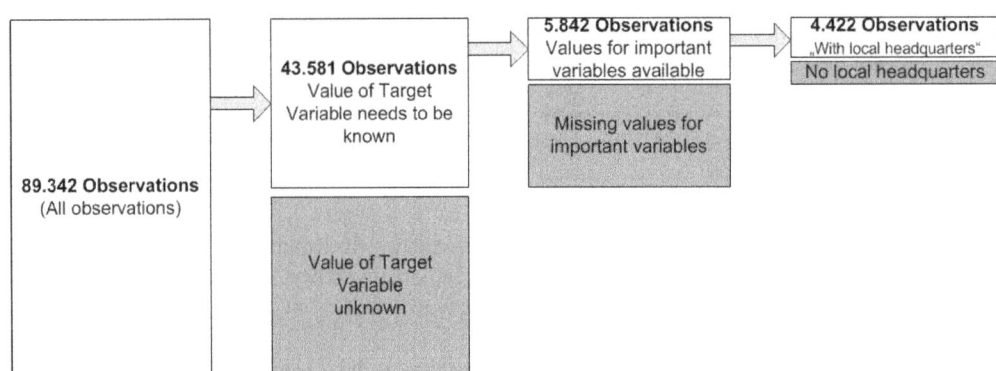

Finally, 4,422 observations are left for the analysis. From these observations, 177 are events (with an event rate of approximately 4%). This is still sufficient to train a model, maybe by including bootstrapping methods. This example shows that a rough initial check about the number of observations and events can be misleading when assessing the data quantity for a given analysis task.

The problem of balancing between including additional variables by losing observations in return can be frequently observed. Decisions about whether to include them or not can only be made individually for each business problem. The important criteria are the expected benefit of including the respective variables and the availability of the data also later in the scoring process. In many cases, the analysis is performed on both sets: one with many observations but only a subset of the complete variables and one with fewer observations but more variables. The results are compared to allow a better assessment of the benefit of each option.

In general, more data are needed and a reduction in the degrees of freedom occurs when the research design attempts to distinguish between and among more treatments, classifications, classes, subclasses, geographies, regions, and time periods. The more complex the model, the more data are required to achieve the same level of precision in the results.

Pre-selected subsets of analysis subjects

Historic data often also suffer from the fact that the operative process has already been applied and they do not reflect the original picture. Consider the following example:

- Suppose you want to analyze the loan defaults in a retail bank. You have to consider that the population of customers who have a loan with this bank has been filtered by the credit approval process in earlier periods. For rejected observations, no data about their ability to pay back the loan are available. In some cases, no data about the rejected applications are available at all.

In credit scoring, for example, methods like "reject inference" are available (compare [11] Siddiqi). This method includes the rejected applicants in model training and weighs their potential default event with the probability of default of the respective model.

This problem is not only true for credit scoring. It also applies in many other areas where the analysis sample is not a random sample:

- In marketing, a response model for a certain product is only built with those households that have been included in a direct mailing. Here, a test group can help to assess the generalizability of the results of the direct mailing group.
- In health care, health profiles are derived from people who have voluntary health checks. This is, however, not necessarily a random sample of the population.

If the application of a model moves outward from the current data, the results will not generalize well and the predicted models are less precise along time periods, geographic regions, or other dimensions and classification variables.

7.6 Conclusion

This chapter has discussed specific information for data quality for analytical models that goes beyond basic data quality criteria in simple analyses. Predictive models provide important findings about the relationships of the inputs to the output(s), and they provide scoring code that can be used by researchers and businesses to score new observations (cases) for the current period and to forecast values for future periods. Predictive models deliver important findings and logic that can be applied beneficially in the business process.

However, they also pose additional requirements, especially on the availability of the data. This has been discussed along with the need for data availability from historic periods and from future periods for scoring.

Most analytical models cannot deal with missing values; thus, a closer look at the completeness and usability of data is necessary to decide whether an analysis can be performed on the basis of the available data.

This chapter has focused primarily on predictive models. Predictive models are a widely used method. The next chapter goes into detail about data quality characteristics of analyses in general.

Chapter 8: Analytics for Data Quality

8.1 Introduction 96
8.2 Correlation: Problem and Benefit 96
    Problem description 96
    Imputing missing values based on other variables 96
    Substituting the effect of unavailable or unusable variables 97
    Multicollinearity or the need for independent variables 97
    Sign inversion 98
    Derived variables from transactional data 98
    Derived variables for customer behavior 99
8.3 Variability 100
    General 100
    Statistical variability and the significance of p-values 100
    Introducing variability 100
    Instability of the business background and definitions 100
    Undescribed variability 101
8.4 Distribution and Sparseness 101
    General 101
    Missing values 101
    Distribution of interval variables 102
    Outliers 102
    Categorical variables and rare events 102
    Grouping sparse categories 102
    Sparse values in time series forecasting 103
    Clustering 103
8.5 Level of Detail 104
    Detailed data and aggregated data 104
    Data structures for analytics 104
    Sampling 104
8.6 Linking Databases 105
    Linking and combining data 105
    Multivariate plausibility checks 105
    Checking parent/child relationships 106
    Project time estimates 106
    Complex regulations 106
8.7 Conclusion 106

8.1 Introduction

Earlier chapters have shown that data quantity is an important criterion that has to be considered. From an analytical point of view, a certain minimum number of observations is necessary. This is also true for simple analyses because only from reports on a representative sample can relevant conclusions be made. This is especially true for analytical methods because the estimation of coefficients, the decision for significance, and the validation of the stability of the models require a sufficient number of observations.

The previous chapter described very specific properties of predictive modeling on data availability and data quality. This chapter extends the previous discussions and provides details on specific properties:

- **Correlation**: the fact that some variables in the data are usually correlated has a strong impact on analytical models, with positive and negative consequences.
- **Variability**: the variability in the data affects the estimation of the parameters in a model and has a strong impact on the stability of the models and the model results.
- **Distribution**: for many analytical methods, the distribution of the data, especially the existence of outliers or the fact that some categories have only sparse frequencies, is of importance.
- **Granularity**: for analytical models, data often need to be available on a more detailed level than for simple reporting tasks.
- **Data linkage**: for many analytical questions, data from different sources need to be linked together. This poses requirements on the ability to access and combine these databases.

8.2 Correlation: Problem and Benefit

Problem description

In practice, different variables in an analysis table are usually not independent from each other, but they are correlated or associated.

Researchers include specific input variables in the model because they are correlated with the output (target variable). For example, as disposable income rises, so do new car purchases, and models that are designed to predict new car sales typically include measures of the current and recent past levels of disposable income.

Strictly speaking, correlation only refers to numeric variables like age in years or weight in kilograms, whereas the correlation between categorical data is called association. An example of association is the relationship between the customers' retail purchase behavior and his life-cycle segment.

Correlation is measured and described by so-called correlation coefficients (for example, the Pearson and the Spearman correlation coefficient). Association is measured, for example, by Gamma, Lambda, or Kappa coefficients. For simplicity, the terms "correlation" and "association" are used as synonyms in this section.

The correlation between different variables in an analysis table is a feature that is relevant for data quality considerations from an analytic perspective. Correlation can have both positive and negative effects on the analysis. The next subsections discuss the advantages and disadvantages of correlation.

Imputing missing values based on other variables

One benefit of correlated data is that missing values or non-existing information for one variable in the data can be derived from other variables. Here, the correlation helps to substitute missing information. In data mining, missing values are often imputed by the most appropriate replacement values. These imputation methods include, for example, tree imputation or regression imputation methods. A predictive model is built, which imputes the missing values by a value that is predicted based on other variables. If the variables in the analysis table were completely independent from each other, no imputation would be possible. Chapter 10 shows examples of how to impute missing values with SAS.

Substituting the effect of unavailable or unusable variables

Also, in predictive modeling itself, like multiple regression or decision trees, the correlation between variables has a positive influence from a business point of view.

- Consider an example where variable A cannot be used in the analysis because it does not exist, there is no trust in the data value itself, or the variable has too many missing values. Usually the predictive model is built without variable A in this case.
- The predictive power of the analytical model will most likely not be as high compared to a hypothetical predictive model that could use variable A.
- If variable A, however, is correlated to some extent with other variables, the reduction of predictive power in the absence of variable A will not be that dramatic because other variables can introduce inherent knowledge about variable A into the model.
- In the case of very high correlation between variables, the loss in predictive power when one variable is absent in a model can result almost in zero.

This example shows that correlation can benefit the analysis. This does not mean that any missing value or unavailable information can be ignored because other variables will compensate for them. Variables are usually higher correlated if they come from the same functional domain and can therefore express the same content better (see also chapter 17).

Consider the following example that is visualized in Table 8.1. Assume there are 9 variables in a data set: 3 variables from each functional domain A, B, and C.

If 3 out of these 9 variables are missing, it makes a difference whether they are missing from the same functional domain or from different domains.

Assume that the content of each domain is relevant for the prediction of the target variable and the variables from the same functional domain are stronger when correlated among each other than variables across different domains. Then a single missing variable in each domain (scenario 1) will not harm the predictive power as much as if all variables from one domain are missing (scenario 2).

Table 8.1: Available variables per functional domain

	Domain A			Domain B			Domain C		
Variable	A1	A2	A3	B1	B2	B3	C1	C2	C3
Scenario 1	OK	OK	Missing	OK	Missing	OK	OK	Missing	OK
Scenario 2	OK	OK	OK	Missing	Missing	Missing	OK	OK	OK

- In scenario 1, the missing variables can usually be imputed with a better accuracy than in scenario 2, where all variables from a domain are missing and no correlated information is available.
- For predictive modeling, the missing variables in scenario 1 might not be needed for the prediction because the other two variables already contribute to the inherent effect of the respective domain.

Multicollinearity or the need for independent variables

The advantages of correlation have been presented in the previous subsections. The disadvantage, however, is, that strong correlation between variables violates the assumption of a linear model; namely, it violates the fact that variables in the design matrix should be independent from each other.

The fact that one input variable can be completely explained by a set of other input variables (that is, it is a linear combination of these variables) is called perfect multicollinearity. In this case the matrix if input variables does not have full rank, and parameters cannot be correctly estimated. Multicollinearity means that an input variable is highly correlated with other variables but does not fully depend on them. Here it may be able to estimate the parameters, but they might be quite unstable in magnitude as more data are added.

Sign inversion

Consider the example where a variable that is positively correlated in univariate analysis receives a negative regression coefficient in multiple regression analysis. This effect is also called sign-inversion. It occurs when the direction of the effect of a variable changes if it is analyzed or quantified together with other variables (for example, in a multiple regression). From a business point of view, inverted signs are often problematic because the logic of a regression model cannot be interpreted from a business point of view.

Sign inversion occurs very frequently if the input variables have a high correlation among each other. Here is a way to explain sign inversion to business people:

- Three friends go to a theater to see a performance of a classical play.
- Each of them like the performance a lot. If each one of them was asked individually, each would provide positive feedback to the act. So their univariate coefficient is positive.
- However, if they are asked together whether they liked the play, the following situation could happen:
    - The first one responds: "It was the best play I have ever seen. The actors were performing so well that I could not stop watching them."
    - The second replies: "The choreography was wonderful. I have never seen a comparable ensemble of scenery, actors, music, and lighting."
    - The third one, even though he liked the play a lot, thought that the first two exaggerated the positive impression of the play to some extent. He wants to correct this impression, so after the first declarations he might say that he generally liked the play, but he thought the choreography was a little bit too exaggerated and the male actor did not really fit the role.
- In the context of the two other opinions, the third opinion moves the results from a positive opinion to a slightly negative one. So, the sign for his opinion changes from positive to negative.

From a strict mathematical point of view, this above situation is a little bit different from the estimation of coefficients in multiple regression because the coefficients are determined simultaneously and not sequential as in the story above.

However, this description gives a good explanation that can be used when speaking to non-analytical and non-mathematical people to explain the inversion of signs of regression coefficients.

Derived variables from transactional data

Different variables that hold the same or similar information but are calculated by different formulas have a very high correlation. For example, if the trend over time is calculated from transactional data, the trend calculated by a regression coefficient and the trend calculated by a ratio between different periods are highly correlated. If one variable is omitted from the model, the predictive power will only decrease slightly because the other variable will provide almost the same explanatory information in the model.

If derived variables are created from other variables, for example, in the case of transactional data that are aggregated for one-row-per-subject data marts, errors in the source data (the transactional data) will distribute the error to all derived variables. In this case, the correlation between the different variables also has the disadvantage that an error in one variable will most likely also affect other variables. The same is true for missing values.

The fact that derived variables from transactional data have a high correlation among each other does not necessarily mean that they can be used to easily replace missing. Because they are derived from the same source variable, their values are, in most cases, either all available or all missing.

Moreover, an additional problem arises with the creation of derived variables from transactional data. In predictive modeling, it is desirable to create a number of derived variables because the predictive power of the model will be maximized and it is not possible to define a priori which aggregation or derived variable is best suited to the analysis. Thus, an analyst usually creates many different derivations. In this case, the number of variables increases, where each variable describes the same feature in a slightly different way.

The analytical model, for example, regression or decision tree, then selects which set of variables are best suited to explain the target. If, however, too many possible explanatory variables are available (which describe the same or a similar feature), the model itself may become very unstable. Unstable in this context means that eliminating a variable from the data or training the model on a different subset of observations or at a later point in time will result in a very different set of explanatory variables.

Derived variables for customer behavior

The following example illustrates this problem. Assume that there are transactional data for each customer. The monthly total product usage is recorded for the last six months. For each customer, months 1, 2, 3, 4, 5, and 6 are available. Based on these data, a number of derived variables can be calculated:

- MEAN: The mean usage over all months.
- MEAN3: The mean usage for the last three months.
- STDDEV: The variability of the usage in terms of the standard deviation of the six months.
- UPDOWN: The variability of the usage in terms of the number of ups and downs over consecutive months.
- TREND_REG: The trend over time in the form of a linear regression coefficient.
- TREND_REG3: The trend over time in the form of the linear regression coefficient for the last three months (and another variable for the last four months, TREND_REG4).
- TREND_DIF3: The trend over time calculated as the difference of months 1, 2, and 3 compared to months 4, 5, and 6.
- TREND_RATIO3: The trend over time calculated as the ratio of months 1, 2, and 3 compared to months 4, 5, and 6.
- TREND_DIF2: The trend over time calculated as the difference between months 1 and 2 compared to months 3, 4, 5, and 6.
- TREND_RATIO2: The trend over time calculated as the ratio of months 1 and 2 compared to months 3, 4, 5, and 6.
- This list could be easily extended to a large number of items.

Assume that the overall product usage and the product usage divided by different usage categories A, B, and C are available. In this case, the three-fold number of derived variables is created.

If all the derived variables are provided to train a predictive model, the variable selection has to be done from a highly correlated set of variables. The derived variables describing the trend for example, will differ only slightly from each other. The resulting set of explanatory variables for the model may look like the following:

- MEAN3, STDDEV, TREND_DIF3_A, TREND_RATIO2_B

If the model is trained for a different time period or for a specific customer segment, the resulting set of explanatory variables may result in the following:

- MEAN, UPDOWN, TREND_RATIO2_A, TREND_DIF2_C

It is very hard to explain, from a business point of view, why the entire set of explanatory variables completely changes if the model is re-trained only a month later. In most cases, the reason is not that the causal relationships change after a period but the model itself is unstable when being trained on a set of highly correlated input variables.

8.3 Variability

General

Variability of the data is a very important feature that has a strong impact on analytical models and the usability of the data for analytics. This section shows these aspects. Only the first aspect covers the pure statistical variability in terms of variances. The other aspects deal with variability (instability) in data definitions and sources.

Data that have no variability might have great precision for one value, but if other input values are used, then the model does not predict well. If additional observations are added that have different input values, then the model parameters might change dramatically.

Data that have low variability also might have great precision over a small range of input values, but again, if other input values are used, then the model might not predict well. If additional observations are added that have a greater range of input values, then the model parameters might be unstable and polynomial terms for nonlinearities might be needed.

Data that have high variability might have low precision over a broad range of input values, but because the model was developed across a broad range of input data, the predicted values should be reliable. If additional observations are added, then the parameter estimates should become more precise, and more subtle long-term trends might be identified.

The issues involve the precision of the parameter estimates, the power of the statistical tests, and the accuracy of the predicted values.

Statistical variability and the significance of *p*-values

The size of data variability has a very strong impact on analytical methods and results. Variability of data is expressed by statistical measures like standard deviation, variance, or interquartile distances. In inferential statistics, for example, in the field of testing of hypotheses, the standard deviation has a direct impact on whether a difference between two groups is considered to be significant.

In the case of comparing the means in a two-group experiment, the absolute difference between the two groups is not a measure of a significant difference. For the statistical test, the means are compared in units of their standard deviation. The means are divided by the standard deviation. If the standard deviation is large, the absolute size of the difference needs to be larger to be recognized as statistically significant (or a larger number of observations need to be tested for the analysis).

Thus, variability of the data, especially the variance of interval variables, has a strong impact on analytical results.

Introducing variability

For predictive modeling, it is often desirable to extract similar data from different data sources. For example, when creating financial rating models of corporate customers, analysts can use official balance sheet data as one important data source. Other data sources, like annual reports, where companies publish their data on a voluntary basis, are also a good data input. The two data sources may differ from a content perspective; however, both sources may be relevant for predictive modeling because they provide different aspects of the data.

In this case, additional variability in terms of heterogeneous data sources is introduced in the data to get a broader picture.

Instability of the business background and definitions

Many analytical models are based on the assumption that the business background, and the respective data and the definition of the content of the data sources, remains stable over time.

- This is important, for example, for time series forecasting where the definitions of the quantities over time (like order or sales number) stay stable over time. Otherwise, they cannot be used in the analysis as a continuous time series.
- In predictive modeling, definitions of categories and data content need to be stable over time because models will be used to score new data in the future.

If business processes, data definitions, and data content are too volatile, analytical models cannot (yet) be built. A high volatility in definitions also often reflects the fact that the business or the functional background is too new to allow the deep application of analytical models. In this case, it is better to use descriptive methods only and to continue to collect data and record the respective definitions over time. Later you can move to using complex analytical methods.

In time series forecasting, if the number of sold or demanded units does not depend on repeatable customer demand patterns but, for example, is only caused by the production of large specific buckets, like the construction of plants, the demand can only be explained for historic periods. The demand cannot be forecast for future periods because the quantity predominantly depends on whether a new large order will be placed.

Undescribed variability

Another aspect of data variability is the fact that patterns or variability are observed that cannot be explained by the available data alone. For example, sales numbers over time are correctly documented in the data, but the causal factors for the variability of the sales numbers are not available in the data. Thus, variability is observed in the data, but it cannot be explained by the data themselves. An analytical model will, thus, be unable to explain or correctly predict future values.

For example, an exceptionally bad winter might explain why sales were so low during that period, but knowing that relationship does not help us predict future sales. One cannot know when the next exceptionally bad winter will occur.

8.4 Distribution and Sparseness

General

For analytical methods, the distribution of the variables to be analyzed plays an important role. The following aspects of the distribution are of interest for analytical methods:

- number and percentage of missing values
- shape of the distribution of interval variables
- existence of outliers
- number of categories and frequency in these categories

While a simple report is also affected by outliers or sparse densities in some categories, the effect of these criteria is higher in analytical methods where parameters are estimated that may be biased, for example, by outliers. The following subsections discuss these points.

Missing values

Compared to simple analyses like reporting, missing values have a much larger impact on analytical models. While in descriptive statistics the number and percentage of missing values are tabulated as an own category, analytical methods like regression or cluster analysis suffer from missing values. In this case, the entire record cannot be used. See also section 4.5.

Analytical methods, however, also provide methods to impute missing values. This includes simple methods like imputing with average values or methods like regression and decision trees that infer the most probable value from other variables of the respective record.

Distribution of interval variables

Analytical methods pose certain restrictions on the distribution of interval variables. Interval variables to be used as target variables in linear regression, for example, should have a distribution that is close to the normal distribution. Many distributions, however, do not have this shape, due to outliers or right-skewed distributions. In this case, transformations are usually applied to transform the variable to the desired distribution. These transformations include methods like log-transformations or square root transformations. If transformations are performed on the target variable, they need to be undone for scoring, prediction, and forecasting to obtain values on the original scale of the target data.

Not only target variables should be normally distributed or close to this distribution. Also, the distribution of input variables should not contain too many outliers or values that skew the distribution. In this case, transformations are usually applied. Outliers bias the parameter estimates and mean that a few observations dominate the parameter value. In many cases, outliers are then filtered or shifted.

Skewed distributions or distributions with outliers are usually not an issue for reporting analyses because the statistical requirements for means and frequencies are not as high.

Outliers

Outliers in the data can have a strong effect on the values of moment-based parameters (for example, on the mean and standard deviation). The effect is also very strong on the estimation of parameters in models like regression analysis. Simple reporting and analytics are both affected in this case. In analytical models, the effect of outliers is more sustainable because the biased parameter estimates are used to forecast values in future periods.

Filtering means that observations that contain an outlier are removed from the analysis set, in order to be able to estimate the parameters with a more "normal" distribution. In this case, however, observations and their corresponding information are removed from the data and decrease data quality. Thus, in this case, often observations with extreme values are not filtered from the data, but the extreme values are shifted to a more central but still high value. In this case, the inherent information from the original data is maintained in the analysis data.

Categorical variables and rare events

In many cases with categorical data, classes exist that have a low frequency when compared to other classes. In the case of reporting analyses, data are then simply reported with the low frequency, or the low-frequency categories are summarized fewer categories. The effect is even stronger in analyses because parameter estimates might be unstable for the categories that have few cases.

Even if two variables have a solid distribution from a univariate point of view, the number of categories increases quickly if the two variables are used in a multivariate way (crossing). In this case, not only the number of categories increases, but also the potential for classes with low frequencies goes up. As already indicated in chapter 4, a large number of categories also means that the models will consume more degrees of freedom.

Grouping sparse categories

If categorical variables have many categories, some of these categories are usually grouped together. The grouping is usually performed from different viewpoints. Here are some examples:

- Similar categories that have few observations are often grouped together.
- In event prediction, categories with a small number of events are usually grouped together. Otherwise, the relative event proportion will depend on too small a number and the resulting parameter estimates for these groups will be unstable.
- Categories with similar event rates can be grouped together in order to reduce the number of categories.
- Grouping is often useful for categorical variables that have many levels, such as postal codes or US ZIP codes.

- Categories are grouped together from a business point of view. Groups with a similar meaning can be grouped together in order to reduce the number of categories. For example, the classification of car insurance policy holders with the groups RURAL, HIGH RURAL, URBAN, and HIGH URBAN can be grouped to RURAL and URBAN.

Sparse values in time series forecasting

Sparse values also affect time series forecasting models (for example, if there are initial periods with zero values or if the time series is analyzed on a too detailed level, where only occasional values remain for each ID).

- In the launch period for a new product, for example, sales might not have started in all branches or regions in the same week. This results in data where, for some subgroups, the initial weeks have a zero value. This phase with zero values will not be repeated later on and is not representative of the behavior of the time series. Simple reporting is not affected. They only show the values over time. In time series forecasting, however, where a logic method is applied for future periods, these initial periods need to be treated in a special way.
    - Either exclude the initial start-up weeks for more stable market analysis or create observations based on sales for each week of product availability. The first weekly observation is the first week that the product was available in that region, even though the date of availability might differ. Both approaches have advantages and disadvantages, and the approach used should be most appropriate to answer the questions of interest.
- For interpretational and analytical reasons, sparse time series are often aggregated to a higher level, because only then can stable forecasts be obtained. If the time series at the most detailed level only has occasional occurrences of values, it is a best practice to aggregate this time series to a higher level. This higher level can be a higher time aggregation level (like weeks instead of days) or a higher level in cross-sectional hierarchies (like product hierarchies). For sparse time series, special methods like intermittent demand models are available in forecasting. This method is implemented in SAS Forecast Server. These methods model the time series by estimating the time until the next value and the expected size of this value.
    - The choice of data collection frequency typically depends on when the data can be collected and the goals of the research design. If there are high costs involved with collecting the data on a daily basis, then it is important to consider if the benefits to be gained from performing the granular daily analysis are worth the costs of doing so. Collecting data on a weekly basis and performing weekly analyses might be cheaper and have the same practical usefulness.

Clustering

An important feature of data distribution is clustering and segmenting observations into groups. These methods are used to define groups of observations in the data that have the same or similar behavior. Usually clustering denotes statistical methods that define a measure of similarity that is used to put observations into the same group. Segmentations are usually performed on categorical or categorized data, like age classes, grouped usage numbers, regions, gender, and categories of product codes. These categories usually allow a simple and direct interpretation of the segments.

While categorical or categorized data are a good basis for segmentation tasks, they are not that usable for statistical clustering. Using categorical and interval variables together will usually cause problems. Categorical variables will, in this case, always dominate the segmentation because the differentiation between their categories will allow a good separation between clusters. The interval variables, even if using them would make sense for clustering of the observations, only play a subordinate role.

For analytical clustering, only interval data should be used. In this case, the distribution of the data plays an important role. Variables should not be very different in terms of their standard deviation because the variables with the highest standard deviation dominate the clustering.

8.5 Level of Detail

Detailed data and aggregated data

For analytics, the mere availability of data is not enough. In order to perform predictions and forecasts on the analysis subject level, data also need to be available on at least that level. If only aggregated data for all or for a subsegment of the analysis subjects exist, differentiated results on a more granular level cannot be created.

For reporting or for simple descriptive statistics that describe a group or all analysis subjects, in many cases it is sufficient to have aggregated data. For example, if the report will differentiate only on a segment level, the availability of aggregated data on a segment level is sufficient.

Consider the following examples, where descriptive statistics can be created but analyses like predictions or forecasts cannot be performed:

- The number of customers who have cancelled in each month is available for private and commercial customers. However, the information is not available for an individual customer, only as an aggregated statistic. Thus, a chart for the cancellation rates per customer group can be created, but it is not possible to build a model that predicts the cancellation probability for each customer.
- In a clinical trial, two therapies were compared based on the number of occurrences of a certain diagnosis (yes/no). For the analysis, only the aggregated information in terms of a 2x2 table is available. Based on these data, you can analyze whether there is a significant difference in the occurrence between the two therapies. However, to detail the analysis for the effect of co-variables like age, gender, or weight, detailed data on the patient level are needed.
- If data have been recorded on a monthly basis only (for example, for the sales numbers of a certain product for the past), it is only possible to report, analyze, and forecast data on a monthly basis. If a more granular time unit will be used, the data need to be collected and stored at this more granular level.
  - Note that PROC EXPAND, which is part of SAS/ETS, allows disaggregating data to a more granular level by interpolating the data and partitioning them into time units. See chapter 9 for more detail.
  - Expanding monthly data to weekly or even daily data, however, does not add information, and a larger number of degrees of freedom by using this data expansion approach cannot be claimed. This problem is especially troublesome if there are few input variables and most or all of them have been expended by using PROC EXPAND. However, the impact is lessened if only one out of many input variables has been expanded.

Data structures for analytics

Analytical methods also require certain data mart structures. Predictive analysis, for example, requires a one-row-per-subject data mart. Association analysis, for example, requires a multiple-row-per-subject data mart, and time series forecasting requires a longitudinal data mart. Compare also [1], Svolba chapters 7–10.

If transactional data are available and reporting and time series forecasting will be performed, the data are aggregated along different dimensions like time, geography, and various codes (product or clinical diagnosis, for example), but only if it is deemed useful.

If the same data will be used in predictive modeling, a one-row-per-subject data mart needs to be created and the data need to be aggregated and/or transposed on an analysis subject level. Thus, before the analysis, data need to be aligned in a timely manner.

Sampling

If a sample is to be drawn from the transactional data, the selection of the observations typically cannot be done by using a simple random selection method. In this case, a clustered sample needs to be drawn from the data. Here, for an analysis subject, for example, an article, all transactional records need to be in the final sample or the analysis subject is not part of the sample at all. Generating a clustered sample is not complicated from a methodological point of view, however, poses requirements from a performance point of view.

8.6 Linking Databases

Linking and combining data

Linking here means that records in different tables are combined based on common keys. Analytics often has a broader view than simple reporting. To answer analytic business questions, data sources often need to be combined. For example, transactional data are not analyzed as they are, but they are enhanced with base data in order to have additional information on the analysis subjects.

If business questions will be answered that consider the business process from a broader context, for example by including different aspects and viewpoints, the data for the analysis must also reflect this by providing information in a more comprehensive way and linked context. A business question can only be detailed and extended, if the underlying data allow this.

Consider the following two examples:

- In normal sales reporting, only sales numbers and sales amounts from the sales database are used. These data are usually aggregated to different analysis levels. If more complex analytical questions will be answered, like the effect of marketing promotions in certain periods or the different sales behavior in the product's launch phase, the respective data need to be joined to the sales data.
- In another example, web log data will be analyzed. Reporting on the number of hits on individual pages or the number of sessions can be done on the web log data alone. However, if information will be linked together to determine how a user navigated through the website, recording the number of hits per page is not enough. It is necessary to link the page hits with the respective user or session identification. In order to achieve this, it is necessary to retrieve the data on which page request belongs to which user or browser session. One way to solve this problem is to enable cookies in the website that can trace the page requests. Only then can a more detailed analysis like web path analysis or web mining be performed.

Multivariate plausibility checks

Analytical methods analyze data at a more detailed level compared to simple reporting. Thus, some features of the data only come to the surface in the se analytical methods.

For example, data in a data warehouse are often considered to be correct as long as they have been analyzed from a restricted or univariate point of view. If, however, data are analyzed at a more detailed level, data quality problems can surface, for example:

- Data from various tables are combined and checked for completeness and consistency across databases.
- Distributions are not only checked from a one-dimensional point of view, but combinations of different variables are analyzed.
- More complex validation rules are checked from a multivariate point of view. For example:
  - From a univariate point of view, the distribution of age classes and the distribution of price plans for a service provider are considered to be correct.
  - If a cross-tabulation of age class and price plan is created, it may appear that certain age,class/price,plan combinations exist that are not offered from the operational process.
- Very often predictive models uncover relationships in the data that offer deep insight into the causal dependencies. These dependencies need to be checked from a business point of view. They may show that impossible combinations of data attributes exist in the database.

Checking parent/child relationships

Assume that there are two tables: a CUSTOMER table and an ACCOUNT table. The CUSTOMER table has 423,323 customers and the ACCOUNT table has 734,913 accounts. So far, only separate analyses on the two tables have been performed: customer reporting and account reporting.

In order to build an analytic data mart, the two tables now need to be joined together. A one-row-per-subject data mart will be created on the customer level. The data from the account data are aggregated (for example, number of accounts per customer, number of transactions, and sum of transaction values per customer). When this aggregation on the customer level is joined with the CUSTOMER table, it shows that there are 9,814 customers without an account and 11,830 accounts that do not belong to any customer.

In the account reporting, this fact was so far not identified because the tables had never been considered in combination. Only the more complex analysis and combinations show the data quality problems.

Advanced analytics compared to simple analysis often requires combining data from different sources and, thus, these types of data quality problems occur more frequently.

Project time estimates

From a project management point of view, these situations can be problematic for analytic projects because the effort to prepare and clean the data can be underestimated.

The perception of the data quality status may be too optimistic, if the experience is only based on simple analyses or descriptive reporting. In these projects, very often data quality problems are uncovered when the analysis, or at least the pre-analysis, has already started.

Complex regulations

Another problem in data quality is that the combination of data sources that is relevant from a business point of view may not be able to be performed from a juristic or regulatory point of view. The situations and legislations differ between countries and industries:

- Customer usage of a website, for example, can technically be retrieved in small enough detail to define how a user scrolls his mouse over a page. From a data security point of view, these data are, in some areas, private and not allowed to be captured.
- Also, information that is gathered from social media sites about the clients of a company can be very informative from a business point of view. But in many areas it is not allowed to combine this type of data with other data sources in the company.

8.7 Conclusion

This chapter has shown that analytical methods have stronger requirements on data quality. These stronger requirements result, on the one hand, from special prerequisites of the methods themselves, like the distribution of variables, the independency of variables, and the variability of the data. This also includes the requirements from the previous chapter to provide historic snapshots of the data.

On the other hand, the features and performance of the analysis itself highlight data quality requirements and problems in the data that would otherwise remain unknown. This includes combining data from different domains, performing multivariate analysis of the data, and linking records across tables. This allows for checking and verifying additional features in the data that would be undiscovered in typical analysis.

Analytical methods, however, not only pose additional requirements on the data, they also allow methods to improve the data quality. These methods are discussed in more detail in chapters 11 and 12 in part II of this book. They include methods like imputing missing values, matching records without unique identifiers, or performing complex checking methods for outliers.

Chapter 9: Process Considerations for Data Quality

9.1 Introduction 108
9.2 Data Relevancy and the Picture of the Real World 108
    Technical data quality and business data quality 108
    Relevancy 108
    Intent of the data retrieval system 109
    Possible consequences 109
    Reformulation of the business questions 110
    Conversion of real-world facts into data 110
    Conclusion 111
9.3 Consequences of Poor Data Quality 111
    Introduction 111
    Analysis projects are not started 111
    Analysis results are not trusted 112
    Analysis projects take longer than expected 112
    Wrong decisions can be made 112
    Loss of company or brand image 112
    Regulatory fines or imprisonment 112
    The desired results are not obtained 113
    No statistical significance is reached 113
    Conclusion 113
    Different consequences for reporting and analytics 113
    Required degree of accuracy 114
9.4 Data Quality Responsibilities 114
    General 114
    Responsible departments 114
    Data quality responsibilities separated from business projects 115
    Process features that trigger good data quality 116
9.5 Data Quality as an Ongoing Process 116
    General 116
    Maintaining the status 116
    Short-term fixing or long-term improvement 117
9.6 Data Quality Monitoring 117
    General 117
    Example KPIs for data quality monitoring 117
    Dimensions for analyzing the data quality 118
    Analysis over time 119
    Outlook 119
9.7 Conclusion 120

9.1 Introduction

As the last chapter in the first part of the book, this chapter deals with **data quality from a process point of view**. The data quality status of a field, record, or table cannot be considered only from a static perspective. It should also be seen as a process over time.

This includes an understanding that all data that are made available for analysis have a history: not a history in terms of a data history from previous periods or a historic snapshot of the data, but a history in terms of the reasons for the existence of the data. Some data may have been collected in the past for a different purpose and are now used for another analysis. Other data have been collected specifically for that analysis. Data availability has already been discussed in chapter 3. The next section focuses on data relevancy and the usability of data for a specific analysis question.

Data and data quality also have a **future** aspect. Again, this is not seen in terms of predicting data values for future periods but as the consequences of data and the data quality status for the future. Section 9.3 focuses on the consequences of bad data quality or of the perception that the data quality is bad.

Achieving and maintaining a certain status of data quality is very crucial, especially if data are collected from systems that are administered by the organization itself. If companies and organizations implement data storage systems in order to run their operational processes or to provide analysis data in data warehouse systems, it is necessary to define responsibilities for checking, improving, and maintaining the data quality. This is discussed in section 9.4.

Section 9.5 specifically discusses the need not only to improve data quality at one point in time but to implement measures to maintain the data quality status over time.

Finally, section 9.6 shows how to monitor data quality with key performance indicators and how to provide a picture of the data quality status.

9.2 Data Relevancy and the Picture of the Real World

Technical data quality and business data quality

In analytic applications, the difference between data quality in technical terms and data quality in business terms becomes more apparent. The fact that data are available and fulfill certain technical data quality requirements does not necessarily mean that the data are usable to answer the business question.

For example, the fact that billing data in euro amounts is within its validation limits and exactly matches the real values reflects the correctness of the data. However, even if the data are 100% correct with respect to the definition, it is not certain whether the definition is suitable for the analysis question of interest.

On the other hand, it is possible that even if the data are not 100% correct on a technical level, the quality level may still be sufficient from a business point of view. The question is whether the quality of the analysis results and the respective impact on a good decision increases sufficiently to justify the effort of data quality improvement in general. The analyst needs to decide whether increasing data correctness from 90% to 95% sufficiently improves the data quality. Thus, quality data for an analysis does not necessarily mean perfect and 100% correct data.

Relevancy

The appropriateness of the data to be able to answer the business question is also an important feature of data quality. Data of good quality from a technical point of view may still be considered as bad quality data, if they do not contain the features that are relevant for the analysis question.

For example, many types of data that are collected automatically by technical systems, like billing data in the telecommunications industry or sales data in the retail industry, or many other types of data that are collected automatically by machines or IT systems, usually have a good degree of technical data quality.

Intent of the data retrieval system

Systems that retrieve or store data are usually built for a particular reason. This leads to the fact that the data that are stored in this system do not only describe the business facts that they measure, but they also describe them in the context of the intention of the data retrieval system.

Thus, certain additional features of the data that are related from a business point of view but are not needed for the initial analysis may not be stored in the system. This means that the system does not mirror the complete picture of the real world but a point of view that is specific to the system. An important paradigm in data retrieval and data provision is the following:

Why and how data are retrieved and stored influence the picture of the real world that is described.

Thus, data that are retrieved and stored with a particular system can have good data quality for a certain analysis, but they might have bad quality for a different analysis. This may happen if the new analysis questions have, for example, the following features:

- requires additional information that has not been collected so far
- needs information that is collected with a different definition
- requires information on a higher quality level
- will be based on data on a more detailed aggregation level

Possible consequences

The intention of the data system as described in chapter 3 can also be seen as a process example for data quality. Usually, business questions trigger the need for data. The non-existence of the required data leads to one of the two actions:

- The business problem is being reformulated (redefined) because the desired data are not available and it is decided to proceed with the analysis based on the available data. In this case, the data quality status has a direct impact on whether the business questions will be analyzed. See more details on this in the next subsection.
- The need to find the answer for a certain business question is stronger than the cost or effort required to improve the data to the expected level. Improvement, in this case, includes actions like:
  - retrieving the data
  - including new data sources
  - surveying additional persons
  - **postponing the analysis** in order include or collect additional data

Postponing an analysis is often done to get more and better quality data. The analysis will then benefit from one of the following:

- Waiting for more data history, like executing the analysis six months later, which means that six additional months of data history are available.
- Waiting for additional analysis subjects. If the analysis is performed at a later point in time and inclusion of analysis subjects is an ongoing process, more observations are available (for example, the inclusion of patients into a clinical trial or the conversion of customers to a new product).
- Waiting for additional events. In many cases in predictive modeling, the number of events is not large enough. If the analysis is performed at a later point in time, more analysis subjects have the chance (or risk) to experience the event (for example, the default to pay back a loan or the recurrence of a disease for patients under treatment).

- Investing time to clean, standardize, de-duplicate, or match the data with other data sources. In the early phases of analyzing data in a company or organization, it may be too early to analyze certain business questions. Effort and time has to be invested to bring the data to a better data quality state.

These examples show that the data quality process and the business questions/analysis process are closely linked together. Based on the importance of each area and the costs (additional costs for new data and the opportunity costs for non-available business analyses), a decision must be made about how to handle a certain data quality status.

Reformulation of the business questions

Insufficient data quality from both a technical and a business point of view often can lead to the need to reformulate the initial business question. If it turns out that the required data are not available or the available data are not relevant or do not meet the technical requirements, the business question may be amended to answer at least a reduced version of the initial question, for example:

- A time series analysis of sales numbers will be performed for each sales branch on the product level. After investigating the available data, it turns out that the company only started saving data on this level of detail 10 months ago. Earlier data have been collected only on an aggregated level per region and product group. As a consequence, only the sales reporting can be performed on the desired level of detail. Time series analysis of historic data can only be performed on a higher level granularity (namely, region and product group).

- A prediction model for fraud events in retail banking will be created. However, for historic periods, no proven fraud cases are available. There are only cases where the strong suspicion of fraud exists. Thus, a predictive model cannot be trained on the fraud yes/no target variable. A pre-analysis is performed to define a surrogate rule that can be used to flag observations that are assumed to be potential fraud cases. These flagged observations are then used as target events for the predictive model. The pre-analysis, for example, can be a link analysis or an anomaly detection on actual data. The predictive model will find out whether this behavior can be predicted.

- For a clinical trial to show the significance of the difference between two treatment groups, no information about the expected level and variability of the outcome is available. Because only about 20 patients per treatment group can be included into the trial over time, there is a risk that an existing difference between the two groups cannot be detected with statistical significance. The available data are thus used for a pre-study analysis to estimate the size and variability of the effect. This information is then used to plan a clinical trial to answer the questions with statistical significance.

The analysis of a business question depends on the available data. It does not depend only on the quality level of the data or on the concept and definition under which the real-world facts have been converted to data. Data and the resulting analysis results, therefore, mirror the circumstances of the data collection process and how the collected data items are defined.

Conversion of real-world facts into data

Note that the analysis results also include the regulatory restrictions under which the data have been collected. The results, therefore, also depend on the definition of how real-world facts shall be collected and converted into measureable data items,

Keep in mind that data quality may be good from a technical point of view and relevant from a business point of view but may still not show the intended "1:1 picture" of the real-world facts for a certain analysis.

This is also considered as the effectiveness of the measuring instrument to mirror the real-world situation. There is usually a shift from the conception of the measurable to the measurability of the concept. This means that, in many cases, the view on the real world is adapted from the way that it can be measured with the instrument. The true concept on how to measure these facts might look different.

The following examples illustrate these points:

- One of the most obvious examples is the interviewer effect and the selection bias in market surveys. As already mentioned in chapter 6, the interviewer and the form used to conduct the interview confound the data that are gathered. In many cases, this effect cannot be isolated from the original data.
- Also the selection of analysis subjects introduces both a bias and a filter.
  - This is obviously true for **planned studies** and trials where the expected population is defined upfront. It is then influenced by the response behavior of different subjects. For example, dissatisfied customers may be more likely to respond to a survey than satisfied customers.
  - This, however, also happens in **observational studies** where the set of analysis subjects is already established. Consider the case of a company's customer database that will be used to analyze the likelihood of customers to respond to a certain new product offering. This population is not an unrestricted sample of all potential customers, like the population of a country, but a prefiltered set of those individuals who have already decided to become customers of this company. The response behavior of this particular set of customers is usually used to describe the general (unrestricted) behavior of all potential customers due to the lack of more appropriate data.
- The collection of data in a questionnaire or standardized data retrieval form acts like a filter in converting the real-world facts into quantifiable data.
  - The set of questions and available response items, as well as their ordering, affects how real-world facts are represented by the data.
  - Even if free text items are made available, where the respondent can (theoretically uninfluenced) add facts, the preprocessing of the free-form entries and the inclusion into structured analysis may still filter this information.
- The amended picture of the real world usually does not occur with such a strong effect where clearly defined values like birth dates, amounts, durations, or clinical parameters are collected.

Conclusion

These reflections show that the collection process of real-world facts may undergo framework requirements that diverge the picture that is represented by the data.

The next section continues discussing the consequences of bad data quality in general.

9.3 Consequences of Poor Data Quality

Introduction

Bad or insufficient data quality affects the execution of the analysis and the application of the results. The suggestion that data may have inferior quality can cause negative consequences, even if, in truth, the quality of the data is good.

The consequences of poor data quality are manifold. They range from delayed projects to unreliable and inaccurate results. In this section, examples of the consequences of poor data quality are listed and discussed. Chapters 16 through 23 show simulation studies for the consequences of poor data quality for predictive modeling and time series forecasting.

Analysis projects are not started

Data perceived to have bad quality can even delay an analysis from starting. No budget is allocated for analyses where the outcome is expected to be unreliable. In many cases, additional resources such as the financial budget items and manpower that are needed to improve the data quality are not made available.

Over time, this may slow down innovation so that no new ideas are being analyzed and no new knowledge domains are being explored. This is not only true for research areas but also for internal company projects, like customer analytics or risk analysis. If there is no trust in the data, the analysis of critical business questions cannot happen.

But not only bad data quality may cause this effect. In many cases, the mere perception that data are bad can delay projects, meaning the respective results are not available for research publications, investments, or business decisions.

Analysis results are not trusted

If data are perceived to have poor quality, regardless of whether this perception is justified or not, the results of the analysis are not trustworthy. The existence of rumors or personal experiences that the data have bad quality will create uncertainty about the analysis results. Thus, a clear picture about the quality status of the data is very important to instill trust in the results.

Analysis projects take longer than expected

Bad data quality usually also causes projects to take more time and effort to complete. If project planning was performed on the assumption of good quality data, discovering that data cannot be used as planned will cause additional data correction efforts. These efforts include identifying the source of the errors, correcting the data, or amending the analysis procedure.

In many cases, no additional time or effort can be allocated to these projects because they have a fixed budget and deadline. Thus, if bad data quality increases the time that is needed for data preparation and data quality checking, the time available for the analysis itself decreases.

In most of these cases, the analysis will take longer than expected. In some cases, only the analysis itself delivers a full picture about the data quality. Analyses on data with inferior data quality means that the true data quality is only surfaced step by step and data items need to be excluded from the analysis. This may result in higher time efforts and cost to perform the analysis or a time delay as the results are only available at a later point in time.

Wrong decisions can be made

Biased data can lead to biased results and, thus, wrong decisions. Random biases in the data usually don't have the same effect as systematic biases. Systematic biases in the data, like different classifications or different value units in different subsets of the data, can lead to systematically wrong results. With these effects in the data, even a careful and exact execution of the statistical analysis will not be able to compensate for the poor data quality and incorrect results will be produced.

If the analyst is not aware that his data are incorrect or biased, the release or publication of the respective analysis results can cause damage. Decisions will be made based on wrong assumptions.

Loss of company or brand image

In the simplest case, customers are annoyed if they get marketing mailings several times because their data appears as duplicates in the marketing database. This, however, leads to a loss of trust in the company's brand if the customers perceive that the company is not able to maintain their data in a qualitative way.

In a more severe case, the public brand image will suffer substantially if a company has to admit that it published wrong sales numbers or sensitive private customer records due to data quality problems.

Regulatory fines or imprisonment

Regulations like the Basel accords for minimum capital requirements in banking or the Solvency directive in insurance require a certain level of transparency in financial and organization risk. In order to comply with these regulations, enterprises need to make sure that they have a clear and consistent picture about their counterparties and risk-related transactions.

The desired results are not obtained

Another consequence of inferior data quality is that the analysis results will not meet expectations. If, for example, the variable person age in years contains too many missing values, the influence of age on buying behavior cannot be studied and no meaningful report of the age distribution over different demographic districts can be created.

Bad data quality in terms of non-availability or incomplete data means that the business questions cannot be answered.

No statistical significance is reached

In the context of hypothesis testing and inferential statistics, bad data quality may mean that a predefined significance level in a research study cannot be met. The analysis result itself is correct by its definition, but from a statistical point of view no significance can be achieved:

- This may be caused by missing values, which reduce the sample size because some observations cannot be used for the analysis. For example, in longitudinal studies, patients may not show up for follow-up visits. Thus, the respective record cannot be used for the analysis. Even if proper sample size planning was performed, the effective sample size for the statistical analysis may fall too low to obtain a significant result.
- This may also be due to an increased variability that is caused by inaccurate measurements or "noise" in the data.

For many of these research studies, the data are not yet available. They are collected during the study (for example, through measurements on patients or surveys on people). This means that in such a study data quality problems are not inherited from a data collection that took place earlier. The data collection just starts with the study itself and effort can be made in the data collection and entry process to ensure that the data have as high a quality as possible.

Conclusion

Considering the possible consequences of inferior data quality, it is obviously an important point to achieve good data quality or at least to get a clear picture of the status of the data quality.

Different consequences for reporting and analytics

The consequences of biases and small deviations in the data are different in simple statistics and in analytics:

- Reports and simple statistics like tabulations are often exposed to a much more detailed comparison of the resulting numbers than other reporting systems, where the same or similar numbers can be retrieved. Data in controlling applications usually have a very high requirement for data correctness. The values in currency amounts and the quantities that are presented need to give an exact picture of the companies' processes. Even relatively small deviations, for example, a couple of euros, need clarification because trust is damaged if the system does not output correct numbers.
- In analytics, comparisons on such a detailed level often do not take place. If data deviate to a negligible extent, such biases in the data are not checked any further. Also, in analytic methods, like predictive models, absolute values do not play such an important role. Derived variables like percentiles or groupings are used for the analysis instead.

On the other hand, data errors have a more sustainable effect in analytics than in reporting. If data for a report contain an error and the error is corrected before the next generation of the report, only one reporting period is affected by this error. If erroneous data are used to train an analytical model, not only are the analytical results biased, but so is the resulting score logic that is used to score observations in future periods. Thus, an error in such a model may affect the outcome in many future periods.

Required degree of accuracy

Investing more effort into improving data quality will usually result in a higher degree of the quality of the data. In many cases, however, data quality improvement efforts have to be balanced with the resulting benefit for the intended use. Quality data do not necessarily need to be 100% perfect. From a business point of view, the researcher has to consider whether the data are more usable and relevant if the correctness increases from 95% to 98%.

Technically, there are limits to the maximum achievable data quality. Laboratory values in health care can only be measured at a certain precision. Increasing this precision, like being able to measure more decimal digits of the parameter value, would be costly and would not necessarily result in usable data.

Another example in data analysis is correctly calculating the variable age in years from the date of birth. Considering leap years and the different number of days per calendar month, an exact calculation of age can be performed. Or the days between the date of birth and the current date can be divided by 365.25 to obtain an almost exact value for the age variable. Depending on the required analysis, you can group the values into age classes for data mining or use an exact value of the current age to approve a certain service at a certain age value.

9.4 Data Quality Responsibilities

General

Clarifying who is responsible for data quality is a very important topic in a company or organization. For the discussion of responsibility, it is necessary to divide between

- research organizations like national statistical offices, institutions for public opinion research, and CROs (contract research organizations)
- companies and organizations that collect the data of their own business processes in a data warehouse

Deciding the responsibility for the first group is very context- and domain specific. Data are often acquired externally, and the data integration and collection process is very much linked to the business and research questions themselves.

Companies and organizations, however, that build data warehouses to store and maintain data about their own processes and interactions with other parties have a different view of data quality. Here, data are considered as an asset that describe the most recent operations of the company for reporting. The data also describe the company's mid- and long-term history that allows company leaders to draw conclusions for future decisions.

This is also known as data warehousing, business intelligence, or business analytics. The following subsections focus on these aspects in this context:

- Who is responsible for the improvement and maintenance of data quality?
- Will data quality be improved as part of a project?
- Is there a difference between transactional and base data?

Responsible departments

Data quality is very often triggered by a business question or analysis. Thus, the business department is the one that has an interest in improving data quality.

However, even if the detection of many data quality problems and the **main interest for data quality** comes from the business department, business should not be the primary task owner that is **responsible for data quality**. It is the responsibility of all data stakeholders to ensure the integrity of the data across all business processes in the organization.

There is a trend for business and IT departments to collaborate on data quality. One of the biggest issues with data quality from a process point of view is that business departments need the data to evaluate, measure, and change behavior, but in general IT departments own the systems.

Consider the following separation of tasks that can be seen in many companies and organizations:

- Business users define the terms that they work with. They set the business definitions of those terms (for example, for trade, net profit, day sales outstanding) and define the rules.
    o What is "Day Sales Outstanding"? What is "Net Profit"? What is the definition of "Trade"?
    o How is it calculated? How important is it? What other attributes is it related to? How do you test the validity of these data (rules)?
    o Data quality solutions like DataFlux have business glossaries that include what data terms are used by the business, their importance, the related items, rules about the data, how to check data validity, owners, and the interested parties.
- IT implements tools to assist in this process and it implements the data validation rules. IT also has the knowledge and experience to implement data quality improvement as an automated production process in the systems.
- Data stewards monitor the data based on the definitions and rules. Data stewards are usually a separate group in the business or IT department.
- Business departments are alerted to poor quality data.

There should be collaboration between business and IT departments because both are responsible for data and the quality of data within an organization. It is a best practice that IT departments and business departments work together to improve data quality. Each party has its strengths and can contribute valuable input to improve the data.

Figure 9.1 shows an example screenshot for monitoring triggers.

Figure 9.1: Monitoring data quality triggers in DataFlux (DataFlux Data Management Platform)

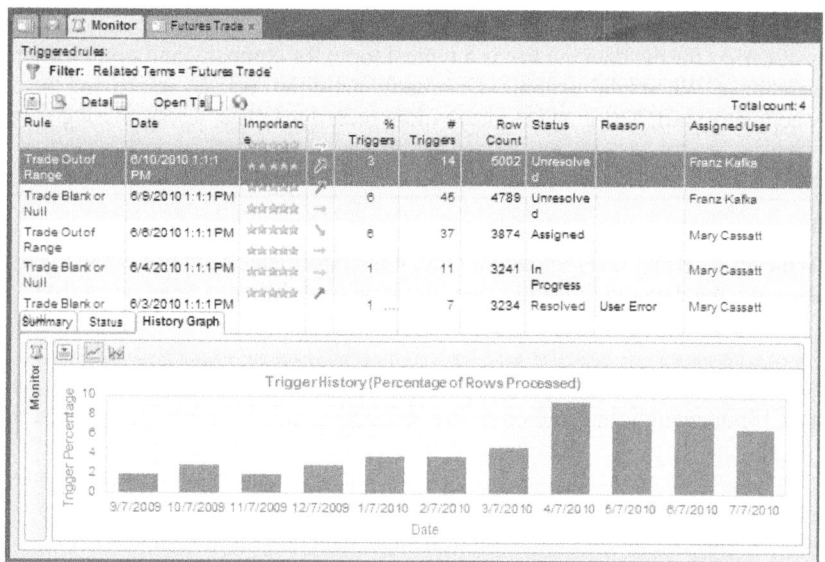

Data quality responsibilities separated from business projects

These steps are usually initiated in the set-up of a data warehouse. But they are also an ongoing process as new analysis questions arise that see the data from different viewpoints. Therefore, projects are often the starting point to check and improve the quality of a certain domain of data in the data warehouse.

Business projects can trigger a consecutive or parallel data quality project and can highlight the need for data quality improvements that need to be resolved before the business project can be completed. Data quality improvement of the source data, should, however, not be part (or the appendix) of a business project in general.

Projects usually span a certain lifetime. If a data quality project is linked with a business project, there is the risk that, after the completion of the business project, not enough ongoing support and focus on data quality will be given. This is especially important because data quality should not be seen as a one-time task but rather as an ongoing requirement. Data quality should be an ongoing operative process rather than part of a business project.

Process features that trigger good data quality

In practice, data quality often also depends on the fact that the

- data are used to control and govern a process
- data are generated and stored by an automated electronic process

If data are used to control and govern a process, the data must be available and in a good quality state; otherwise, the business process cannot be performed. On the other hand, if data are collected and maintained but there is no operational urgency for the data, the data are usually only updated occasionally. This only becomes evident if the data are needed for analysis and the analysis identifies many missing values, free-text entries instead of code, and outdated versions of the data.

Data that are not collected manually but through automatic and electronic transfer are usually more complete and up-to-date than data that need to be manually updated. Often, transactional data like sales numbers, bookings, and orders have better data quality in the data warehouse system because they are collected from rigid operational systems that facilitate the customer order and sales process. Attributes in the customer or product database that need to be maintained manually are often not very accurate or complete.

9.5 Data Quality as an Ongoing Process

General

Similarly, the ongoing process of caring for the data quality is a typical topic for company and organizational data that are stored in a data warehouse. When talking about data quality activities, usually improving the data quality status is understood by company staff. The goal is to have more record matches, fewer missing values, or fewer outliers.

Maintaining the status

However, even if a company achieves a certain level of data quality, it is, unfortunately, not possible to consider the data quality topic solved. Usually, the data quality status does not remain at the same level when data quality activities are stopped.

Data quality actions can be seen in the same context as the need to maintain a house regularly, service a car, or take care of your personal health. If not enough time and effort are invested in this, the quality decreases, for example:

- For customer databases, it is important to perform record matching to eliminate or join duplicate records. This is, however, an ongoing project because, over time, new records are created in the database that are potential duplicates. Thus, even when a duplicate-free version of the database has been created at some point, the ending of data quality actions will cause an increase in duplicate records and a decline in data quality.
- At the start of a forecasting project in the retail industry, extreme and implausible values are manually checked and corrected in order to have a correct version of the data. Based on these data, automatic forecasting for future period is performed. The forecasts at this project stage will reflect the corrected version of historic data and result in the corresponding forecast. Usually time series analyses are

reforecast in regular intervals (for example, by month) in order to include more recent data. If the data that are collected for the new months will not receive the same data quality treatment, their quality and correctness will decline over time and the resulting forecasting quality will decline as well.

In summary, the data quality process is not only an ongoing process in terms of successive improvements to the data, but it is also ongoing in terms of the demand for regular data quality actions to maintain the data quality status at the same level.

Thus, **data quality improvement** is designed to bring data to a certain data quality level (for example, by reducing the number of missing values, achieving a higher correctness of the data, having fewer unexplained outliers, and allowing a shorter time delay for the provision of the data).

Data quality maintenance is designed to maintain the quality at a certain level. In many cases, this means that a certain quality level has been reached and the quality state of the system will be retained for future periods. This includes, for example, that record matching logic is updated due to new conditions. Also, the lists of possible values may need to be updated as new tariff plans or new products and product categories come into play.

Short-term fixing or long-term improvement

If data quality problems arise in an analysis or in a project, there are two options you can take. One is to stop the project, thoroughly improve the data quality, and only then rerun the analysis. The other is to provide a data quality fix to the data as part of the project and proceed with the project.

In practice, the decision depends on the importance of completing the project on time, the severity of the data quality problem, and the ability to provide a quick fix. If possible, data quality problems are fixed as quickly as possible as part of the project, and the project can eventually be completed within the expected timeframe.

The important fact is not who performs the data quality improvement. It is more important to make sure that the measures to improve data quality are not only part of this particular project. Otherwise other departments, projects, or analyses could not benefit from the improvement and the respective logic.

If the data correction logic is not applied centrally, each data consumer will have to improve the data quality on their own. Thus, over time, much more effort is invested into quick-fix solutions than into central data improvement.

The fact that an analysis cannot be performed right now because of a lack of historic data or snapshots should not stop data collection for this domain. It should rather be a trigger to start data collection right now to make sure that historic data and historic snapshots are available in future periods.

9.6 Data Quality Monitoring

General

In order to understand the quality status of the data, data quality features are defined that can be measured and represented over time. The selection of key performance indicators (KPIs) to describe the data quality features is domain-specific. It depends on the specific goals that you want to achieve when profiling the data quality.

Some of the KPIs are suitable for a quick initial profiling of the status of the available data (for example, by providing the number of missing values or the number of observations). Other KPIs only make sense if data quality characteristics are compared over time.

Example KPIs for data quality monitoring

The following list contains example key performance indicators (KPIs) that can be used to profile the quality status of the data. Note that this list is not a complete list of all possible KPIs. Also not all KPIs make sense in all circumstances.

- Number of duplicate records in the database
- Number of parent records without child records
- Number of child records without parent records
- Number of observations without values for mandatory identification variables
- Number of observations
- Number of events
- Number of missing values for key variables of the master data, like address data, telephone numbers, or e-mail addresses
- Number of missing values per variable
- Number of observations with at least one missing value
- Number of observations that cannot be scored
- Number of categories for categorical variables
- Number of new categories compared to the number of categories in previous months for a particular variable
- Number of observations with new categories in at least one of the input variables
- Descriptive statistics that describe the distribution of an interval variable
- Deviation of the distribution of an interval variable compared with a previous month
- Number of observations with outliers
- Number of variables that are available compared to a predefined list of variables
- Number of variables that are included in the final analytical model
- Number of time series that can be forecast with nontrivial models
- Deviations between the two data versions from double data entry
- Length of data history for transactional or time series data

Dimensions for analyzing the data quality

In some cases, the overall picture does not provide enough insight, and it is necessary to detail the analysis by subcategories. These subcategories then allow the researcher to get a more detailed picture and to compare data quality characteristics between different categories.

The most important dimensions are:

- time
- region, like district or country
- functional domain, like subdivision in companies
- staff members, differentiating between the people who are mainly responsible for providing the data (like call center agents or data entry personnel)

These items are only a suggestion of possible dimensions. The most important dimension for subgrouping the analysis is undoubtedly time.

Example screenshots are shown in Figure 9.2.

Figure 9.2: Monitoring data quality KPIs in DataFlux (DataFlux Data Management Platform)

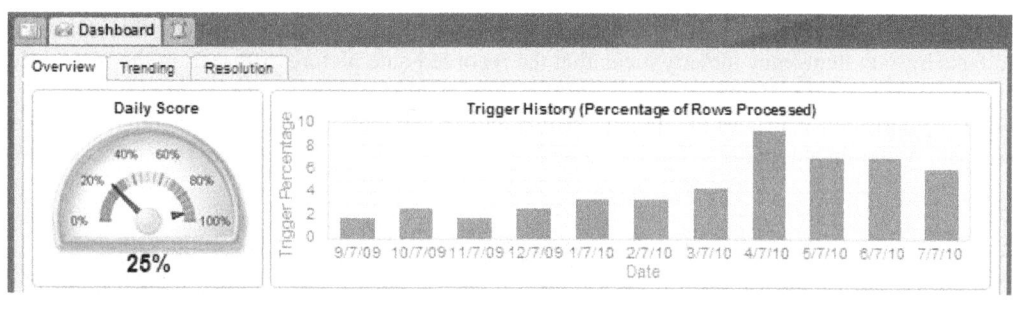

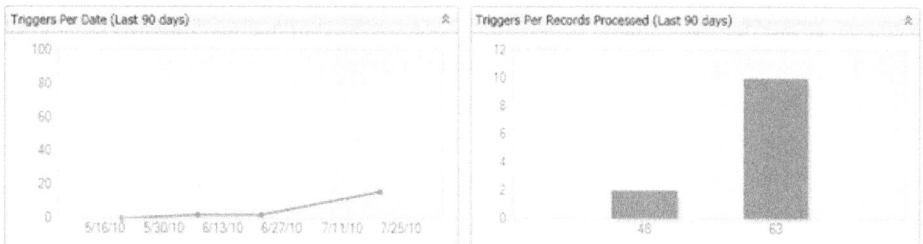

Analysis over time

Data quality should not only be analyzed and reported at a single point in time, but it should be monitored over time. Only then can the upward or downward trends of data quality characteristics be seen. Also, insight into the variability of the data quality status or the dependence on seasonal effects can be gained (see also the lower left chart in Figure 9.2).

The time dimension is also important in a so called "vintage analysis" to see whether customer, that are acquired in different calendar months or through different marketing campaigns, show a different behavior in completing the applications forms or providing personal data.

For this reason, a quality monitoring system should not only show static information, but it should display information over time with line charts. Only then is a comparison with previous periods possible. A comparison over time allows you to judge whether a data quality status at a particular point in time is problematic or not.

Many data quality problems also start small; only a few records are not matched when two tables are joined. If in a database with 6 million customers, 27 records do not match, from an analytical point of view, there is no reason to worry because there is still a large enough set of records to run the analysis. The problem, however, is the underlying reason for the non-match. If in the current month 27 records do not match, it is very likely that in the next month a larger number will not match (for example, 103). In the next month, there are 181 non-matches. From a static point of view, 181 non-matches are still not a serious problem; however, the increase over time shows the trend and allows you to assess the size of the problem for the future.

Outlook

Part II of this book introduces different capabilities of analytical methods to profile the data quality status. Specific methods (for example, to obtain deep insight into the structure of missing values) are presented in these chapters.

9.7 Conclusion

This chapter has shown aspects of data quality from a process point of view. It has shown that the representation of real-world facts by data items may already mean that the recorded data picture differs from the real world.

The chapter discussed the consequences of bad data quality and shown that data quality is an ongoing process, not a one-shot task. A certain amount of effort has to be invested just to keep data at their current data quality level.

Another section discussed the fact that data quality should not be seen as part of a project but rather as a separate operational task in the IT department. Analysis projects, however, can definitely be the trigger for data quality projects.

Finally, data quality characteristics that are suitable for profiling and monitoring the data quality status over time were presented and the importance of comparing the characteristics over time highlighted.

Part II: Data Quality—Profiling and Improvement

Introduction
This second part of the book shows methods you can use to profile and improve the quality of the data.

Profiling
The methods for profiling focus primarily on advanced features of the data like:

- the structure of missing values in a one-row-per-subject data mart
- the structure of missing values in a time series data mart
- the fact that observations in time series data are missing
- the detection of complex outliers like multivariate outliers or outliers in time series data
- the detection of duplicate records in the data based on matching algorithms

Methods for simple data profiling and validation are only briefly mentioned and the reader is directed to the respective references.

Improvement
The chapters also show how to improve the data quality status, for example:

- inserting missing records in time series data sets
- imputing missing values for both one-row-per-subject data marts and time series data marts
- generating match keys for record matching and de-duplication
- calculating individual imputation and correction values that are based on other variables of the respective analysis subject

SAS focus
These chapters focus on SAS solutions that provide these methods. Methods from classical SAS software like Base SAS, SAS/STAT, and SAS/ETS are shown as well as methods in SAS Enterprise Miner, SAS Text Miner, and SAS Forecast Server as well as features of advanced solutions like DataFlux.

These methods are presented as examples with full SAS code. For some specific tasks, SAS macros are presented. The code for these macros can be found in the appendixes. All SAS resources (programs, macros, and data sets) that are presented here can be downloaded from the sascommunity.org page available at http://www.sascommunity.org/wiki/Data_Quality_for_Analytics.

Overview of the chapters
Part II has five chapters:

- Chapter 10, "**Profiling and Imputation of Missing Values**," deals with missing values in general.
- Chapter 11, "**Profiling and Replacement of Missing Data in a Time Series**," discusses the topic of missing values and records in time series data.
- Chapter 12, "**Data Quality Control across Related Tables**," shows methods to profile data quality in tables that are related (for example, through a one-to-many relationship).

- Chapter 13, "**Data Quality with Analytics**," covers the capabilities of analytics for data quality.
- Chapter 14, "**Data Quality Profiling and Improvement with SAS Analytic Tools**," connects the capabilities of the SAS offering with the requirements of data quality.

Chapter 10: Profiling and Imputation of Missing Values

10.1 Introduction 124
More than simple missing value reporting 124
Profiling missing values 124
Imputing missing values 124
SAS programs 124

10.2 Simple Profiling of Missing Values 125
General 125
Counting missing values with PROC MEANS 125
Using a macro for general profiling of missing values 125

10.3 Profiling the Structure of Missing Values 126
Introduction 126
The %MV_PROFILE_CHAIN macro 127
Data to illustrate the %MV_PROFILE_CHAIN macro 128
Simple reports based on the missing value profile chain 128
Advanced reports based on the missing value profile chain 130
Usage examples 132
Usage information for the %MV_PROFILING macro 132

10.4 Univariate Imputation of Missing Values with PROC STANDARD 133
General 133
Replacement values for the entire table 133
Replacement values for subgroups 134

10.5 Replacing Missing Values with the Impute Node in SAS Enterprise Miner 135
Overview 135
Available methods for interval variables 136
Available methods for categorical variables 137
References 137

10.6 Performing Multiple Imputation with PROC MI 137
Single versus multiple imputation 137
Example data for multiple imputation and analysis 138
Performing multiple imputation with PROC MI 138
Analyze data with PROC LOGISTIC 139
Combine results with PROC MIANALYZE 140

10.7 Conclusion 140
The SAS offering 140
Business and domain expertise 140

10.1 Introduction

More than simple missing value reporting

This chapter deals with methods to profile and impute missing values in one-row-per-subject analysis tables. The profiling of missing values can start with the simple counting of observations that have a missing value, as shown in section 10.2.

Many data cleaning and data quality tools and solutions provide profiling reports that show the number of missing values, with frequencies and percentages, in tables and other graphical representations. This information is important to get a picture of the number of valid observations that meet the definition and can be used for the analysis. Actually, it is not difficult from an analysis point of view to query and aggregate this information.

For analytical purposes, but also for data cleaning and data profiling in general, it is often important to know which variables are usually missing together. This allows the analyst to find structures and associations in the data that may reflect patterns that mirror specifics of the data collection, data retrieval or data entry process. Most importantly, the display and analysis of such patterns also provides an indication whether the missing values occur randomly or whether systematic combinations can be found.

The definition of a missing value in this context is an observation that has a NULL value (or a predefined missing value identifier) for a particular variable. Checking for observations that are missing is not the focus in this chapter. It is shown in the next chapter in the context of time series data.

Profiling missing values

The focus of this chapter is to introduce more advanced methods to profile missing values. This is shown in section 10.3. Not only are the number of missing values analyzed, but this chapters looks at the inherent structure of missing values across different variables.

For each observation, a characteristic pattern is created that describes the distribution of missing values across variables. This pattern, which is described in the form of tabulations and a tile chart, offers firsthand insight into the structure of missing values. Additionally, cluster analysis and principal components are used to create more insight into frequent co-existences of missing values across variables. They also show the "closeness" and relationships between variables based on the fact that missing values occur together for these variables.

Imputing missing values

The final sections of this chapter show different methods used to impute missing values:

- Section 10.4 shows how simple imputations can be performed (for example, by replacing missing observations with the mean of the existing observations).
- Section 10.5 shows more advanced methods to impute missing values that are available in the Impute node in SAS Enterprise Miner.
- While these methods only impute a single value for a missing value, section 10.6 also shows how multiple imputations for missing values can be performed and analyzed.

SAS programs

To profile missing values, this chapter introduces the macros %COUNT_MV (count missing values) and %MV_PROFILING (missing value profiling). The code for these macros is shown in appendix A. All macros, SAS programs, and example data sets that are used in this chapter can be downloaded from the source, as shown in the introduction at the beginning of this book.

10.2 Simple Profiling of Missing Values

General

Simple profiling of missing values is not difficult from an analytical point of view. The number of missing values for each variable in an analysis table can easily be profiled with basic methods of descriptive statistics.

Counting missing values with PROC MEANS

A simple and quick method to check the number of missing values in a table is to use PROC MEANS with the NMISS option:

```
proc means data = hmeq nmiss;
run;
```

Note that only variables with a numeric format can be analyzed with this method. If no variables are specified in the VAR statement, all numeric variables in the data set are used for the analysis. An alternative would be to use the VAR statement in the MEANS procedure:

```
Var mortdue value yoj;
```

The output that is generated from this procedure is shown in Output 10.1.

Output 10.1: Output of the MEANS procedure for counting the number of missing observations

```
             N
Variable   Miss
─────────────────
MORTDUE    518
VALUE      112
YOJ        515
```

Using a macro for general profiling of missing values

In the case of categorical variables, PROC FREQ can be used to count the number of missing values for each variable. However, each variable requires a separate TABLES statement, and the output is shown in a separate table.

A macro can help to automate the calculation of the number of missing values for categorical variables. This macro can be used not only for character variables but for all variable types. Additionally, the percentage of missing values is calculated for each variable. The syntax and the description of the macro can be found in appendix A. Here is an example:.

```
%count_mv(data=hmeq, vars = job reason yoj value);
```

Note that JOB and REASON are categorical variables and YOJ and VALUE are interval variables. The output is shown in Output 10.2.

Output 10.2: Output of missing value profiling with the %COUNT_MV macro

```
           Number    Proportion_
Variable   Missing   Missing       N

Job        279       0.05          5960
reason     252       0.04          5960
yoj        515       0.09          5960
value      112       0.02          5960
```

The following parameters can be specified with the macro:

DATA
    Name of the input data set.

VARS
    List of variables in the data set to be analyzed. Variable names must be separated by a blank.

10.3 Profiling the Structure of Missing Values

Introduction

The previous section introduced the most basic methods for profiling missing values. This section presents a method for profiling the structure of missing values in a data mart, which not only counts the missing values for each variable but also displays a profile of the missing values that occur together. This allows the analyst to learn which variables in the data mart are correlated by the fact that they have missing values for the same observations. Profiles provide insight on whether values are missing at random or systematically. For example, this allows an analyst to see whether variables from specific data sources or domains have a high frequency of missing values.

Such a method is implemented in JMP as "Missing Data Pattern" (see also [12], p. 68). Example output is shown in Outputs 10.3 and 10.4.

Output 10.3: Table output of JMP Missing Data Pattern

	Count	Number of columns missing	Patterns	BAD	JOB	DEROG	DELINQ	CLAGE	NINQ	CLNO
1	4924	0	0000000	0	0	0	0	0	0	0
2	52	1	0000010	0	0	0	0	0	1	0
3	64	1	0000100	0	0	0	0	1	0	0
4	21	1	0001000	0	0	0	1	0	0	0
5	56	2	0001010	0	0	0	1	0	1	0
6	174	1	0010000	0	0	1	0	0	0	0
7	19	2	0010010	0	0	1	0	0	1	0
8	132	2	0011000	0	0	1	1	0	0	0
9	149	3	0011010	0	0	1	1	0	1	0
10	90	5	0011111	0	0	1	1	1	1	1
11	113	1	0100000	0	1	0	0	0	0	0
12	22	2	0100100	0	1	0	0	1	0	0
13	12	3	0110010	0	1	1	0	0	1	0
14	132	6	0111111	0	1	1	1	1	1	1

Output 10.4: Graphical output of JMP Missing Data Pattern

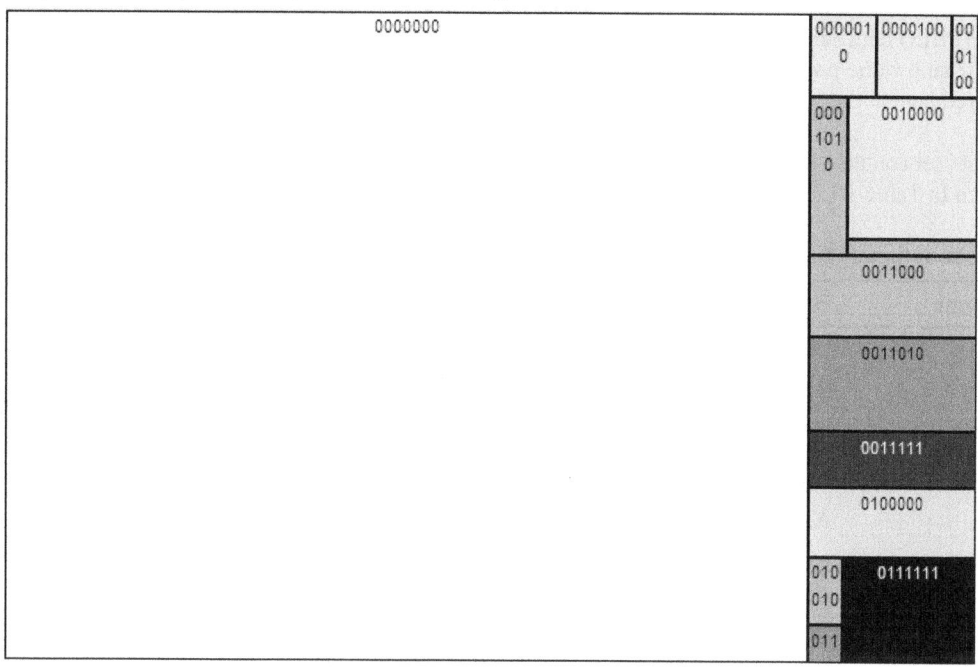

This section introduces the %MV_PROFILE_CHAIN macro, which provides the same functionality in SAS. Additionally, methods of cluster and principal components analysis are applied to provide more insight into the inherent multivariate structure of missing values. Unlike the JMP implementation shown, the SAS macro presented here calculates the total number of missing values per observation.

The %MV_PROFILE_CHAIN macro

The missing value profile chain (%MV_PROFILE_CHAIN) macro defines a segmentation of the observations based on its missing value structures. For each variable, a dummy variable is created that indicates whether a value is missing or not. This set of indicator variables is then concatenated into a string. Table 10.1 illustrates the definition of the profile chain.

Table 10.1: Definition of the missing value profile chain

Age	Gender	Income	MV_Profile_Chain
68	M	.	001_1
.	F	.	101_2
.	M	3040	100_1

Note that after the profile chain itself, the total number of missing values per observation is concatenated separated by a '_'. This approach shows the number of missing values for each profile and is very useful when larger numbers of variables are analyzed.

Using the %MV_PROFILE_CHAIN macro has many advantages:

- It provides insight into the structure of missing values across variables.
- It identifies the frequency and accumulation of missing value patterns.
- It shows information efficiently. Instead of looking at multiple graphs and reports, an analyst can get a quick overview of the missing value structure.

Data to illustrate the %MV_PROFILE_CHAIN macro

Data from the HMEQ data set, which is frequently used in SAS Enterprise Miner examples and demos, illustrate the missing value profile chain macros. The data set is available together with the programs and macros presented here at the location described in the introduction of this book.

The HMEQ data set contains histories of credit customers with different types of home equity loans. The variables shown in Table 10.2 were used to analyze the missing values.

Table 10.2: Description of the variables used in the %MV_PROFILE_CHAIN macro

Variable Name	Description
BAD	Defaulting or repaying the loan
DELINQ	Number of delinquent trade lines
DEROG	Number of major derogatory reports
NINC	Number of recent credit inquires
CLAGE	Age (in months) of the oldest trade line
CLNO	Number of trade lines
JOB	Current job, six categories

Simple reports based on the missing value profile chain

Output 10.5 shows a lookup table that displays the MV_PROFILE assigned to each variable. This identifies which digit of the %MV_PROFILE_CHAIN macro is used for which variables and eases navigation through the MV_PROFILE_CHAIN report.

Output 10.5: MV_PROFILE pattern for each variable

Variable	MV_PROFILE
BAD	1000000
DELINQ	0100000
DEROG	0010000
NINQ	0001000
CLAGE	0000100
CLNO	0000010
JOB	0000001

The order of the 0s and 1s in the %MV_PROFILE_CHAIN macro depends on the order of the variables names in the &VARS macro variable. If the &VARS parameter is omitted in the macro, all variables of the data set are used in alphabetical order for the %MV_PROFILE_CHAIN macro by default. The macro variable ORDER=POS can be used to force the order to match the order in the data set.

Output 10.6 shows the output for the frequencies of the different patterns in the %MV_PROFILE_CHAIN macro.

Output 10.6: Output of the %MV_PROFILE_CHAIN macro for the HMEQ data set

MV_PROFILE_CHAIN	Frequency	Percent	Cumulative Frequency	Cumulative Percent
0000000_0	4924	82.62	4924	82.62
0010000_1	174	2.92	5098	85.54
0111000_3	149	2.50	5247	88.04
0110000_2	132	2.21	5379	90.25
0111111_6	132	2.21	5511	92.47
0000001_1	113	1.90	5624	94.36
0111110_5	90	1.51	5714	95.87
0000100_1	64	1.07	5778	96.95
0101000_2	56	0.94	5834	97.89
0001000_1	52	0.87	5886	98.76
0000101_2	22	0.37	5908	99.13
0100000_1	21	0.35	5929	99.48
0011000_2	19	0.32	5948	99.80
0011001_3	12	0.20	5960	100.00

First, the output shows that there are 82.6% complete records without missing values. The next line shows that there are 174 observations (2.9%) that have a missing value for the third variable (in this case, variable DEROG). The next entries in the table reveal that in 2.5% of the cases three variables are missing together (namely, variables DELINQ, DEROG, and NINQ) (the second, third, and fourth variables) and in 2.2% of the cases variables DELING and DEROG are missing. This provides first-hand insight not only into the frequency but also into the structure of the missing values.

This information is formatted as both a table and a tile chart, as shown in Output 10.7. A tile chart splits a rectangle that represents the population into smaller rectangles, where the area of each rectangle is proportional to the frequency of the respective category. In this case, this gives a graphical representation of the proportion of the complete records and the records with missing values. Additionally, it can easily be seen that the top four patterns with missing values make up more than 50% of all missing value patterns.

Output 10.7: Tile chart for the frequency of the %MV_PROFILE_CHAIN macro

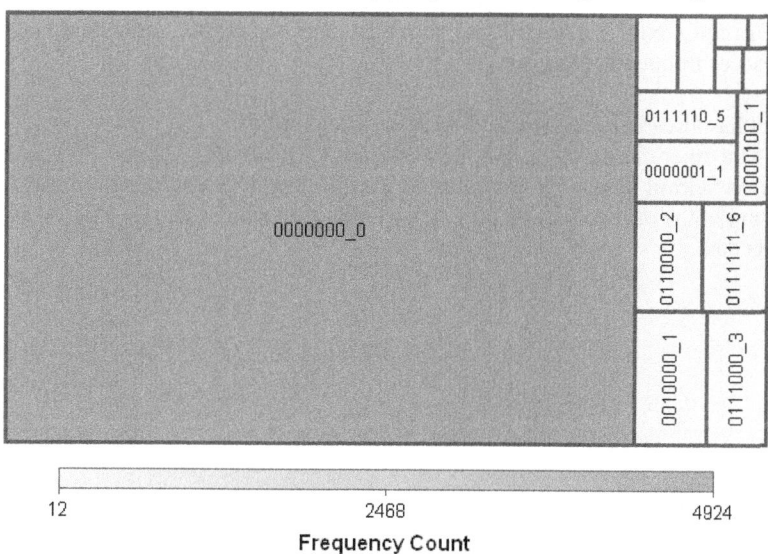

The next two output examples show the distribution of the number of missing values per record (Output 10.8) and a ranked list of missing values per variable (Output 10.9).

Output 10.8: Distribution of the number of missing values per record

N_MV	Frequency	Percent	Cumulative Frequency	Cumulative Percent
0	4924	82.62	4924	82.62
1	424	7.11	5348	89.73
2	229	3.84	5577	93.57
3	161	2.70	5738	96.28
5	90	1.51	5828	97.79
6	132	2.21	5960	100.00

Output 10.9: Ranked list of missing values per variable

Variable	Missing_Abs	Missing_Rel	MV_PROFILE
DEROG	708	11.88%	0010000
DELINQ	580	9.73%	0100000
NINQ	510	8.56%	0001000
CLAGE	308	5.17%	0000100
JOB	279	4.68%	0000001
CLNO	222	3.72%	0000010
BAD	0	0.00%	1000000

Output 10.8 shows that there are 82.6% complete records and that in 6.4% of the cases three or more variables have a missing value. Looking at Output 10.9, you can see that target variable BAD does not include any missing values. Combining this information with Output 10.7 lets us conclude that in 2.2% (pattern 0111111_6) of the cases all input variables are missing.

Advanced reports based on the missing value profile chain

In the next step, the structure and patterns of the missing values are analyzed in more detail. In order to do this, a variable clustering and a principal component analysis are applied to binary indicator variables that were created earlier. The binary indicator variables are assigned a value of 1 if the respective variable has a missing value and 0 otherwise. For this analysis, all observations without missing values are excluded from the analysis because they do not provide any additional insight.

The tree plot in Output 10.10 displays the result of the variable clustering. Two main clusters with three variables each are produced. The DELINQ, NINQ, and DEROG variables form one cluster, indicating that missing values for these variables occur frequently together.

A preliminary indication of this fact has already been seen from the distribution of the %MV_PROFILE_CHAIN macro. The other cluster comprises the variables CLAGE, CLNO, and JOB. From a business point of view, this analysis is helpful because it identifies which missing value patterns occur together. Reasons for this pattern may be found in the data collection process or in the data availability in general, as well as in the business background of the data.

Output 10.10: Tree plot for missing values based on variable clusters

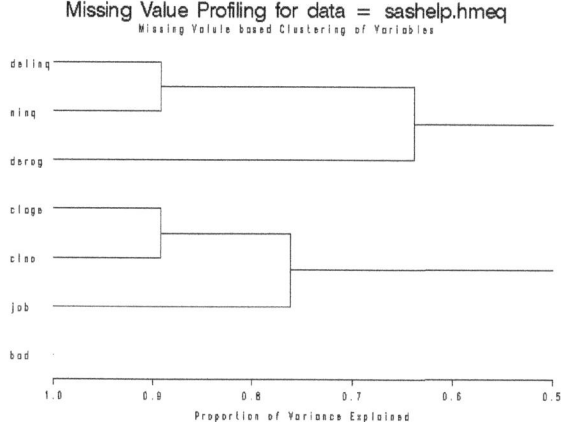

Output 10.11 shows a similar picture, where the closeness of the variables is derived from the first and second principal component of the missing value variable indicators.

Output 10.11: Principal component plot based on missing value indicators

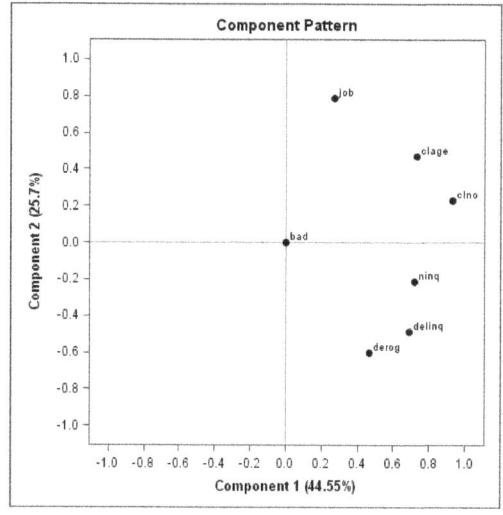

Note that the proportion of the Eigenvalues for each component is shown in brackets on the respective axis. More details about the proportion of several Eigenvalues can be obtained from Output 10.12. The macro provides a parameter NCOMP to control the number of principal components used for the principal component plots. The default value is 2 (resulting in one plot).

Output 10.12: Proportion and cumulative proportion of the Eigenvalue per principal component

	Eigenvalue	Difference	Proportion	Cumulative
1	2.67320227	1.13123746	0.4455	0.4455
2	1.54196481	0.81059024	0.2570	0.7025
3	0.73137457	0.15854384	0.1219	0.8244
4	0.57283073	0.18179027	0.0955	0.9199
5	0.39104046	0.30145331	0.0652	0.9851
6	0.08958715	0.08958715	0.0149	1.0000
7	0.00000000		0.0000	1.0000

Usage examples

Consider the following usage examples for the %MV_PROFILING macro:

Profile all variables from data set HMEQ:

    %MV_Profiling (data=hmeq, vars = _all_)

Profile the variables JOB, REASON, DEBTINC, and VALUE from data set HMEQ:

    %MV_Profiling (data=hmeq, vars = job reason debtinc value);

Sample the data for better performance to 10%:

    %MV_Profiling (data=hmeq, vars = _all_, sample=0.1);

Using only variables starting with "D," turn off creating the tile chart, the variable clustering, and the principal component analysis:

    % MV_Profiling (data=hmeq,vars = d:,ods = NO, princomp=NO);

Usage information for the %MV_PROFILING macro

In order to profile the structure of missing values and to create these results, the SAS macro %MV_PROFILING has been programmed. The macro requires Base SAS, SAS/STAT, and SAS/GRAPH software. The full code for this macro is documented in appendix A.

Note the following:

- In order to produce the principal component chart, ODS HTML has to be activated for the macro. You can do this either by
  - submitting the statement ODS HTML; before the macro and the statement ODS HTML CLOSE; after the macro
  - running the macro within SAS Enterprise Guide.
- In order to produce the tile chart, the graphical device has to be set to ACTIVEX or JAVA in addition to ODS HTML. This can be done either by submitting GOPTIONS DEVICE = JAVA; or GOPTIONS DEVICE = ACTIVEX; in the SAS code or by specifying ActiveX or JAVA as graphic format in the SAS Enterprise Guide options.
- The list of variables can also be quoted by a character followed by a colon ":". This allows you to specify all variables that start with the same character(s). If no variables match the quoted string, the macro stops executing.

The following parameters can be specified with the %MV_PROFILING macro:

DATA
Name of the input data set.

VARS
List of variables in the data set to be analyzed. Variable names are separated by a blank. If _ALL_ is specified, all variables in the data set are used. The default value is _ALL_. A wildcard notation with a colon ":" can also be used (for example, DEMO: causes all variables starting with "DEMO" to be used).

ORDER
Only relevant when VARS=_ALL_ is specified. Possible values are POS or ALPHA. The default value is ALPHA. ALPHA means that the variables of the data set are used in alphabetical order for the %MV_VALUE_CHAIN macro. POS means that the variables are used in the same order as in the data set for the %MV_VALUE_CHAIN macro.

ODS
Specifying ODS = YES creates the tile chart, the VARCLUS tree plot, and the principal component analysis. Default = YES.

VARCLUS
Specifying VARCLUS = YES creates the VARCLUS tree plot. Default = YES.

PRINCOMP
Specifying PRINCOMP=YES creates the principal component analysis. Default = YES.

NCOMP
Defines the number of principal components that are used for the plots of components. NCOMP = 2 creates one plot for the first and second components. NCOMP = 3 creates three plots: one for the first and second components, one for the first and third components, and one for the second and third components, and so forth.

SAMPLE
Specifies a sample proportion to allow you to run missing value profiling on a sample rather than on the entire data set. Default = 1 (100%), no sampling. Values from 0 to 1 are valid.

SEED
Seed value that is used for sampling. Default = 18,419.

10.4 Univariate Imputation of Missing Values with PROC STANDARD

General
PROC STANDARD is part of Base SAS and allows you to standardize the data. For example, data can be normalized to a mean of 0 and a standard deviation of 1 or they can be standardized to a specific mean and standard deviation. The procedure also provides a REPLACE option, which replaces observations with missing values by its mean. Only numeric variables can be used for replacement.

Replacement values for the entire table
Consider the following data of credit customers that is extracted from the HMEQ data set, which can be downloaded from the link named in the introduction.

Table 10.3: Credit customer data

ID	BAD	MORTDUE	VALUE	JOB
1	1	25860	39025	Other
2	1	70053	68400	Other
3	1	13500	16700	Other
4	1			
5	0	97800	112000	Office
6	1	30548	40320	Other
7	1	48649	57037	Other
8	1	28502	43034	Other
9	1	32700	46740	Other
10	1		62250	Sales
11	1	22608		
12	1	20627	29800	Office
13	1	45000	55000	Other
14	0	64536	87400	Mgr
15	1	71000	83850	Other
16	1	24280	34687	Other
17	1	90957	102600	Mgr
18	1	23030		
19	1	28192	40150	Other
20	0	102370	120953	Office
21	1	37626	46200	Other
22	1	50000	73395	ProfExe
23	1	28000	40800	Mgr
24	1	18000		Mgr
25	1		17180	Other

For some observations, values for MORTDUE (amount due on existing mortgage), VALUE (value of current property), and JOB (job type) are missing. In order to replace the values of the numeric variables MORTDUE and VALUE with the mean of the existing values, the following code can be used:

```
proc standard data = CreditData
 out = CreditData_Replaced
 replace;
 var mortdue value;
run;
```

Note that the REPLACE option must be specified in order to force a replacement of missing values.

Table 10.4 shows the data with the replaced values. Note that for each variable the same imputation value is imputed for all observations.

Table 10.4: Credit customer data with replacement values

ID	BAD	MORTDUE	VALUE	JOB
1	1	25860	39025	Other
2	1	70053	68400	Other
3	1	13500	16700	Other
4	1	73760.8172	101776.049	
5	0	97800	112000	Office
6	1	30548	40320	Other
7	1	48649	57037	Other
8	1	28502	43034	Other
9	1	32700	46740	Other
10	1	73760.8172	62250	Sales
11	1	22608	101776.049	
12	1	20627	29800	Office

Replacement values for subgroups

In order to perform a subgroup-specific replacement of missing values (for example, by job type), you can run PROC STANDARD with a BY statement. The data must be sorted in advance using the BY variable.

```
proc sort data = CreditData
 out = CreditData_Sorted;
 by job;
run;

proc standard data = CreditData_Sorted
 out = CreditData_Replaced_ByJob
 replace;
 var mortdue value;
 by job;
run;
```

After sorting the data by ID, the result of the replaced data is shown in Table 10.5.

Table 10.5: Credit customer data with replacement values grouped by job type

ID	BAD	MORTDUE	VALUE	JOB
1	1	25860	39025	Other
2	1	70053	68400	Other
3	1	13500	16700	Other
4	1	62687.3094	83701.9179	
5	0	97800	112000	Office
6	1	30548	40320	Other
7	1	48649	57037	Other
8	1	28502	43034	Other
9	1	32700	46740	Other
10	1	82266.1818	62250	Sales
11	1	22608	83701.9179	
12	1	20627	29800	Office
13	1	45000	55000	Other
14	0	64536	87400	Mgr
15	1	71000	83850	Other
16	1	24280	34687	Other
17	1	90957	102600	Mgr
18	1	23030	83701.9179	
19	1	28192	40150	Other
20	0	102370	120953	Office
21	1	37626	46200	Other
22	1	50000	73395	ProfExe
23	1	28000	40800	Mgr
24	1	18000	108628.263	Mgr
25	1	59337.1479	17180	Other

Now individual replacement values for MORTDUE, depending on different job types (MISSING, 'Sales' and 'Other') for ID 4, 10, and 25 are used.

10.5 Replacing Missing Values with the Impute Node in SAS Enterprise Miner

Overview

SAS Enterprise Miner provides a rich set of methods for automated imputation of missing values with the Impute node. The Impute node allows you to generate imputation values for numeric as well as for categorical variables. An example of the node usage can be seen in Figure 10.1.

The Impute node is part of the MODIFY group of nodes in SAS Enterprise Miner. MODIFY refers to the modify step in data mining SEMMA (Sample, Explore, Modify, Model, Assess) methodology.

Figure 10.1: Example of the Impute node

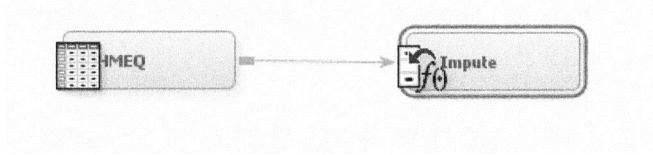

The node parameters are controlled by a property sheet, where imputation methods for all interval or categorical variables can be specified globally on a data set level or individually on a variable level. Figure 10.2 shows an example of the property sheet. Separate methods can be specified for input or target variables.

Figure 10.2: Property sheet to specify the imputation method globally for the data set

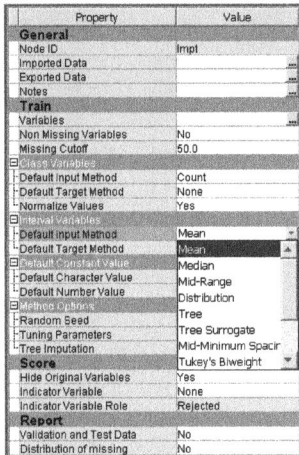

Successor nodes of the Impute node receive the input data with the imputed values in new variables starting with the prefix "IMP_" in their name. Additionally, the original variables with the nonimputed values are available in the data and can be dropped if desired. The Impute node also allows you to create an indicator variable for each variable in the data set that indicates whether a missing value has been imputed or not.

Note that the imputation is not only performed for the training data partition but also for the validation and test data partition, if they exist.

Available methods for interval variables

For interval data, the available imputation methods are listed in Table 10.5.

Table 10.5: Imputation methods for interval data

Method	Description
Mean	Arithmetic mean of the non-missing observations
Median	Median of the non-missing observations
Distribution	Replacement values are calculated on the random percentiles of the variable's distribution of the non-missing values. This method does not change the distribution of the data very much.
Tree	A decision tree for each variable is run that predicts the imputation value based on other values in the data set. The variables that are used as input variables for the tree can be selected. This method produces more accurate imputation values, but it is also more time-consuming.
Tree Surrogate	Same method as the tree method. Additionally, surrogate splitting rules are created when an input variable is missing.
Midrange	(Minimum+Maximum)/2
Mid-Minimum	A certain proportion of the data is trimmed, by default the lower and upper 5%. From the remaining data, the MIDRANGE is calculated as the imputation method.
Tukey's Biweight	Tukey's biweight robust M-estimator
Huber's	Huber's robust M-estimator
Andrew's Wave	Andrew's Wave robust M-estimator
Default Constant	Default constant to replace the missing values

Note that Tukey's biweight, Huber's, and Andrew's wave are robust M-estimators of location. Common estimators such as the sum of squared residuals can become unstable when they use outlier data points and can distort the resulting estimators. M-estimators try to reduce the effect of outliers by using substitute functions.

Available methods for categorical variables

For categorical variables, the available imputation methods are listed in Table 10.6.

Table 10.6: Imputation methods for categorical data

Method	Description
Count	Replace the missing value with the category that has the highest frequency.
Default Constant	Default constant to replace the missing values.
Distribution	Replacement values are calculated on the random percentiles of the variable's distribution of the non-missing values. This method does not change the distribution of the data very much.
Tree	A decision tree for each variable is run that predicts the imputation value based on other values in the data set. The variables that are used as input variables for the tree can be selected. This method produces more accurate imputation values, but it is also more time-consuming.
Tree Surrogate	Same method as the tree method. Additionally, surrogate splitting rules are created when an input variable is missing.

Note that the tree method and the tree surrogate method, which are the two most advanced methods, are available for both categorical and numeric variables.

References

For more details on the imputation methods, refer to the node reference help in SAS Enterprise Miner.

Chapter 18 documents a simulation study where the Impute node is used in a SAS Enterprise Miner process flow.

10.6 Performing Multiple Imputation with PROC MI

Single versus multiple imputation

The methods that have been presented in the preceding sections have one thing in common: they impute a missing value with a single value. The definition of the imputed value varies from a predefined value, a calculated value, a random value, and a predicted value to a smoothed value.

In the subsequent analysis, however, in all cases the imputed value is treated like a real value. These methods are also called single imputation methods. Single imputation does not, however, reflect the uncertainty about the predictions of the unknown missing values.

Methods of multiple imputations, in contrast, do not impute a missing value with a single value but with a set of potential imputation values. This accommodates any uncertainty about the true value to impute. Thus, the result of multiple imputations is not only a single input data table for the analysis but a set of input data tables. The consecutive analysis is then performed for each data table by standard procedures for complete data, and the analysis results are combined.

Multiple imputation inference involves three distinct phases:

1. The missing data are filled in *m* times to generate *m* complete data sets. This step is performed by PROC MI, which is part of SAS/STAT.
2. The *m* complete data sets are analyzed using standard statistical analyses (for example, PROC LOGISTIC or PROC REG).
3. The results from the *m* complete data sets are combined to produce inferential results. This can be performed with PROC MIANALYZE, which is also part of SAS/STAT.

In many cases, as few as three to five imputations are adequate in multiple imputation (compare [13], Rubin 1996).

Example data for multiple imputation and analysis

To illustrate using PROC MI and PROC MIANALYZE for multiple imputation and the respective analysis, a short example of logistic regression is shown. This example uses the credit scoring customers data from the HMEQ data set. Only a few of the original input variables are kept, as well as the target variable BAD (binary variable: 0 for customers who did not default, 1 for customers who defaulted).

```
data CreditData;
  format ID 8.;
  set hmeq;
  id = _N_;
  mortdue = mortdue / 1000;
  value = value / 1000;
  keep id bad clage mortdue value;
run;
```

Performing multiple imputation with PROC MI

Next, PROC MI is run to create a set of imputation data tables for the analysis:

```
proc mi data = CreditData
  out = CreditData_MI;
  var clage mortdue value;
run;
```

This results in five imputation data tables (default setting) that are concatenated into one SAS data set. Each imputation is identified by the _IMPUTATION_ variable.

The MI procedure also analyzes the structure of missing value patterns and the distribution of existing values. The respective output is shown in Outputs 10.13 and 10.14.

Output 10.13: Structure of missing value pattern

		Missing Data Patterns				Group Means		
Group	CLAGE	MORTDUE	VALUE	Freq	Percent	CLAGE	MORTDUE	VALUE
1	X	X	X	5160	86.58	179.417210	73.954943	105.421195
2	X	X	.	79	1.33	163.058402	76.557911	
3	X	.	X	399	6.69	188.539638		64.857240
4	X	.	.	14	0.23	152.660936		
5	.	X	X	197	3.31		68.734712	93.917668
6	.	X	.	6	0.10		35.008333	
7	.	.	X	92	1.54			74.273321
8	O	O	O	13	0.22			

This example shows a row for each missing data pattern. The Xs describe the pattern and indicate whether a value is available for a certain variable. For each missing data pattern, the group means are shown for the non-missing variables.

Output 10.14: Distribution of the existing values (parameter estimates)

Variable	Mean	Std Error	95% Confidence Limits		DF	Minimum	Maximum	Mu0	t for H0: Mean=Mu0	Pr > |t|
CLAGE	179.609813	1.160835	177.3296	181.8900	552.69	179.410268	180.126102	0	154.72	<.0001
MORTDUE	71.469891	0.608992	70.2759	72.6639	3509.6	71.385812	71.578019	0	117.36	<.0001
VALUE	101.851175	0.746861	100.3871	103.3153	5873.2	101.799920	101.876744	0	136.37	<.0001

Table 10.7 shows an extract of table CREDITDATA_MI of the first imputation. Note that for ID 4, no values have been imputed because all the variable values are missing for this observation.

Table 10.7: Extract of CREDITDATA_MI data

	Imputation	ID	BAD	MORTDUE	VALUE	CLAGE
1	1	1	1	25.86	39.025	94.3666667
2	1	2	1	70.053	68.4	121.833333
3	1	3	1	13.5	16.7	149.466667
4	1	4	1			
5	1	5	0	97.8	112	93.3333333
6	1	6	1	30.548	40.32	101.466002
7	1	7	1	48.649	57.037	77.1
8	1	8	1	28.502	43.034	88.7660299
9	1	9	1	32.7	46.74	216.933333
10	1	10	1	11.9023372	62.25	115.8
11	1	11	1	22.608	53.3440974	68.5008852
12	1	12	1	20.627	29.8	122.533333
13	1	13	1	45	55	86.0666667
14	1	14	0	64.536	87.4	147.133333
15	1	15	1	71	83.85	123
16	1	16	1	24.28	34.687	300.866667
17	1	17	1	90.957	102.6	122.9

Analyze data with PROC LOGISTIC

These data are now analyzed with PROC LOGISTIC for each imputation:

```
proc logistic data = CreditData_MI;
  model bad(event='1') = clage mortdue value / covb ;
  ods output ParameterEstimates = Estimates
  Covb = CovMatrix;
  by _Imputation_;
run;
```

Note that the parameter estimates as well as the covariance matrix are output using the ODS OUTPUT statement and the analysis is run "BY _IMPUTATION_." Table 10.8 shows the parameter estimates for the first two imputations.

Table 10.8 Parameter estimates from PROC LOGISTIC

Imputation	Variable	DF	Estimate	StdErr	WaldChiSq	ProbChiSq
1	Intercept	1	-0.5032	0.0902	31.1290	<.0001
1	CLAGE	1	-0.00543	0.000452	144.4298	<.0001
1	MORTDUE	1	-0.00181	0.00162	1.2567	0.2623
1	VALUE	1	0.00151	0.00129	1.3737	0.2412
2	Intercept	1	-0.4899	0.0900	29.6004	<.0001
2	CLAGE	1	-0.00554	0.000451	150.5512	<.0001
2	MORTDUE	1	-0.00273	0.00164	2.7610	0.0966
2	VALUE	1	0.00219	0.00130	2.8150	0.0934

Combine results with PROC MIANALYZE

Next, PROC MIANALYZE is used to combine the results from the individual analyses based on the imputation:

```
proc mianalyze parms=Estimates covb=CovMatrix;
 modeleffects clage mortdue value;
run;
```

Output 10.15 shows the results of PROC MIANALYZE in the form of the combined parameter estimates.

Output 10.15: Results from PROC MIANALYZE

Parameter	Estimate	Std Error	95% Confidence Limits	
intercept	-0.477818	0.095274	-0.66513	-0.29051
clage	-0.005585	0.000492	-0.00656	-0.00461
mortdue	-0.001634	0.001807	-0.00522	0.00195
value	0.001386	0.001417	-0.00141	0.00419

This example shows users how to perform multiple imputations and how to combine the respective results. For more details on multiple imputation and on the options and parameters of the MI and MIANALYZE procedures, refer to [14] *SAS/STAT Users Guide*.

10.7 Conclusion

The SAS offering

This chapter has shown methods for profiling missing values in SAS and how to replace these missing values. SAS offers not only simple descriptive methods to profile the frequency and percentage of missing values per variable. It also offers advanced analytical methods to provide more insight into the structure and co-existence of missing values.

For the imputation of missing values, users can select from simple averages or frequency based methods up to a complete set of imputation techniques that is integrated into the data mining process flow environment of SAS Enterprise Miner. In addition, SAS offers methods for multiple missing value imputations as well as the respective analysis capability of multiple imputations.

Business and domain expertise

The technical tools for profiling and imputing missing values need to be accompanied by business and domain expertise. Deciding whether to impute missing values and selecting the method to do so cannot be based solely on technical characteristics. For example: Deciding whether to impute a variable with a high percentage of missing values or to drop this variable from the data set has to be based on a business point of view. The methods presented here can form the basis for those decisions.

The important point is that the capabilities of SAS do not constrain analysts when selecting possible actions. The analyst can choose the method that best fits his business problem.

Chapter 11: Profiling and Replacement of Missing Data in a Time Series

11.1 Introduction 141
   General 141
   SAS programs 142
11.2 Profiling the Structure of Missing Values for Time Series Data 142
   Missing values in a time series 142
   Example data 142
   The TS_PROFILE_CHAIN 143
   Reports for profiling a time series 143
   Macro %PROFILE_TS_MV 146
11.3 Checking and Assuring the Contiguity of Time Series Data 147
   Difference between transactional data and time series data 147
   Example of a non-contiguous time series 148
   Checking and assuring contiguity 149
   PROC TIMESERIES 149
   Macro implementation and usage 150
11.4 Replacing Missing Values in Time Series Data with PROC TIMESERIES 152
   Overview 152
   Functionality of PROC TIMESERIES 152
   Examples 153
   Changing zero values to missing values 154
11.5 Interpolating Missing Values in Time Series Data with PROC EXPAND 155
   Introduction 155
   Statistical methods in PROC EXPAND 155
   Using PROC EXPAND 156
11.6 Conclusion 156
   General 156
   Available methods 157
   Business knowledge 157

11.1 Introduction

General

This chapter focuses on missing values and observations in time series data. Different from one-row-per-subject observations, where each observation can be considered independently, observations in time series data need to be considered as sequence groups of observations.

Whether a value is missing in time series data not only depends on the existence of a NULL value for a certain record. Missing values, for example, include records where a value is missing as a whole for a certain period. Thus, the chapter is titled, in part, "Missing Data" instead of "Missing Values" to accommodate both cases.

As time series data are arranged in sequence groups of observations, different methods to profile and impute missing values apply and are shown in the next subsections:

- Section 11.2 shows a method for profiling the missing value pattern in time series data. Here a missing value is defined as a NULL value for an existing record.
- Consequently, section 11.3 focuses on missing records in time series data. Methods to profile noncontiguous time series and to insert the missing records are introduced.
- Section 11.4 shows methods to insert missing records with PROC TIMESERIES and additional features of this procedure for the treatment of missing and zero values.
- Finally, section 11.5 shows how PROC EXPAND can impute missing values in a time series.

SAS programs

For advanced profiling in the imputation of missing values in a time series, this chapter introduces the macros %MV_TS_PROFILING (missing value in time series profiling) and %MISS_OBS_TS (missing observation in time series data).

All macros, SAS programs, and example data sets that are used in this chapter can be downloaded from the source referenced in the introduction at the beginning of this book.

11.2 Profiling the Structure of Missing Values for Time Series Data

Missing values in a time series

This section shows a method for profiling the structure of missing values for time series data. In time series analysis, it is relevant to know whether the individual time series has leading, trailing, or embedded missing or zero values. A missing value here is defined as a NULL value in a particular record of time series data. Do not confuse it with missing records in time series data that are discussed in the next section.

Counting the number of missing values for each time series and comparing it with the number of observations identifies the proportion of missing values. However, it makes a difference whether these values occur within the series or at the beginning or end of the series. It also makes a difference whether the missing values occur in blocks or are distributed over different periods.

Missing or zero values at the beginning or end of a series may result when no data are available before or after a certain point in time (for example, when a product has not been launched before a certain date or is not sold anymore after a certain date).

In order to gain insight into the structure and pattern of available data in time series, a macro is introduced that creates a profile for each time series to reveal sequences of non-missing, zero, and missing values. This puts the analyst in the position to quickly check his data to decide whether additional data preprocessing is required and how data should be used in the analysis.

Example data

For small data sets, an overview can also be obtained by looking at the table view of the data. The example in Table 11.1 shows that there are 6 zero values at the beginning for series OB2 and there are embedded zero values for series OC3. From a business point of view, you need to decide whether the zero values are real zero quantities or whether they are missing values. For larger data sets with longer time series data or more cross-sections, however, it is often impossible to get an overview just by browsing the table.

Table 11.1: Sample Data

monyear	OB1	OB2	OC1	OC2	OC3	OC4
2004.01	3386	0	3542	11419	1055	1597
2004.02	2977	0	5664	15439	4701	1648
2004.03	3268	0	6211	15712	4808	1744
2004.04	2847	0	3714	11418	8143	867
2004.05	2891	0	3179	11731	5274	457
2004.06	2741	0	6302	21551	4835	278
2004.07	2855	4392	4425	8950	0	102
2004.08	2660	4684	3202	5019	0	155
2004.09	3421	4156	4309	10535	2344	180
2004.10	3480	6791	5137	18292	4355	1377
2004.11	3191	6405	3842	17580	2048	1297
2004.12	3291	5743	5236	16839	7438	485
2005.01	3817	7728	3016	3579	3326	2661
2005.02	4250	6207	4756	8070	6764	1904
2005.03	3842	6883	5209	11408	3307	768
2005.04	3361	7167	5334	9984	392	485
2005.05	3245	8305	4402	12543	465	487
2005.06	4807	8029	3895	9667	0	383
2005.07	3321	7070	3387	14373	0	396
2005.08	2024	7743	4444	17088	0	304
2005.09	1722	6887	6734	14669	0	382
2005.10	3329	4579	6926	13750	24	405
2005.11	3840	6617	7049	15778	2113	608
2005.12	4986	6356	7156	8105	5352	712
2006.01	4365	6327	5358	7762	1092	815
2006.02	4460	7676	8188	7832	358	817
2006.03	4595	5621	5417	8028	358	851
2006.04	2280	4691	4687	12913	2602	519
2006.05	3364	6189	3860	11066	7955	274
2006.06	3950	9424	2928	6855	3314	0
2006.07	3595	5548	9619	16547	5100	1974
2006.08	3881	5956	4355	6096	2762	1475

The TS_PROFILE_CHAIN

In order to profile time series data, for each time series the TS_PROFILE_CHAIN is calculated, which represents a sequence of 0-, X-, and 1 values. If the time series has a non-missing value for the respective period, a 1-value is inserted; for a zero value, a 0 is inserted; and missing values are indicated by an X value.

The following short example illustrates this. A time series has the values 0, 0, 28, 26, 0, MISSING, 42, 12, MISSING, 43, and 53. The respective TS_PROFILE_CHAIN is 00110X11X11, which illustrates the pattern of the time series.

In addition, a TS_PROFILE_CHAIN_UNIQUE is created that only contains a value if the sequence changes from one state to another. For this example, this results in 010X1X1 and shows that one or more non-missing values are followed by one or more missing values, and so on.

It is possible to specify the values that will be considered missing and result as X-values in the chain.

Reports for profiling a time series

Reports for profiling time series data are shown using a data set with 46 time series values. In Output 11.1, a frequencies table lists the TS_PROFILE_CHAIN in order of descending frequencies:

- The top two entries are patterns without non-missing values: 18 time series with a length of 54 months and 17 time series with a length of 60 months.
- Five time series have 6 leading 0 values. After this period, the time series contains only non-missing and non-zero values. This may be due to the fact that the product was only launched in month 7, but data have already been recorded for the earlier period. Here the 0 values should be interpreted as missing values. See section 11.4 for a method on how to convert these observations to missing values with PROC TIMESERIES.
- Then there are a number of unique patterns, where missing or zero values are embedded at the beginning, end, or somewhere in between. For example, the fifth pattern in this list has to be judged from a business point of view on whether the embedded zero values are real zero quantities or whether they are unknown values and should be replaced with missing values. See section 11.5 on how to replace missing values in time series data.

Note that at the end of the TS_PROFILE_CHAIN pattern, the total number of observations and non-missing values is concatenated. For example, <pattern>_60_0 reads as "Time series has 60 observations and no missing or zero values."

Output 11.1: Frequencies of TS_PROFILE_CHAIN

TS_Profile_Chain	Frequency	Percent
111111111111111111111111111111111111111111111111111111_54_0	18	39.13
111111111111111111111111111111111111111111111111111111111111_60_0	17	36.96
000000111111111111111111111111111111111111111111111111111111_60_0	5	10.87
111111001111111100001111111111111111111111111111111000001_60_0	1	2.17
111111111111111111111111100000000000011111111111111111111111_60_0	1	2.17
1111111111111111111111111100111111111111111111111111111111111_60_0	1	2.17
11111111111111111111111111111111111111111111111111111_53_0	1	2.17
1111111111111111111111111X111111111111XX1X1XX11111XXXX_60_10	1	2.17
1111XX1111111111111111111111111111111X1X11XXXX11111XX111_60_10	1	2.17

The next table shows a more compressed view of the TS_PROFILE_CHAIN. Here, only transitions between Xs, 0s, and 1s are listed:

- There are 36 time series that have no missing value.
- Pattern 0 indicates that there are 5 time series (as described above) that start with zero values. After that, they only have non-zero values.
- Pattern 101 means that there are 2 time series that start with one or more non- missing values. One or more missing values are embedded, and non-missing values are at the end.
- These two tables provide a quick overview of the patterns in the time series and support decisions about performing further steps in data preparation or using the time series data for the analysis.

Output 11.2: Frequencies of TS_PROFILE_CHAIN_UNIQUE

TS_Profile_Chain_Unique	Frequency	Percent
1	36	78.26
01	5	10.87
101	2	4.35
1010101	1	2.17
1X1X1X1X1X	1	2.17
1X1X1X1X1X1	1	2.17

The macro also creates a data set, MV_PROFILE_TS, that contains one row for each time series (see Output 11.3). Beside the time series ID, this table contains the TS_PROFILE_CHAIN string and the TS_PROFILE_CHAIN_UNIQUE string and allows you to filter the time series IDs based on their missing data patterns.

Output 11.3: Data Set MV_PROFILE_TS

	_id	TS_Profile_Chain	TS_Profile_Chai	N	NMiss
1	OA1	111111111111111111111111111111111111111111111111111111111111_60_0	1	60	0
2	OA2	111111111111111111111111111111111111111111111111111111111111_60_0	1	60	0
3	OA3	111111111111111111111111111111111111111111111111111111111111_60_0	1	60	0
4	OB1	111111111111111111111111111111111111111111111111111111111111_60_0	1	60	0
5	OB2	000000111111111111111111111111111111111111111111111111111111_60_0	01	60	0
6	OB21	000000111111111111111111111111111111111111111111111111111111_60_0	01	60	0
7	OB22	000000111111111111111111111111111111111111111111111111111111_60_0	01	60	0
8	OB23	000000111111111111111111111111111111111111111111111111111111_60_0	01	60	0
9	OB24	000000111111111111111111111111111111111111111111111111111111_60_0	01	60	0
10	OC1	111111111111111111111111111111111111111111111111111111111111_60_0	1	60	0
11	OC2	111111111111111111111111111111111111111111111111111111111111_60_0	1	60	0
12	OC3	111111100111111110000111111111111111111111111111000001_60_0	1010101	60	0
13	OC4	1111111111111111111111111111X11111111111XX1X1XX11111XXXX_60_10	1X1X1X1X1X	60	10
14	OC5	1111XX111111111111111111111111111111111X1X11XXXX11111XX111_60_10	1X1X1X1X1X1	60	10

Output 11.4 and Output 11.5 show the distribution of the time series lengths and the distribution of the number of missing values.

Output 11.4: Distribution of the length of the time series

N	Frequency	Percent	Cumulative Frequency	Cumulative Percent
53	1	2.17	1	2.17
54	18	39.13	19	41.30
60	27	58.70	46	100.00

Output 11.5: Distribution of the number of missing values

NMiss	Frequency	Percent	Cumulative Frequency	Cumulative Percent
0	44	95.65	44	95.65
10	2	4.35	46	100.00

In order to display the patterns of the time series data graphically, a profile plot, as shown in Output 11.6, is created. The horizontal axis of this plot corresponds with the overall time window of the data. For each time series, a horizontal line is plotted and the line is interrupted at those time IDs where a missing or zero value is found in the data. In this plot, zero and missing values are treated the same for simplicity.

The plot provides an intuitive visual overview of the data. The plot is designed to provide a high-level overview of the data clusters. The plot groups the time series data based on the similarity of the patterns.

It shows the interruptions in the time series and how they form patterns. There are different blocks of time series data with different start dates.

Output 11.6: Profile plot for the structure of the time series

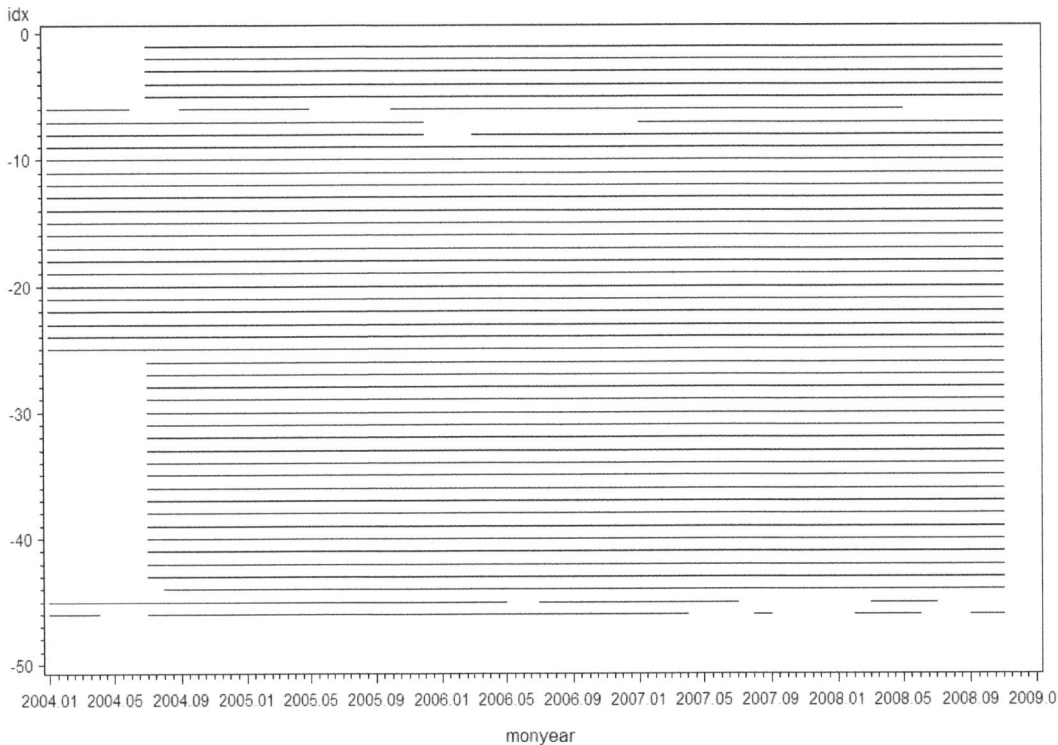

Note that when large numbers of time series are plotted, the run time of the macro is excessive. The plotting behavior can be controlled with macro options PLOT and NMAX_TS.

Macro %PROFILE_TS_MV

In order to profile the structure of missing values in a time series and to create the results shown here, the SAS macro %PROFILE_TS_MV has been programmed. The macro code is documented in appendix A.

Note the prerequisites for using the macro:

- Each time series needs to have an &ID variable. If a time series with the same &ID occurs more than once in the data (for example, per sales region or sales channel), the respective cross-sectional dimensions variable needs to be specified with the &CROSS macro variable. The macro then concatenates these variables to a new ID variable.
- As a best practice, use GIF or JPG as graphical devices, especially if the input data have hundreds of time series. Otherwise, if JAVA or ACTIVEX is used, creating the profile plot is very time consuming. The statement GOPTIONS DEVICE = GIF can be used before the macro call to change the graphics device to GIF.

The following parameters can be specified with the macro:

DATA
   Name of the input data set (mandatory).

ID
   Name of the time series ID variable (mandatory).

CROSS
List of variables for the cross-sectional dimensions (optional). The values of these variables are concatenated first to a string and then to the ID variable in order to allow the analysis of the same time series ID for different subgroups.

DATE
Name of the time ID variable (mandatory).

VALUE
Name of the variable that holds the values of the time series (mandatory).

MV
List of values to be considered as missing. Values need to be specified in brackets, separated by a comma. Examples are mv=(.) and mv=(.,9). The default value is (.). Note that this list is used with an IN operator in the DATA step.

ZV
List of values to be considered as zero. Values need to be specified in brackets, separated by a comma. Examples are mv=(**0**) and mv=(**0,-1**). The default value is (0). Note that this list is used with an IN operator in the DATA step.

PLOT
A profile plot is only produced if PLOT is set to YES. Default = YES. This option should be set to NO if your data contain more than 500 time series to avoid excessive run times.

NMAX_TS
Depending on the number of time series in the data, this option controls whether a profile plot will be produced. The default number is 100. Thus, if your data contain more than 100 time series, a profile plot will not be produced and you will need to set the value higher. In this case, the following message in the log file is produced:

```
---------------------------------------------------------------------------
Number of time series = 1026, is higher than nmax_ts value(= 100).
No plot has been created. Reset parameter NMAX_TS to a value of at least 1026
---------------------------------------------------------------------------
```

W
Defines the thickness of the lines used in the profile plot to represent one time series.

11.3 Checking and Assuring the Contiguity of Time Series Data

Difference between transactional data and time series data

In order to explain the necessity for checking and correcting the completeness of time series data, it is important to differentiate between *transactional data* and *time series data*.

- Transactional data are time-stamped data that are collected with no particular frequency over time. Examples of transactional data include Internet data like website visits or point-of-sale data in retail.
- Time series data, on the other hand, are time-stamped data that are collected at a particular frequency over time, like web hits per hour or sales amounts per day.

Tables 11.2 and 11.3 show examples of transactional data and time series data, which have been aggregated on an hourly basis.

Table 11.2: Transactional data from a web server

	Session Identifier	requested_file
1	43d0a4da826149b5 2002-02-17 08:38:12	/Home.jsp
2	43d0a4da826149b5 2002-02-17 08:38:12	/Cookie_Check.jsp
3	43d0a4da826149b5 2002-02-17 08:38:12	/Home.jsp
4	43d0a4da826149b5 2002-02-17 08:38:12	/Corporate_Relations.jsp
5	43d0a4da826149b5 2002-02-17 08:38:12	/Retail_Store.jsp
6	43d0a4da826149b5 2002-02-17 08:38:12	/Store/Store_Locations.jsp
7	43d639ebce6c73d8 2002-02-17 23:43:16	/Home.jsp
8	43d639ebce6c73d8 2002-02-17 23:43:16	/Cookie_Check.jsp
9	43d639ebce6c73d8 2002-02-17 23:43:16	/Home.jsp
10	43d639ebce6c73d8 2002-02-17 23:43:16	/Department.jsp
11	43d639ebce6c73d8 2002-02-17 23:43:16	/Department.jsp
12	43bb8704bb370e09 2002-02-17 13:44:04	/Home.jsp
13	43bb8704bb370e09 2002-02-17 13:44:04	/Home.jsp
14	43bb8704bb370e09 2002-02-17 13:44:04	/Subcategory.jsp
15	43bb8704bb370e09 2002-02-17 13:44:04	/Product.jsp
16	43bb8704bb370e09 2002-02-17 13:44:04	/Department.jsp
17	43bb8704bb370e09 2002-02-17 13:44:04	/Product.jsp
18	43bb8704bb370e09 2002-02-17 13:44:04	/Department.jsp

Table 11.3: Aggregated web server data per hour (time series data)

	Time	NumberOfReqestedFiles
1	1:00:00	116
2	2:00:00	93
3	3:00:00	17
4	4:00:00	158
5	6:00:00	30
6	7:00:00	66
7	8:00:00	210
8	9:00:00	130
9	10:00:00	143
10	11:00:00	298
11	12:00:00	239
12	13:00:00	145

Note that in Table 11.3, there is no record for time 5:00:00 because no web server traffic occurred during that time. For time series analysis, however, the records need to be equally spaced at the time axis. The existence of all consecutive periods is called contiguity. Time series that are not contiguous can cause problems in analyses because the time series methods will not work properly.

Example of a non-contiguous time series

From the SASHELP.AIR data, approximately 10% of the records have been deleted with the following statements:

```
data air_missing;
 set sashelp.air;
 if uniform(12345) < 0.1 then delete;
run;
```

From Table 11.4, you can see that now some records are missing (for example, there is no record for month JULY 1949).

Table 11.4: Extract of the AIR_MISSING data

	DATE	AIR
1	JAN49	112
2	FEB49	118
3	MAR49	132
4	APR49	129
5	MAY49	121
6	JUN49	135
7	AUG49	148
8	SEP49	136

Next, a time series analysis with PROC ESM is executed:

```
proc esm data = air_missing;
  id date interval = month;
  forecast air;
run;
```

PROC ESM checks the data and recognizes that the specified ID variable DATE is not contiguous with the specified INTERVAL MONTH. Thus, the following warning is displayed in the SAS log:

```
WARNING: 1 omitted observations have been detected before observation
 number 7 according to the INTERVAL=MONTH option and the ID
 values. The current ID is DATE=AUG1949, the previous ID was
 DATE=JUN1949.
```

Similar warnings are displayed when the data are analyzed with PROC FORECAST or PROC ARIMA.

Checking and assuring contiguity

From this example, you can see that it is important to check time series data for contiguity before running the analysis. If a noncontiguity is detected, the missing records need to be inserted into the data. Note that the point here is not that missing values will be imputed with another value, but that missing records will be inserted into the data, most likely with a value of 0.

Inserting a record with a zero value is plausible when, for example, data such as those shown in Tables 11.8 and 11.9 are considered. Because there was no web server traffic at hour 5, no records are found in the transactional data for this time interval and, thus, the respective record is not created in the time series data. If a record is inserted, the value that makes most sense for this interval is 0 because it represents no traffic.

There are various ways to check for missing records and to insert those records. Implementations include DATA step programs that check for the time interval between two records as well as the creation of a master table that holds a contiguous set of records, which is then merged with the original data to find records for the missing time intervals.

These implementations, however, can become very complicated for time series data with cross-sectional dimensions. In this case, it is necessary to check the contiguity for all possible cross-sectional dimensions that also may have different start and end points.

PROC TIMESERIES

A very smooth implementation of this task can be achieved with PROC TIMESERIES, which provides the following functionalities:

- Aggregates, aligns, and describes transactional or time series data
- Converts transactional data into time-stamped data
- Performs seasonal decomposition of the data

- Checks for and inserts missing records
- Handles and converts missing and zero values

A typical PROC TIMESERIES example is shown here:

```
proc timeseries data = air_missing
  out = TIMEID_INSERTED;
  id date interval = MONTH setmiss=0;
  var air;
run;
```

The ID statement sets the time variable and the interval spacing. Here, the SETMISS option defines which values to insert for the VAR variable AIR if a missing record is detected. Missing records are identified automatically based on the ID variable and the spacing.

Macro implementation and usage

A macro has been coded to help you use PROC TIMESERIES to check for and insert missing records.

This macro can run in two modes: CHECK mode and INSERT mode. CHECK mode checks for the existence of missing records and INSERT mode inserts the missing records.

Running this macro in CHECK mode for the AIR_MISSING data with the following syntax generates the records in the TIMEID_MISSING data set (see Table 11.5):

```
%check_timeid(data = air_missing,
  mode = CHECK,
  timeid = date,
  value = air);
```

Table 11.5: Content of data set TIMEID_MISSING

	DATE	AIR
1	JUL1949	-123456789.1
2	APR1951	-123456789.1
3	AUG1951	-123456789.1
4	APR1952	-123456789.1
5	MAR1953	-123456789.1
6	MAY1954	-123456789.1
7	DEC1955	-123456789.1
8	NOV1956	-123456789.1
9	NOV1957	-123456789.1
10	DEC1958	-123456789.1
11	MAR1960	-123456789.1
12	MAY1960	-123456789.1

If there were no missing records from a contiguity perspective, the data set would have been empty. Note that the displayed values for the AIR variable for these records are only an arbitrary value that is used to check for contiguity in the data. They are not inserted into the original table.

Running the macro in INSERT mode with the following syntax creates an output data set called TIMEID_INSERTED (see Table 11.6):

```
%check_timeid(data = air_missing,
  mode = INSERT,
  timeid = date,
  value = air);
```

Table 11.6: Content of data set TIMEID_INSERTED

	DATE	AIR
1	JAN1949	112
2	FEB1949	118
3	MAR1949	132
4	APR1949	129
5	MAY1949	121
6	JUN1949	135
7	JUL1949	0
8	AUG1949	148
9	SEP1949	136
10	OCT1949	119

For month JUL1949, a record has been inserted and the value of AIR has been set to 0. This data set now fulfills the requirements of contiguity.

Note that when using the macro, you can specify a list of variables that receive 0 values for the inserted records. And the macro can also be run with one or more BY variables to insert records if the data are structured with cross-sectional dimensions.

For example, if there are missing records in the SASHELP.PRDSALE data (or in a copy of the data set called PRDSALE_MISSING), these records can be inserted for each BY group with the following statements:

```
%check_timeid(data = prdsale_missing,
 out = prdsale_insert,
 mode = INSERT,
 timeid = month,
 value = actual,
 by = country region division prodtype product);
```

You can see from the syntax that a list of variables defining the cross-sectional dimensions has been specified with the BY variable.

Note the following from the syntax:

- PROC TIMESERIES is used to check for missing records and to insert the missing records.
- Inserting the missing records is done using typical PROC TIMESERIES syntax. An insert value different from 0 can be specified with the INSERTVALUE macro variable.
- Checking for missing records is also done with PROC TIMESERIES:
  o Here, records with an arbitrary value of -123456789.123456789 are inserted if a record is missing. The inserted value is then used to identify the inserted records.
  o If the default value of -123456789.123456789 conflicts with real data, a different value using the CHECKDUMMYVALUE variable can be specified.
- Cross-sectional dimensions are treated with the BY statement in PROC TIMESERIES.
- If BY variables are specified with the macro, the input data are sorted for these variables, which may take some time for large data sets.

The following variables can be specified with the macro:

DATA
Name of the input data set that holds the time series data (mandatory).

OUT
Name of the output data set that holds the original data plus the inserted records. Default = TIMEID_INSERTED.

OUT_CHECK
Name of the output data set that holds missing records.
Default = TIMEID_MISSING.

MODE
Defines whether the macro will run in CHECK or INSERT mode. Valid values are CHECK or INSERT. Default = INSERT.

TIMEID
Name of the time ID variable (mandatory).

INTERVAL
Time interval of the time series data. This interval is used to check for contiguity.

VALUE
Name(s) of the value variable(s) that receives the inserted value (mandatory). Note that a list of variables or the global variable _NUMERIC_ can be specified here to denote all numeric variables in the data set.

INSERTVALUE
Value that will be inserted into the value variable for missing records. Default = 0.

BY
Optional variable that contains the names of the variables that define the cross-sectional dimensions.

CHECKDUMMYVALUE
Arbitrary value that is inserted into the data in CHECK mode to identify the missing records. Default = 123456789.123456789.

11.4 Replacing Missing Values in Time Series Data with PROC TIMESERIES

Overview
Using PROC TIMESERIES has already been demonstrated in the previous section. This section discusses using the procedure to replace missing values in time series data. Missing values in this context mean that the missing values occur explicitly in time series data where the value for a certain time period is missing, which is different from the implicit missing values discussed in the previous section.

Functionality of PROC TIMESERIES
PROC TIMESERIES allows you to replace missing values by using one of the replacement methods listed in Table 11.7. These methods are controlled with the option SETMISS. For details, refer to the documentation of PROC TIMESERIES, section ID statement, SETMISS option. Table 11.7 lists the possible values for this option.

Table 11.7: Possible values for the SETMISS option

Option value	Missing values are set to
Number	Any number. (for example, 0 to replace missing values with zero)
MISSING	Missing
AVERAGE	Average value of the time series
MEDIAN	Median value of the time series
MINIMUM	Minimum value of the time series
MAXIMUM	Maximum value of the time series
FIRST	First non-missing value
LAST	Last non-missing value
PREVIOUS	Previous non-missing value
NEXT	Next non-missing value

Examples

In order to demonstrate how to use PROC TIMESERIES to impute missing values, artificial missing values are inserted in the SASHELP.AIR data set. A new variable, AIR_MV, is created as a copy of the AIR variable and approximately 15% of the missing values are inserted:

```
data air_missing;
  set sashelp.air;
  if uniform(12345) < 0.15 then air_mv = .;
  else air_mv = air;
run;
```

Table 11.8 shows the first 17 rows of table AIR_MISSING with missing values for April and July 1949 and March 1950. Note that the AIR variable is only kept for explanatory purposes. In a real-world situation, variable AIR would not exist, and the missing values found in AIR_MV would need to be replaced for the analysis.

Table 11.8: Data set AIR_MISSING with missing values in variable AIR_MV

	DATE	AIR	air_mv
1	JAN49	112	112
2	FEB49	118	118
3	MAR49	132	132
4	APR49	129	129
5	MAY49	121	.
6	JUN49	135	135
7	JUL49	148	.
8	AUG49	148	148
9	SEP49	136	136
10	OCT49	119	119
11	NOV49	104	.
12	DEC49	118	118
13	JAN50	115	115
14	FEB50	126	126
15	MAR50	141	141

PROC TIMESERIES can now be used to set missing values to zero with the following code:

```
proc timeseries data = air_missing
  out = air_setmissing_zero;
  id date interval =month setmiss=0;
  var air_MV;
run;
```

Another example shows how to replace missing values with the previous value in the time series:

```
proc timeseries data = air_missing
   out = air_setmissing_previous;
   id date interval =month setmiss=PREVIOUS;
   var air_MV;
run;
```

A third example shows how to replace missing values with the mean value in the time series:

```
proc timeseries data = air_missing
   out = air_setmissing_mean;
   id date interval =month setmiss=MEAN;
   var air_MV;
run;
```

In Table 11.9, the output data sets of these three examples have been joined together for illustrative purposes.

Table 11.9: Results of replacing values with PROC TIMESERIES

	DATE	AIR	air_mv	air_mv_zero	air_mv_previous	air_mv_mean
1	JAN49	112	112	112	112	112
2	FEB49	118	118	118	118	118
3	MAR49	132	132	132	132	132
4	APR49	129	129	129	129	129
5	MAY49	121		0	129	284.54385965
6	JUN49	135	135	135	135	135
7	JUL49	148		0	135	284.54385965
8	AUG49	148	148	148	148	148
9	SEP49	136	136	136	136	136
10	OCT49	119	119	119	119	119
11	NOV49	104		0	119	284.54385965
12	DEC49	118	118	118	118	118
13	JAN50	115	115	115	115	115
14	FEB50	126	126	126	126	126
15	MAR50	141	141	141	141	141

Note that you can also specify a BY statement with PROC TIMESERIES in order to replace missing values by BY group.

Changing zero values to missing values

PROC TIMESERIES also offers the ability to change zero values to missing values. This is important if the data show zero values, which, for the analysis, however, are interpreted as missing values (for example, the introduction of new products or the retirement of products in the retail or fashion industry). Here, the zero value represents different phases of the product lifecycle rather than a comparable quantity.

Table 11.10 shows the data before PROC TIMESERIES is applied, and Table 11.11 shows the data after PROC TIMESERIES has set the leading and trailing zeros to missing. Note that beside ZEROMISS=BOTH, which replaces both leading and trailing zeros, LEFT and RIGHT can also be used as option values.

```
proc timeseries data=sales_original out=sales_corrected;
   id date interval=month zeromiss=both;
   var sales;
run;
```

Table 11.10: Data before the replacement by PROC TIMESERIES (not all records shown)

	DATE	sales
1	JAN49	0
2	FEB49	0
3	MAR49	0
4	APR49	0
5	MAY49	0
6	JUN49	0
7	JUL49	148
8	AUG49	148
9	SEP49	136
10	OCT49	119
11	NOV49	104
12	DEC49	118
13	JAN50	115

Table 11.11: Data after the replacement by PROC TIMESERIES (not all records shown)

	DATE	sales
1	JAN1949	.
2	FEB1949	.
3	MAR1949	.
4	APR1949	.
5	MAY1949	.
6	JUN1949	.
7	JUL1949	148
8	AUG1949	148
9	SEP1949	136
10	OCT1949	119
11	NOV1949	104
12	DEC1949	118
13	JAN1950	115

SAS Forecast Studio also offers this functionality. For details, see chapter 13.

11.5 Interpolating Missing Values in Time Series Data with PROC EXPAND

Introduction

The EXPAND procedure is part of SAS/ETS software. It allows you to convert time series data from one sampling interval to another and to interpolate missing values. Time series data can be collapsed from a higher frequency interval to a lower frequency interval or expanded from a lower frequency interval to a higher frequency interval. Monthly data can be aggregated into annual data, and quarterly data can be interpolated from an annual series. For examples, compare [1], Svolba, p. 250f.

Aperiodic time series that are observed at non-regular points in time can be converted into periodic estimates. For example, events that are documented at random time points can be interpolated to form weekly average event rates.

This section focuses on interpolating missing values with PROC EXPAND.

Statistical methods in PROC EXPAND

By default, PROC EXPAND fits cubic spline curves to the non-missing values of the input variable to form continuous approximations of the time series. The respective output series are then generated from the spline approximations. Different from commonly used natural spline methods, which use zero second-derivate endpoint constraints, the SPLINE method in PROC EXPAND uses the not-a-knot method by default, which has proven to be more appropriate for time series data. For more details see the SAS documentation for PROC EXPAND, chapter "Conversion Methods" [15].

Other methods to interpolate missing values with PROC EXPAND include:

- the JOIN method, where interpolations are calculated based on the successive straight line collection between the non-missing points.
- the STEP method, which is a discontinuous piecewise constant curve where the value is interpolated from the most recent non-missing value.

Using PROC EXPAND

This example demonstrates how to use PROC EXPAND based on the data in Table 11.8.

With the following SAS statements, a new variable, AIR_EXPAND, is produced that contains the non-missing values of AIR_MV as well as interpolated values for those months where AIR_MV is missing:

```
proc expand data = air_missing out = air_expand;
 convert air_mv=air_expand;
 id date;
run;
```

Table 11.12 shows the first 17 rows of data set AIR_EXPAND. The interpolated values are stored in variable AIR_EXPAND.

Table 11.12: Data set AIR_EXPAND with interpolated values

	date	air	air_mv	air_expand
1	JAN49	112	112	112
2	FEB49	118	118	118
3	MAR49	132	132	132
4	APR49	129	129	129
5	MAY49	121	.	128.29783049
6	JUN49	135	135	135
7	JUL49	148	.	144.73734152
8	AUG49	148	148	148
9	SEP49	136	136	136
10	OCT49	119	119	119
11	NOV49	104	.	116.19900978
12	DEC49	118	118	118
13	JAN50	115	115	115
14	FEB50	126	126	126
15	MAR50	141	141	141
16	APR50	135	135	135
17	MAY50	125	125	125

PROC EXPAND can also be used with a BY statement if the input data contain different time series in a cross-sectional format that are identified by one or more BY variables. Assuming that the SASHELP.AIR data set contains a COUNTRY variable that separates the number of airline passengers by country, the previous code example would appear as follows:

```
proc expand data = air_missing out = air_expand;
 convert air_mv=air_expand;
 id date;
 by country;
run;
```

11.6 Conclusion

General

This chapter has shown considerations for profiling and replacing missing values in time series data. It has been highlighted that time series data have specifics that require different methods than one-row-per-subject data marts for profiling the missing value structure and the recognition and handling of missing values. These features are important for data quality and for the decision whether the data are usable for analysis and correctly mirror the true value.

Available methods

When profiling missing values, a pattern string consisting of 0, 1, and X values has been introduced that allows you to gain insight into the completeness and structure of time series data. Also, this chapter demonstrated how to detect and replace missing records in a time series.

When treating missing values in a time series, the SAS procedures PROC TIMESERIES and PROC EXPAND have been introduced to replace zero values with missing values and vice-versa and to impute missing values.

Business knowledge

In terms of business knowledge, the same is true as for missing values in the previous chapter: the available methods need to be combined with business and domain knowledge in order to apply the results for good data quality.

For example, deciding whether to treat a missing value as a zero value or vice-versa has to be based on business knowledge.

Also, for time series data, the important point is that the capabilities of SAS allow the analyst to choose that method to profile and impute missing values that best fits his business problem and does not constrain his selection of possible actions.

Chapter 12: Data Quality Control across Related Tables

12.1 Introduction 159
  General 159
  Relational model 159
  Overview 160
12.2 Completeness and Plausibility Checks 160
  General 160
  Completeness check of records 160
  Plausibility check of records 161
12.3 Implementation in SAS 161
  Merging tables 161
  Other methods in SAS 162
12.4 Using a SAS Hash for Data Quality Control 162
  Example data 162
  Completeness control in parent-child relationships 163
  Plausibility checks in parent-child relationships 164
12.5 Conclusion 165

12.1 Introduction

General

The focus of this chapter is data quality control across different tables. In relational databases, individual tables are connected by so-called relationships. Relationships are represented by the appropriate columns or additional tables.

This chapter shows methods to control and validate the completeness and correctness of data in relational tables. A very frequent case is the so-called 1:n relationship or parent-child relationship. Parent-child relationships are found frequently in analytic data marts. For every record in the parent table, a number of related child records in the child table exits. Examples include

- a customer who has different accounts
- a patient who has repeated observations over time
- a mobile phone line that has call record data for each month

Relational model

One-to-many relationships are represented by a foreign key column. In this case, the "many" table holds the primary key of the "one" table as a foreign key. In the example of customers and accounts, the "many" table ACCOUNT holds the customer ID for the corresponding CUSTOMER.

Figure 12.1 shows the physical data model for the customer account example.

Figure 12.1: Example of a 1:n relationship in a relational database

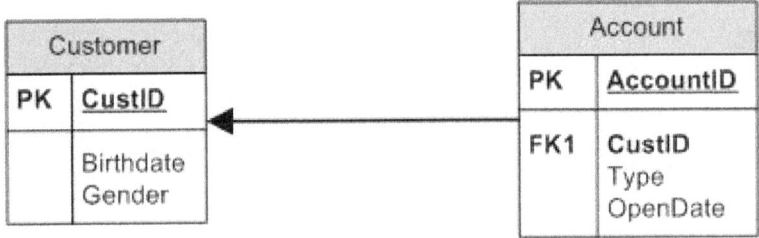

Here, CustID is stored in the ACCOUNT table as a foreign key. This column is important to join the tables together in order to create an analysis table.

Overview

This chapter discusses this relationship and the need to check for completeness and plausibility across tables. SAS methods are shown that allow you to combine different tables. This is possible by joining the two tables; however, here an additional method, the use of a HASH, is shown and discussed.

12.2 Completeness and Plausibility Checks

General

If the parent table and the child table are considered as individual tables, the data in them apply to data validation and profiling as in any other table. Because the parent and child tables are related, additional features can be profiled from the data. These features include

- completeness checks of records
- plausibility checks across tables

Completeness check of records

A completeness check of records in the context of a relational table means that the existence of a certain record in a table requires the existence of a respective record in another table, for example:

- Typically, the rule that a child record can only exist if a respective parent record exists needs to be followed. Otherwise, the child record would be considered a lost child. If a child record has no respective parent record, it is ignored in many analyses because it could create implausible records. Here, the number of child records in the aggregation does not equal the number of child records that belong to a parent record.
- On the other hand, depending on business rules and processes, a parent record could exist that does not have a child record or at least it does not yet have a child record.
    - Consider the case where a customer is registered in the customer database of an insurance company, but the customer does not yet have an insurance policy.
    - Another example is a patient who is part of a clinical trial and who has not yet had an examination.
- This situation is also a generalization of the check for contiguity of a time series, which was presented in section 11.3. Here, the completeness check has to verify that for each virtual parent time interval, a child record in the time series must exist.

Usually, a relational database system can verify and enforce these criteria by so-called integrity constraints, which, for example, make sure that no account record can exist without a corresponding customer record. Thus, in operational systems, there should be no child records without parent records.

Plausibility check of records

A correctness check of records extends the concept of the completeness check because it verifies not only the existence of the records but also the content of the record and whether certain conditions or constraints are met.

- Plausibility checks in this respect can include rules such as a customer account must not be older than the customer relationship itself. In other words, the customer cannot open an account before he becomes a customer.
- Another example is that a certain product may only be sold to a customer who has reached a certain minimum age. In this case, the relationship between the CUSTOMER table and the PRODUCTS table needs to be checked.
- In the medical history of a patient, a record in the PREGNANCY_TEST_RESULTS table should not be found for male patients.

12.3 Implementation in SAS

Merging tables

From an implementation point of view, there are different methods to validate parent-child requirements.

One group of methods physically merges the parent and child records to detect missing records. In the case of a completeness check, an outer join is performed in SAS, and the records without counterparts in the other table are output.

In SAS, an outer join can be performed by a SAS DATA step or by a SQL merge. Examples of this method can be found in Ron Cody [9] chapter 6.

Here is an example of a SAS DATA step:

```
data Cust_Acct_Matches
     Cust_Only
     Acct_Only;
merge customer (in=in_cust)
      accounts (in=in_acct);
by custid;
if in_cust and in_acct then output Cust_Acct_Matches;
else if in_cust then output Cust_Only;
else if in_acct then output Acct_Only;
run;
```

The SAS DATA step can create multiple output files, and the records are directed into the appropriate table.

The log may show the following notes, indicating that there were 4,049 records in the accounts table that did not match any customer record.

```
NOTE: There were 21766 observations read from the data set
 WORK.CUSTOMER.
NOTE: There were 37971 observations read from the data set
 WORK.ACCOUNTS.
NOTE: The data set WORK.CUST_ACCT_MATCHES has 33922
 observations and 9 variables.
NOTE: The data set WORK.CUST_ONLY has 0 observations and 9 variables.
NOTE: The data set WORK.ACCT_ONLY has 4049 observations and 9
 variables.
```

The advantage of this method is that the content of the two tables is combined and available for plausibility checks between different attributes. Most important, and different from the following methods, completeness checks can also be performed in the parent table.

Other methods in SAS

The other method available in SAS is not to physically merge the tables but to use other features of the SAS language to apply the content of the parent table to the child table. In this case, the parent table is treated like a lookup table and a **SAS format** or a **SAS hash** is defined based on this table. This SAS format or SAS hash is then applied to the child table, and the respective features are checked.

The advantages of these methods are that no physical merged output table is created and performance is much faster. In the case of a large child table (for example, transactions for hundreds of thousands of customers), the merging of the tables can consume both time and storage.

The disadvantage is that the checks can only be performed in one direction, going from parent to child:

- Does the child record have a corresponding parent record?
- Is the value of the child record plausible with respect to the parent record?

Physically merging the tables allows you to check in both directions (as shown earlier). The following section shows how to use a SAS hash in more detail.

12.4 Using a SAS Hash for Data Quality Control

Example data

The method presented here uses SAS hash tables for data quality control. For illustration, sample data of CUSTOMERS and ACCOUNTS are used.

Table 12.1: The CUSTOMER table

CustID	Birthdate	CustomerSince	Gender	MaritalStatus
1000002	26DEC1958	01JAN2000	Male	Married
1000005	25JUN1947	01APR1999	Male	Single
1000006	10DEC1945	01SEP1996	Female	Married
1000007	02JUN1934	01SEP1997	Male	Married
1000008	15DEC1957	01JAN1996	Male	Single
1000009	11MAR1959	01JUL2001	Male	Single
1000014	23AUG1952	01MAY1996	Male	Single
1000015	12MAY1959	01FEB1999	Male	Single
1000016	11FEB1967	01FEB2001	Male	Married
1000019	30DEC1936	01JAN2002	Male	Single
1000021	12AUG1959	01OCT2003	Male	Divorce
1000022	10JUL1961	01NOV1993	Male	Married
1000026	20JAN1972	01DEC1995	Male	Married
1000027	26APR1969	01MAR2000	Female	Married

Table 12.2: The ACCOUNTS table

AccountID	CustID	Type	Opendate	Balance
10001	1000002	Saving Account	10JUN2000	1852.63
10002	1000002	Loan	03NOV2001	1134.29
10003	1000005	Saving Account	01APR1999	2022.45
10004	1000006	Funds	01SEP1996	1602.60
10005	1000007	Funds	10AUG1997	1387.38
10006	1000007	Saving Account	01SEP1997	2099.45
10007	1000008	Funds	01JAN1996	1112.56
10008	1000009	Saving Account	02NOV2003	1179.21
10010	1000009	Loan	01JUL2001	1638.46
10009	1000009	Funds	22JUL2003	1647.35
10011	1000014	Saving Account	23JUL1998	1510.90
10012	1000014	Saving Account	01MAY1996	815.40
10013	1000015	Saving Account	01FEB1999	1734.42
10015	1000016	Saving Account	15JUN2003	990.68
10014	1000016	Saving Account	04JUN2003	1092.98
10016	1000018	Funds	18MAR2004	1256.87
10017	1000019	Funds	28NOV2001	663.83
10018	1000019	Loan	14APR2003	2146.13
10019	1000021	Loan	03AUG2005	406.54
10020	1000022	Saving Account	01NOV1993	2849.44
10021	1000026	Saving Account	25SEP1996	2629.77
10022	1000027	Funds	20MAR2001	1870.15
10023	1000028	Funds	01SEP2003	1863.66

Completeness control in parent-child relationships

In order to check the completeness across tables, you need to load the list of customer IDs (CUSTID) from the CUSTOMER table into a hash. This hash is then used to validate whether a parent record exists in the CUSTOMER table for each ACCOUNT record. Based on Tables 12.1 and 12.2, the customer ID 100018 exists in the ACCOUNTS table but not in the CUSTOMER table:

```
DATA accounts_no_parent;
  format AccountId CustID 8. Type $20. Opendate date9.;
  *** Define the Hash;
  if _n_ = 1 then do;
  declare hash customer(dataset: "customer");
  customer.definekey('custid');
  customer.definedone();
  call missing(custid);
  end;
  *** Now SET the accounts table;
  SET accounts;
  *** Check for each record whether the CUSTID exists in the parent table;
  if customer.find() ne 0 then output;
RUN;
```

Note from the code that

- Before the SET statement of the ACCOUNTS table, a hash CUSTOMER is defined based on the CUSTOMER table. This hash only has a KEY column CUSTID.

- The CUSTOMER.FIND() method is used to check whether the CUSTID can be found in the hash. The CUSTOMER.FIND() method returns a 0 if a record is found.

If you apply this code to the ACCOUNTS table, the following output table results, which list the CUSTID 1000018 record:

Table 12.3: ACCOUNTS with no parent in the CUSTOMER table

AccountID	CustID	Type	Opendate	Balance
10016	1000018	Funds	18MAR2004	1256.87
10032	1000034	Funds	02SEP1993	640.61
10033	1000034	Loan	01MAR1992	1005.01
10034	1000034	Loan	29JUN1992	2152.68
10084	1000090	Saving Account	23FEB1999	1286.25
10137	1000154	Saving Account	01APR2001	1965.27
10203	1000211	Saving Account	02JAN2001	807.60
10202	1000211	Saving Account	09NOV2000	1403.36
10209	1000215	Funds	01MAY1999	861.07
10266	1000270	Loan	01FEB2001	752.05
10267	1000270	Saving Account	08MAR2001	1403.01
10299	1000302	Saving Account	12JUN2000	1412.57
10379	1000382	Saving Account	21AUG2003	2555.36
10426	1000438	Loan	24MAY2003	1144.40
10425	1000438	Funds	01MAR2002	1502.80
10424	1000438	Saving Account	01MAR2002	2201.35

Plausibility checks in parent-child relationships

In order to check plausibility between the CUSTOMER and ACCOUNTS tables, the CUSTOMERSINCE and OPENDATE variables are compared. A business rule states that an account cannot be opened before the start of the customer relationship. The following code is used to compare the values with a hash table:

```
DATA accounts_opendate_check;
  format AccountId CustID 8. Type $20.
  Opendate CustomerSince date9.;
  *** Define the Hash;
  if _n_ = 1 then do;
  declare hash customer(dataset: "customer");
  customer.definekey('custid');
  customer.definedata('customersince');
  customer.definedone();
  call missing(custid, customersince);
  end;
  *** Now SET the accounts table;
  SET accounts;
  *** Call the HASH and check the integrity rule;
  rc = customer.find();
  if opendate < customersince then output;
  drop rc;
RUN;
```

Note from the code that

- Before the SET statement of the ACCOUNTS table, a hash CUSTOMER is defined based on the CUSTOMER table. This hash contains the KEY column CUSTID and the CUSTOMERSINCE variable.
- The hash is queried with the CUSTOMER.FIND() method to find the CUSTOMERSINCE value for the respective CUSTID. Note that this method has to be called in an expression (for example, by writing the result into an arbitrary variable return code).
- After this call, the lookup value from the hash is available in the SAS DATA step and can be used for data quality checks.

If you apply this code to the ACCOUNTS table, the following output table results. It shows that the records for CUSTID 100007 and 100019 display an OPENDATE that occurred before the CUSTOMERSINCE date.

Table 12.4: ACCOUNTS with OPENDATE before the CUSTOMERSINCE date

AccountID	CustID	Type	Opendate	CustomerSince	Balance
10005	1000007	Funds	10AUG1997	01SEP1997	1387.38
10017	1000019	Funds	28NOV2001	01JAN2002	663.83
10059	1000063	Funds	01OCT1999	01DEC1999	858.76
10063	1000072	Loan	02DEC1988	01FEB1989	1453.06
10079	1000086	Loan	14FEB2000	01APR2000	1143.54
10106	1000121	Funds	30APR2000	01JUL2000	1221.50
10131	1000146	Saving Account	03FEB1999	01APR1999	2133.63
10134	1000150	Saving Account	02DEC1999	01JAN2000	2375.63
10143	1000161	Saving Account	05MAY2001	01JUN2001	464.33
10166	1000177	Funds	08MAR1987	01MAY1987	1674.00
10183	1000194	Funds	10JUL2000	01SEP2000	1042.63
10186	1000196	Loan	29FEB1996	01MAY1996	1256.90

12.5 Conclusion

This chapter has shown a specific case of data quality control, the check for completeness or plausibility across different tables. Methods in SAS, especially the method using a SAS hash, have been presented.

In analytics, such data quality problems often surface when data from different tables are combined to achieve a more complete picture of the customer (for example, for predictive modeling where a one-row-per-subject data mart is created from different tables).

Chapter 13: Data Quality with Analytics

13.1 Introduction 168
13.2 Benefit of Analytics in General 168
    Outlier detection 168
    Missing value imputation 168
    Data standardization and de-duplication 168
    Handling data quantity 168
    Analytic transformation of input variables 169
    Variable selection 169
    Assessment of model quality and what-if analyses 169
13.3 Classical Outlier Detection 169
    Ways to define validation limits 169
    Purpose of outlier detection 169
    Statistical methods 170
    Implementation 170
    Outlier detection with analytic methods 170
13.4 Outlier Detection with Predictive Modeling 171
    General 171
    Methods in SAS 171
    Example of clinical trial data 171
    Extension of this method 174
13.5 Outlier Detection in Time Series Analysis 174
    General 174
    Time series models 174
    Outlier detection with ARIMA(X) models 174
    Decomposition and smoothing of time series 175
13.6 Outlier Detection with Cluster Analysis 176
    General 176
    Conclusion 177
13.7 Recognition of Duplicates 177
    General 177
    Contribution of analytics 177
13.8 Other Examples of Data Profiling 178
    General 178
    Benford's law for checking data 178
13.9 Conclusion 179

13.1 Introduction

This chapter shows how data quality can benefit from analytics. Chapters 3 through 9 in the first part of this book have shown that analytics can pose special requirements on data quality. In return, this chapter shows that analytics also provide special capabilities to data quality.

These capabilities include profiling and improving the data quality.

Chapters 10 and 11 have already shown an important capability of analytics handling missing values. This chapter focuses on additional features of analytics for data quality. This chapter focuses on detecting outliers, which includes classical outlier detection and advanced outlier detection with predictive models, time series models, and cluster analysis.

Missing value imputation and complex outlier detection are the two main capabilities of analytics in data quality control; however, there are also additional features that are discussed in this context.

13.2 Benefit of Analytics in General

The following points offer an overview of the typical features of analytics that are relevant for data quality.

Outlier detection

- **Simple profiling of univariate data**. Analytics play an important role in detecting outliers based on statistical measures like standard deviations or quantiles.
- **Outlier detection with** methods of **cluster analysis** and distance metrics. These methods allow you to identify outliers in the data from a multivariate viewpoint.
- **Individual outlier detection with predictive models and time series methods**. These methods allow you to calculate validation limits and optimal correction values on an individual basis. As an overall average might introduce unwanted bias into the analysis, a within-group average might be a better choice for replacement.
- Analytics not only provide methods for profiling and identifying outliers and non-plausible values, they also can provide a **suggestion for a most probable value** that should be entered here.

Missing value imputation

- Analytics can deliver **replacement values for missing values**. Chapters 10 and 11 have shown how missing values in one-row-per-subject data marts and time series data marts can be imputed.
- These imputation methods range from average-based imputation values to analysis subject individual imputation values, which are based on analytic methods like decision trees or spline interpolations for time series.

Data standardization and de-duplication

- **Identification and elimination of duplicates** in a database where no unique key for the analysis subjects is available can be based on statistical methods that describe the similarity between records.
- These methods provide a measure of the closeness and similarity between records that is based on information like addresses, names, phone numbers, and account numbers, among others.

Handling data quantity

- Analytics allow you to **plan the optimal number of observations** for a controlled experiment with sample size and power calculation methods.
- In the case of small samples or small numbers of events in predictive modeling, methods for **modeling rare events** are provided.

- In the case of time series forecasting, so-called **intermittent demand models** are provided that model time series with only occasional non-zero quantities.

Analytic transformation of input variables

- Analytical methods are used to **transform variables to a distribution** that is suited for the respective analysis method. Log and square root transformations are, for example, used to transfer right-skewed data to a normal distribution.
- For variables with many categories, analytics provide methods to **combine categories**. Here, the combination logic for these categories depends on the number of observations in each category and the relationship to the target variables. Examples of these methods include decision trees or weight of evidence calculations.
- **Text mining** allows you to convert free-form text into structured information that can then be processed by analytical methods.

Variable selection

- Various methods for **variable selection** allow you to identify the subset of variables that have a strong relationship with the target variable in predictive modeling. These methods include simple metrics like R-square and advanced metrics like LARS, LASSO, compare also [16].
- Many analytical methods allow different methods of variable selection in the respective analysis model itself. Consider, for example, the forward, backward, and stepwise model selection in regression.

Assessment of model quality and what-if analyses

- Analytical tools are often designed to assist in model creation and validation. In predictive modeling, for example, it is often important to get a quick initial insight into the predictive power of the available data (this is also referred to as **Rapid Predictive Modeling**).
- These tools also provide measures to assess model quality very quickly and features for what-if-analyses.
- What-if analyses are especially useful to determine the importance of variables or groups of variables. Here, the consequences on the predictive power will be estimated if certain variables are not available.

13.3 Classical Outlier Detection

Ways to define validation limits

The purpose of outlier detection is to identify those observations that are considered to be outside of the typical value range for a variable. The definition of the typical range can be either done on

- statistical measures of the variable itself
- rules from a business- or domain-specific point of view

If business rules are available to define upper or lower limits, these are usually applied. For example, for the age in years variable, it usually makes more sense to define domain-specific limits than to calculate the limits based on the mean and standard deviation of the age values.

If, from a business- or domain-specific point of view, no validation limits can be defined or if the respective definition requires too much effort or is unfeasible, statistical validation limits can be used to define the acceptable range for the values of an observation.

Purpose of outlier detection

The purpose of defining these limits and checking these values is mainly to make sure that only valid and plausible observations are used for the analysis. From a statistical point of view, however, for some methods it is also advisable to identify observations as outliers even if their values are plausible and correct. Many analytical methods, for example, work on the assumption that the interval variables have a distribution that is close to the normal distribution:

- For example, in a distribution of the usage duration of a certain service, most values are around 120 hours. The distribution has two extreme outliers with 2,430 hours and 4,302 hours. From a business point of view, these values are correct and plausible; for the analysis, however, they are most likely filtered or shifted in order to achieve a more well-shaped and central distribution.

If outliers are identified in data, there are two strategies to deal with them:

- Observations with outliers can be excluded from the analysis. This is also called filtering.
- Alternatively, the values that are considered as outliers can be replaced by appropriate replacement values. Often, these replacement values are located closer to the center of the distribution. This is called shifting values.

Statistical methods

Statistical outlier detection uses information from the whole observations database to calculate validation limits. These limits are then applied for each observation independently. Statistical methods to define outliers include the following:

- Statistical measures like the mean of the distribution +/- the standard deviation multiplied by a factor:
  - Here the calculation of the mean and the standard deviation can be performed on the entire available data or on a central subset of the data in order to avoid biasing the calculation of the central metrics by outliers.
  - The determination of the number of standard deviations is domain-specific. Three standard deviations, for example, results in approximately 1% of outliers for normally distributed data. SAS Enterprise Miner, for example, uses three standard deviations in its default settings.
- Quantiles like the 1% and the 99% quantiles.
- Special forms of trimmed means, robust estimators, or other quantile-based measures like the 1.5-fold interquartile distance added to the third quartile and subtracted from the first quartile. Note that this definition is also used by many statistical software packages for outliers in box plots.

Implementation

These methods can easily be calculated in SAS and applied to data marts. SAS offers a wide range of functions and analytical procedures to calculate virtually any statistical validation range. Compare Cody [9] for coding examples of how standard deviation and percentile-based limits can be calculated.

SAS/QC software also provides methods for statistical quality control. Here, PROC SHEWHART provides many features to define outliers based on statistical measures and patterns. These methods can be used for data quality control as well. Compare also Svolba [17], for examples of how to use methods of statistical quality control for clinical trials, which also is related to data quality control.

Outlier detection with analytic methods

The following sections show examples of outlier detection with more advanced analytical methods, which include outlier detection in

- predictive modeling
- time series analysis
- cluster analysis

13.4 Outlier Detection with Predictive Modeling

General

An extension of the method described earlier is to determine the validation limits not from the variable itself in a univariate way but by inferring a most likely value based on relationships to other variables. Such other variables can be attributes like age, sex, region, or others that are used to individualize the validation limits.

The relationship between the variable of interest and the potential predictor variables is determined by predictive modeling. The predictive model uses a set of base variables as input variables to predict the variable of interest as the target variable. In this method, an expected value (reference value) for the variable is calculated based on the values of the input variables.

For each analysis subject, the individualization can calculate an individual reference value. If only a few categorical variables are used, the reference value corresponds to the defined segments. This can also be understood as a peer group approach, where similar analysis subjects are assigned the same reference value.

This reference value can then be compared against the actual value. The deviation between the reference value and the actual value is then used to judge whether the value is considered as outlier.

The reference value is also used to define the individual acceptance ranges. The following methods can be applied:

- using limits from a business point of view in absolute numbers
- using limits from a business point of view, which are calculated as absolute or relative differences between the values
- calculating the upper and lower limits based on the standard deviation for the predicted value or the deviation

Whether a value is defined as an outlier is based on the reference value, which itself is based on the actual values of variables of the respective observation.

Methods in SAS

Analytical methods to calculate individual validation limits usually include the following:

- Linear models like a linear regression or a general linear model. The advantage of this method is that for the prediction and the residual value, the standard deviation is calculated as well. This method can be performed in SAS Enterprise Miner with different regression nodes and in SAS/STAT software, for example, with the REG procedure or the GLM procedure.
- Decision trees, which can automatically detect interactions in the data and provide very specific reference values. However, the set of prediction (expected) values is not continuous. This method is available in SAS Enterprise Miner. See also [3] Schubert.

Example of clinical trial data

In order to illustrate the concept, a data set from a clinical trial is used. The following variables are collected for each patient:

- age (4 groups)
- sex
- weight (3 groups)
- melanoma Stage (2 stages)
- trial center ID (8 centers)

172 *Data Quality for Analytics Using SAS*

For each patient, data are collected over time at different visits. For illustrative purposes, the cholesterol value is used to show how a predictive model based on these variables can help calculate individual reference values. Because all input variables are categorical, the maximum number of different reference values is 4 x 2 x 3 x 2 x 8 = 384. In the example data that follows, the 403 patients fall into 181 different categories with respect to the variable grouping. For these 403 patients, 3,154 measurements have been made over time. An excerpt of the analysis data is shown in Table 13.1.

Table 13.1: Excerpt of the analysis data

PATNR	Age_Grp	SEX	Weight_Grp	STAGE	CENTERNR	VisitDate	CHOL
232	19-30	0	80+	1	1	16SEP1999	237.00
232	19-30	0	80+	1	1	08OCT1999	193.00
232	19-30	0	80+	1	1	24DEC1999	194.00
232	19-30	0	80+	1	1	23FEB2000	205.00
232	19-30	0	80+	1	1	24MAY2000	217.00
232	19-30	0	80+	1	1	14MAR2001	211.00
232	19-30	0	80+	1	1	20JUN2001	223.00
232	19-30	0	80+	1	1	19SEP2001	218.00
191	60+	0	80+	2	1	11AUG1999	139.00
191	60+	0	80+	2	1	06OCT1999	166.00
191	60+	0	80+	2	1	10NOV1999	166.00
605	46-60	1	66 -	1	4	29DEC1999	169.00
605	46-60	1	66 -	1	4	01MAR2000	188.00
605	46-60	1	66 -	1	4	06SEP2000	185.00
605	46-60	1	66 -	1	4	06DEC2000	158.00
605	46-60	1	66 -	1	4	13JUN2001	174.00
225	46-60	0	80+	1	3	23SEP1999	211.00
225	46-60	0	80+	1	3	29OCT1999	185.00
225	46-60	0	80+	1	3	21MAR2000	193.00
225	46-60	0	80+	1	3	05SEP2000	203.00

Based on this data, a predictive model is created with PROC GLM that uses CHOL as the target variable (y) and the other variables (except PATNR and VISITDATE) as categorical input variables:

```
proc glm data=labor_chol_data;
 class sex centernr stage age_grp weight_grp ;
 model chol = age_grp sex weight_grp centernr stage;
 output out=pred_chol p=reference r=residual
  stdi=stdi stdr=stdr stdp=stdp;
run;
quit;
```

Table 13.2 shows the output table, PRED_CHOL. In addition to the input variables, this table also contains the

- predicted value for CHOL (variable REFERENCE)
- residual between the predicted value and the actual value
- standard deviations of the individual, predicted, and residual values

Table 13.2: Output table (PRED_CHOL) with the results of PROC GLM

PATNR	Age_Grp	SEX	Weight_Grp	STAGE	CENTERNR	VisitDate	CHOL	reference	residual	stdi	stdr	stdp
144	46-60	0	80+	1	2	24AUG2001	232.00	205.87601735	26.123982652	36.494115985	36.400139366	1.8507235971
144	46-60	0	80+	1	2	02NOV2001	220.00	205.87601735	14.123982652	36.494115985	36.400139366	1.8507235971
144	46-60	0	80+	1	2	07FEB2002	244.00	205.87601735	38.123982652	36.494115985	36.400139366	1.8507235971
144	46-60	0	80+	1	2	03MAY2002	241.00	205.87601735	35.123982652	36.494115985	36.400139366	1.8507235971
144	46-60	0	80+	1	2	04AUG2002	218.00	205.87601735	12.123982652	36.494115985	36.400139366	1.8507235971
144	46-60	0	80+	1	2	08NOV2002	236.00	205.87601735	30.123982652	36.494115985	36.400139366	1.8507235971
144	46-60	0	80+	1	2	03JAN2003	253.00	205.87601735	47.123982652	36.494115985	36.400139366	1.8507235971
144	46-60	0	80+	1	2	08MAY2003	230.00	205.87601735	24.123982652	36.494115985	36.400139366	1.8507235971
144	46-60	0	80+	1	2	08AUG2003	230.00	205.87601735	24.123982652	36.494115985	36.400139366	1.8507235971
146	60+	0	80+	1	3	05NOV1999	276.00	202.50040776	73.499592235	36.563135081	36.330810622	2.9090008993
146	60+	0	80+	1	6	28NOV2000	276.00	203.511552	72.488447999	36.538977428	36.355106601	2.5887347805
146	60+	0	80+	1	6	28DEC2000	230.00	203.511552	26.488447999	36.538977428	36.355106601	2.5887347805
146	60+	0	80+	1	6	01FEB2001	231.00	203.511552	27.488447999	36.538977428	36.355106601	2.5887347805
146	60+	0	80+	1	6	01MAR2001	230.00	203.511552	26.488447999	36.538977428	36.355106601	2.5887347805
146	60+	0	80+	1	6	31MAY2001	190.00	203.511552	-13.511552	36.538977428	36.355106601	2.5887347805
146	60+	0	80+	1	6	06SEP2001	191.00	203.511552	-12.511552	36.538977428	36.355106601	2.5887347805
146	60+	0	80+	1	6	13DEC2001	220.00	203.511552	16.488447999	36.538977428	36.355106601	2.5887347805

The predicted value (reference value) or the residual values can now be used together with the standard deviations to calculate individual upper and lower validation limits.

To illustrate using the individual reference values, the following sets of validation limits have been defined:

- Overall validation limits UPPER and LOWER have been defined based on the overall mean of 198.34 and the standard deviation of 38.38, with a range of +/-2 standard deviations.
- Individual validation limits UPPER_I and LOWER_I have been defined based on the individually predicted values (REFERENCE) and a range of +/-2 standard deviations.

The results for selected patients are represented graphically in Output 13.1.

Output 13.1: Scatterplot for patient cholesterol with different validation limits

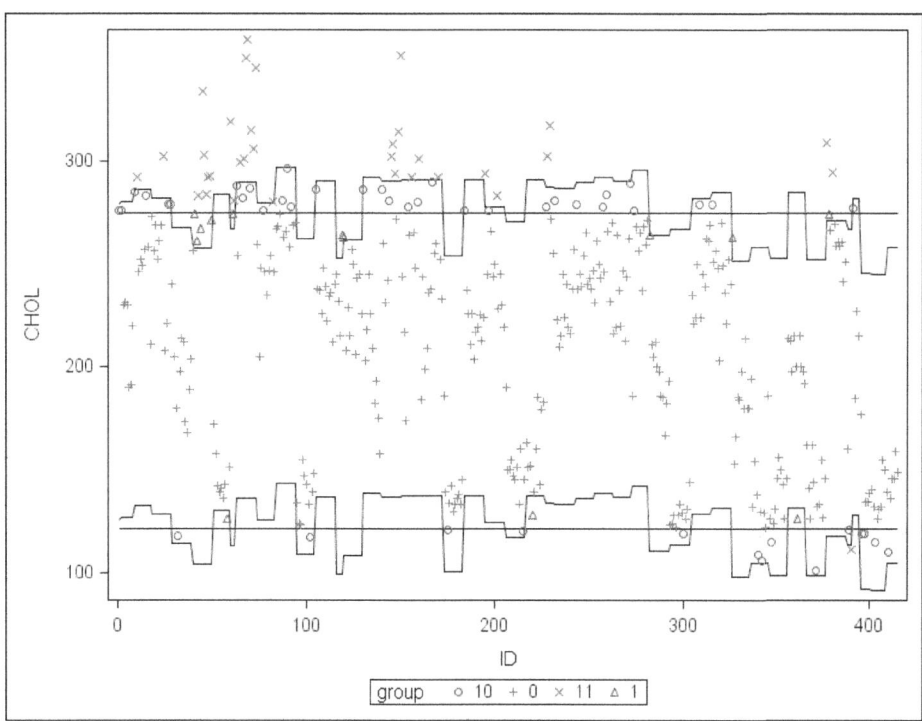

Note the following:

- The graph shows two straight horizontal lines, which are the overall validation limits.
- The two moving black solid lines represent the individual validation limits.
- The cholesterol values are represented on the y-axis, and the ID variable represents an artificial enumeration of the observations.
    o Observations that are within both validation limits, both the general and individual ones, are represented by a plus sign (+).
    o Observations that are outside the general validation limits but within the individual limits are represented as circles (o). These observations are now counted as regular observations if the individual limits are applied.
    o Observations that have not been considered outliers with general validation but are considered outliers when applying the individual validation limits are represented as triangles.
    o Crosses (x) mark those observations that are outliers in both cases.

- On an individual basis, the limits provide much more specific information about the data. The scatterplot was created using PROC SGPLOT with the following code:

```
proc sgplot data = pred_chol;
 series x=id y=upper;
 series x=id y=lower;
 step x=id y=upper_i/justify=center;
 step x=id y=lower_i/justify=center;
 scatter x=id y=chol / group = group MARKERATTRS=(size=8);
run;
quit;
```

Note the simplicity of the code that creates a very detailed and illustrative picture.

Extension of this method

This method can also be extended with a time dimension. Here, a possible time trend of the values is considered.

13.5 Outlier Detection in Time Series Analysis

General

Deciding whether a specific value is an outlier, in many cases, is not only based on the value itself. An example of predictive modeling has been shown in the previous subsection. This method can be extended to include time series information. For example, if it can be assumed that the values underlie seasonal patterns, it makes sense to differentiate the expected values by season (for example, calendar month).

Time series models

If the data are taken from a process over time, it is advisable to apply time series models, which include typical time series elements like seasons, trends, cycles, and shifts.

In this case, modeling is based on the historic time series. Deciding whether a value is an outlier is based on the deviation of the forecasted value in the time series (and not only on static limits).

This can account for the fact that, for example, a sales number of 251,000 pieces could be an outlier in February but would be a normal value in December. The dynamic validation intervals potentially result in less misclassifications.

Outlier detection with ARIMA(X) models

The detection of outliers is a very important topic in time series analysis. ARIMA(X) models can be used to filter the effect of detected outliers. The process, in this case, is as follows:

- An initial time series model is fit based on the available data.
- Those observations that are considered as outliers based on their deviations from the predicted values are flagged.
- A new model is fit that also includes the dummy variables for the detected outliers.
- Based on the new model, new observations might be considered as outliers, and the process repeats itself.

SAS Forecast Studio provides this functionality for ARIMA models. You can set the option in the DIAGNOSTICS tab of the FORECAST SETTINGS. Figure 13.1 shows an example.

Figure 13.1: Outlier detection setting in SAS Forecast Studio

Outlier detection(ARIMA models only):
- ☑ Detect outliers | 1
- Significance level: | 0.01
- Maximum percentage of series that can be outliers: | 2

Output 13.2 shows the model's resulting graph in SAS Forecast Studio. Here, the historic observations are shown as circles and the forecast is shown as a solid line. It appears that some observations have unusually high values.

The ARIMAX model that has been built for these data with the detect outlier option identifies the observations in February 2007 and October 2008 as outliers.

Output 13.2: Line plot of a time series in SAS Forecast Studio

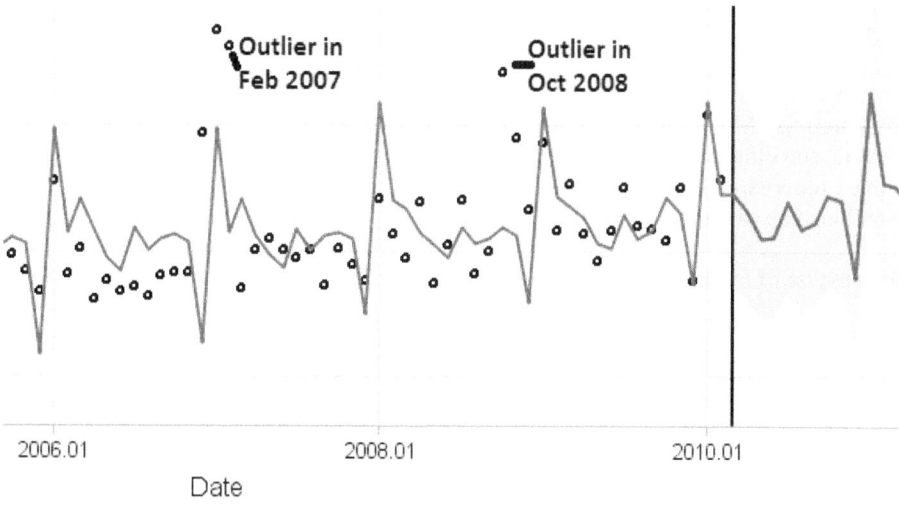

The graph annotates the observations that are considered as outliers. It can be seen that the value in January 2007 (in the upper left next to the outlier in February 2007) is not flagged as an outlier, although the value is larger. In January, values are higher in general so such a value is not seen as out of range, while in February the expected values are usually much lower.

Similar to the predictive model that was shown in the previous section, analytic models here allow you to make a more intelligent judgment about the outlier status of data values.

Decomposition and smoothing of time series

Time series methods can also be used to decompose or to smooth a time series:

- Decomposing means that values of the time series are corrected for a possible existing trend, seasonal effect, cycle, or other influence described by the co-variables.
  - The resulting series for the irregular component shows the course of the values over time after eliminating known effects.
  - Remaining shifts or peaks in the time series can then be interpreted from a business point of view to potentially detect the influence of production changes or external market effects, for example.
- Smoothing means that the values are averaged over a time window or seasons. It reduces the variability in the values of the time series. This results in a time series with less noise and allows a clearer picture on trends.

13.6 Outlier Detection with Cluster Analysis

General

Another method for profiling data is multivariate outlier detection. Different from the situation of univariate outlier detection, where a rule is defined on the individual value only, multivariate outlier detection identifies observations as outliers in a multivariate sense. Rules are based on the combination of the values of two or more variables.

For example, statistical cluster analysis can identify outliers that would go undetected in a univariate analysis. From a univariate point of view, each variable may be within its validation limits. From a multivariate point of view, however, combinations of variable values are detected as unusual and flagged as outliers.

The advantage of this method is that outliers are found that would otherwise stay hidden. On the other hand, the definition of an outlier in this method is not as straightforward as the definition under univariate rules and validation limits.

This method not only detects outliers; it also can identify suspicious cases in fraud detection, see also [3] Schubert.

Output 13.3 shows SAS Enterprise Miner results of clustering insurance customers based on variables like age, income, car Bluebook value, and other factors. All observations are grouped into six clusters. Cluster 6 is a rather small cluster, with 61 observations (from 10,303). From a univariate perspective, the within-cluster average values for age and income can be calculated.

Output 13.3: Clustering results of insurance customers

Segment Id ▼	Frequency of Cluster	Age	Income
6	61	44.19672	62171.06
5	1722	46.30081	133086.8
4	3265	38.83691	39514.4
3	1868	41.47185	43807.37
2	2810	52.6306	54005.21
1	577	47.37674	66922.92

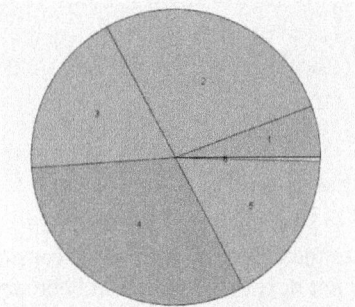

From a multivariate point of view, however, as can be seen in the cluster proximities chart in Output 13.4, cluster 6 is rather distant from the other clusters. Unlike cluster 5, which has 1,722 observations, this cluster is a rather small one. Cluster 6, therefore, may be a start for the examination of the respective observations. Let's check whether business reasons exist for why these observations are clustered together.

Output 13.4: Cluster proximities chart of the insurance customer clustering

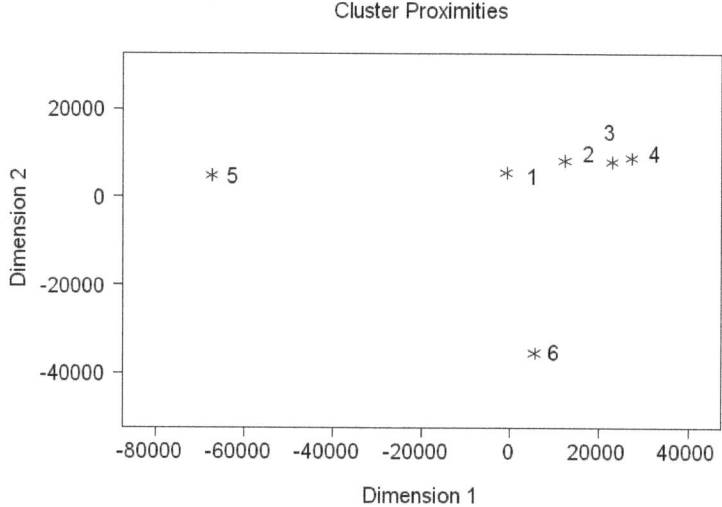

Conclusion

Analytic methods have been shown to help profile data from an advanced perspective and to identify relationships in the data. In data quality profiling, these methods can be applied to identify a subset of observations that differ from the others. You then need to decide whether this combination is considered to be within usual business interpretations or whether it should be further investigated for possible data biases or errors.

13.7 Recognition of Duplicates

General

The recognition of duplicates is an important topic in data quality control. Records in a table or in related tables are checked for duplicates. This task is simple as long as unique keys for the records are available. In the absence of unique keys, however, different attributes of the records like addresses, names, bank accounts, or phone numbers are used to create so-called surrogate keys to identify the records.

Surrogate keys are, for example, needed in the following cases:

- Two companies merge into a single company. Merging the customer database probably is not possible based on the customer IDs because the two companies have their own individual customer IDs.
- A company acquires external data that contains additional attributes on the business customers of the company. No common ID is available, however, to join the external data to the internal database of business customers.

Recognizing duplicates is usually preceded by a standardization of the values in these fields. The standardized version can then be used to calculate how similar (or close) the respective records are to each other. Records that have a high similarity may be candidates for duplicates.

Contribution of analytics

Analytics provide methods to define the similarity between records. These similarity measures can be based on fuzzy matching algorithms. Fuzzy matching methods are based on fuzzy sets and mirror the concept of degrees of group membership. An observation is not only assigned to a single set but to multiple sets with a respective probability. Other similarity methods include, for example, Euclidean distances or clustering methods like *k*-means or hierarchical clustering.

13.8 Other Examples of Data Profiling

General

Even if the data do not contain any missing values and all the values fall within the predefined ranges, the data may still not be useful. Data may be falsified or artificially created. A subdiscipline in fraud detection, forensic data analysis, analyzes the process of data creation and checks whether the data contain unusual patterns.

Thus, there is a link between data quality checks and detection of data anomalies.

Benford's law for checking data

One feature that is frequently checked in the fraud detection context is the distribution of the first (largest) digits of numbers. In 1938, Frank Benford [18] stated, based on the work of Simon Newcomb, that the numbers from many real-life situations follow a specific logarithmic distribution. This phenomenon is known as Benford's law, which is expressed as follows:

$$P(d) = \log_{10} (1-1/d).$$

The probability of the largest digits equaling d decreases when d increases. A leading digit of 1 has a probability of 30.1%, while a leading digit of 2 has a probability of 17.6% (see Table 13.3).

Table 13.3: Benford probabilities for digits 1 through 10

Digit	prob
1	0.30103
2	0.17609
3	0.12494
4	0.09691
5	0.07918
6	0.06695
7	0.05799
8	0.05115
9	0.04576
10	0.04139

Output 13.5 compares the distribution of the observed first digits on income information for customers (dark bars) with the expected percentage following Benford's law (light bars).

Output 13.5: Bar chart for frequency of the first digit of income values

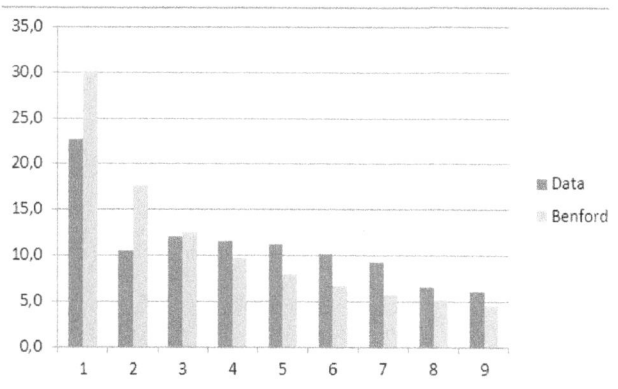

The true distribution does not follow the expected distribution. Analytical methods like the Chi^2-test can be applied to assess the deviation of the observed data from the expected distribution.

13.9 Conclusion

This chapter has presented an overview of different analytical methods that help to improve data quality or that can be used in the data quality process. Some of these topics are discussed in more detail in the next chapter, which addresses the use of different SAS analytic tools for data quality control.

This chapter's main focus was on how analytics can help to identify outliers that might go undetected otherwise. These outliers could have an impact on the intended analysis. Methods have been presented for calculating individual reference values and retrieving individual validation limits. Benefits of using methods like predictive modeling or time series analysis, which allow more advanced outlier detection, have also been described.

Using cluster analysis for outlier detection enables the analyst to identify complex outlier patterns that would remain undetected in a univariate analysis.

Chapter 14: Data Quality Profiling and Improvement with SAS Analytic Tools

14.1 Introduction 182
14.2 SAS Enterprise Miner 182
    Short description of SAS Enterprise Miner 182
    Data quality correction 183
    Assessing the importance of variables 184
    Gaining insight into data relationships 184
    Modeling features for data quality 184
    Quick assessment of model quality and what-if analyses 185
    Handling small data quantities 185
    Text mining 185
    Features for modeling and scoring in SAS Enterprise Miner 186
14.3 SAS Model Manager 187
14.4 SAS/STAT Software 187
14.5 SAS Forecast Server and SAS Forecast Studio 188
    Short description of SAS Forecast Server 188
    Data preprocessing 188
    Outlier detection 189
    Model output data quality 189
    Data quantity 190
14.6 SAS/ETS Software 190
14.7 Base SAS 191
14.8 JMP 191
    General 191
    Detecting complex relationships and data quality problems with JMP 191
    Missing data pattern 193
    Sample size and power calculation 195
14.9 DataFlux Data Management Platform 195
    Short description of DataFlux Data Management Platform 195
    Data profiling 196
    Data standardization and record matching 196
    Defining a data quality process flow 197
14.10 Conclusion 198

14.1 Introduction

SAS is a powerful software package for data integration, data management, analytics, and reporting. It includes a wide range of analytical features across its offerings and allows you to answer analytical questions for a large number of domains. It is, therefore, not surprising that many features of SAS are perfectly suited to profile and improve data quality.

The previous chapters introduced the general capabilities of analytics for data quality. Methods for profiling and imputing missing values and detecting outliers have been presented. In some cases, the use of SAS tools has been showcased (for example, replacing missing values with the Impute node in SAS Enterprise Miner, using PROC MI in SAS/STAT software, or detecting outliers with SAS Forecast Server).

This chapter structures the data quality capabilities of SAS from a tools perspective. For each of the following SAS products, relevant features for data quality for analytics are illustrated:

- SAS Enterprise Miner and SAS Text Miner
- SAS Model Manager
- SAS/STAT software
- SAS Forecast Server
- SAS/ETS software
- Base SAS
- JMP
- DataFlux Data Management Platform

For completeness, the features that have already been mentioned and discussed in previous chapters are also briefly shown here.

This chapter not only highlights the particular features of the SAS tools. It also gives an overview of the capabilities of analytic software for data quality matters in general.

14.2 SAS Enterprise Miner

Short description of SAS Enterprise Miner

SAS Enterprise Miner is the flagship tool for data mining. It represents a comprehensive set of data mining algorithms like decision trees, neural networks, and association analyses, to name a few. The user interface (UI) of SAS Enterprise Miner allows you to define very flexible data mining process flows in a highly interactive drag-and-drop environment.

SAS Enterprise Miner follows the SEMMA methodology, a data mining methodology that has been designed by SAS based on best practices. SEMMA segments the data mining process into the following steps: Sampling, Explore, Modify, Model, and Assessment. Figure 14.1 shows an example of the SAS Enterprise Miner UI.

Figure 14.1: SAS Enterprise Miner UI

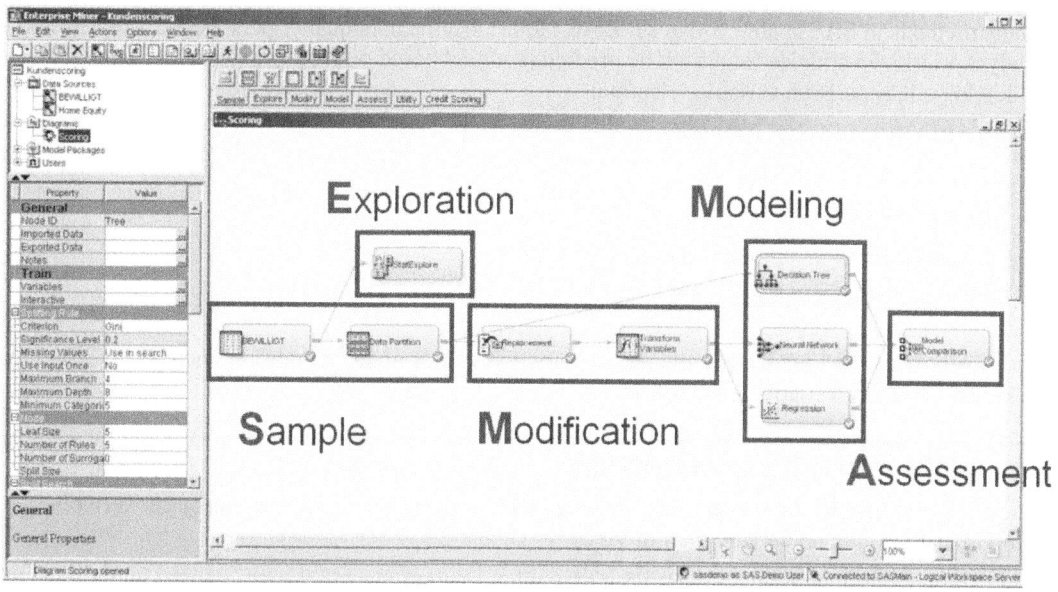

Data quality correction

SAS Enterprise Miner provides a number of tools for data quality improvement. Except for the filter node, which is in the SAMPLE tab, all the respective nodes are found in the MODIFY tab. The most important nodes for data quality correction are:

- The TRANSFORM VARIABLES node allows you to transform the distribution of variables into a more appropriate shape for the analysis. These methods, for example, include LOG or square root transformations.
- The REPLACEMENT node
    - shifts outliers to a more centered value.
    - combines categories with the replacement editor. Figure 14.2 shows an example.

Figure 14.2: Replacement editor in SAS Enterprise Miner

Replacement Editor

Name	Level	Frequency	Type	Char Raw V...	Num Raw V...	Replacemen...
AnzahlKinder	0	9070	N		0.0	
AnzahlKinder	1	804	N		1.0	
AnzahlKinder	2	351	N		2.0	
AnzahlKinder	3	74	N		3.0	
AnzahlKinder	4	4	N		4.0	
AnzahlKinder	_UNKNOWN_	.	N		.	_DEFAULT_
Ausbildung	High School	2952	C	High School	.	
Ausbildung	Bachelors	2824	C	Bachelors	.	
Ausbildung	Masters	2078	C	Masters	.	
Ausbildung	<High School	1515	C	<High School	.	
Ausbildung	PhD	934	C	PhD	.	
Ausbildung	_UNKNOWN_	.	C		.	_DEFAULT_
AutoTyp	SUV	2883	C	SUV	.	
AutoTyp	Sedan	2694	C	Sedan	.	
AutoTyp	Pickup	1772	C	Pickup	.	
AutoTyp	Sports Car	1179	C	Sports Car	.	
AutoTyp	Van	922	C	Van	.	
AutoTyp	Panel Truck	853	C	Panel Truck	.	
AutoTyp	_UNKNOWN_	.	C		.	_DEFAULT_
AutoVerwen...	Private	6513	C	Private	.	
AutoVerwen...	Commercial	3790	C	Commercial	.	
AutoVerwen...	_UNKNOWN_	.	C		.	_DEFAULT_

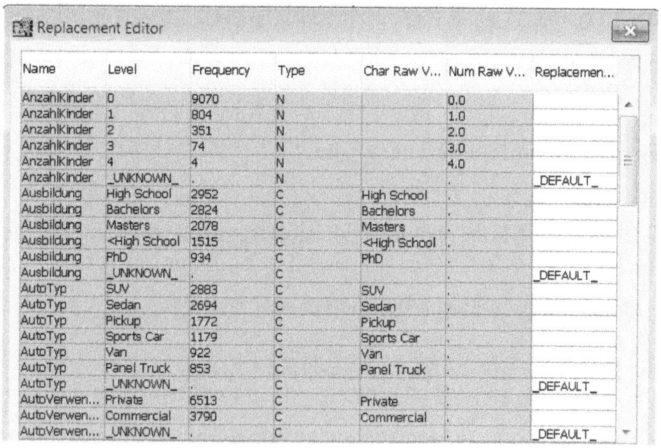

- With the IMPUTE node, missing values can be replaced. Section 10.5 details the available methods to impute missing values for interval and categorical variables.
- The FILTER node filters outliers from the data. The available methods include standard deviations from the mean, percentiles, mean absolute deviations, and others.

Assessing the importance of variables

- Many methods in SAS enable you to select variables. From a data quality perspective, this lets an analyst prioritize variables for a specific predictive model based on their predictive power. The following nodes perform variable selection:
    - VARIABLE SELECTION node
    - TREE node
    - REGRESSION node
    - LARS (least angle regression) node
    - PLS (partial least squares) node

- The VARIABLE SELECTION node and the REGRESSION node, in addition, include the selection of the most important variable interactions.

Gaining insight into data relationships

- The STAT EXPLORE node identifies attributes with a strong relationship to the target variable (importance) from a univariate point of view.
- The explore facility in the VARIABLES dialog allows you to visually investigate the importance of different variables.
- The VARIABLE SELECTION node also facilitates the grouping of distinct values of categorical variables based on their relationship to the target variable. This reduces the number of correlated input variables before the modeling process. And it also provides insight into the structure and relationship of the data.
- With the ASSOCIATION node, association and sequence analyses can be performed. These methods can also be used to analyze the occurrence of combinations and sequences of different attributes of analysis subjects.
    - With this node, relationships in the data can be uncovered and rules can be profiled (for example, the fact that two attributes must not occur together in the data or must occur in a certain sequences). For example, in the event history of an account, the WITHDRAW_MONEY event must not occur before the ACCOUNT_OPENING event.
    - Note that the association node requires a multiple-row-per-subject data mart structure. Refer to [1] Svolba, section 14.5, for more details.

Modeling features for data quality

SAS Enterprise Miner provides many features to build predictive and other types of models. For data quality, the following are of interest:

- Creating a predictive model for the determination of a reference value.
    - This method has been shown in section 13.4.
    - In addition to linear models, SAS Enterprise Miner provides decision trees for nonlinear segment-based definitions of reference values.
    - It is also possible to define segments manually with the interactive tree facility.

- Cluster methods are available to discover multivariate outliers (see section 13.6).

Quick assessment of model quality and what-if analyses

An initial assessment of model quality, along with the predictive power of the available data for the analysis, is critical. SAS Enterprise Miner supports this need in many ways, as follows:

- The model nodes, for example, the REGRESSION node and the TREE node, provide a variety of assessment measures, such as lift curves and error measures. This gives you the first information about the suitability of the data for the business question and the predictability of the target variable with the available data.

- The ASSESSMENT node can compare different models or model variants. Models with different sets of input variables can be compared against each other, and the importance of different variables can be estimated. This also allows a cost-benefit analysis of adding variables to the model by comparing their impact on the uplift in the predictive power of the model to the cost of including this variable in the model.

- The process flow environment of SAS Enterprise Miner can simulate different scenarios of data availability and preparation and assess the consequences on the model output. This provides insight into the impact of additional variables and data quality efforts on final model accuracy. Figure 14.3 shows a SAS Enterprise Miner process flow.

Figure 14.3: SAS Enterprise Miner process flow simulating the impact of different predictor variable subsets

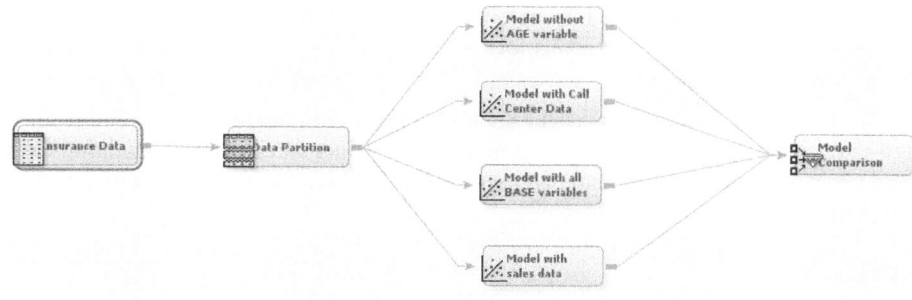

- The SAS Enterprise Miner solution also includes the SAS Rapid Predictive Modeler. This enables business analysts to quickly build predictive models guided by a wizard, without needing to access the SAS Enterprise Miner user interface. This allows you to check the importance of different variables easily and quickly.

Handling small data quantities

In order to handle data with few observations or few events in predictive modeling, SAS Enterprise Miner provides

- the RULE INDUCTION node to model rare events.
- a facility to perform bootstrapping. The START GROUPS and END GROUPS nodes allow iterating and resampling the data to base the analysis on different samples.

Text mining

- SAS Text Miner can process free-form text and convert the unstructured content of text fields and documents into structured information that can be used in the analysis.

- Text mining, together with the character functions in PERL regular expressions (which are part of the SAS language), facilitates the conversion of bad quality unstructured text into usable structured text. See also chapter 9's discussion of converting real-world facts into data.

Features for modeling and scoring in SAS Enterprise Miner

SAS Enterprise Miner provides many features for immediate data quality checks:

- By default, the number of maximum possible categories for a nominal variable is set to 512. This high number helps prevent the usage of categorical variables with a large number of distinct values. Often, these variables can be grouped intelligently or excluded altogether because of the excessive number of degrees of freedom.
- The score code that is produced in SAS Enterprise Miner creates a _WARN_ variable, which is empty if the scoring code did not encounter a problem for a record. The following scoring problems are defined and flagged in the _WARN_ variable:
  - If a variable contains a missing value and no imputation algorithm is defined for that variable, the record cannot be scored. In this case, the _WARN_ variable contains an M to indicate the missing values.
  - If a categorical variable contains a new value that was not present during model training, no coefficient is available for this category in the score code and no score can be produced. In this case, a U is written to the _WARN_ variable to indicate an unknown category.
- The score node also outputs the list of variables that are used for the final model. This is important to streamline the process of data preparation for model scoring from a data quality point of view. For the variables in a SAS Enterprise Miner data source, additional statistical metadata can be generated (click the Compute Summary button in the Data Source Wizard). An example is shown in Figure 14.4.
  - These statistical metadata provide information about the distribution, the percentage of missing values, and the number of distinct levels for each category variable at a very early stage in the analysis process.
  - This helps to set the variable roles for the analysis accordingly and to plan the required preparation steps. For example, a variable with a large number of distinct values requires additional data preparation steps as well as a variable with missing values and variables with a highly skewed distribution.

Figure 14.4: Variables metadata

Name	Number of Levels	Percent Missing	Minimum	Maximum	Mean	Standard Deviation	Skewness	Kurtosis
AGE	.	0.067941	16	81	44.83664	8.606374	-0.03435	-0.08103
BIRTH	513	0	.	.	.	.	.	.
BLUEBOOK	.	0	1500	69740	15660.37	8428.481	0.76924	0.652174
CAR_TYPE	6	0	.	.	.	.	.	.
CAR_USE	2	0	.	.	.	.	.	.
CLM_AMT	.	0	0	123247.1	1511.119	4725.047	9.296862	135.4885
CLM_DATE	512	87.98463	.	.	.	.	.	.
CLM_FLAG	2	0	.	.	.	.	.	.
CLM_FREQ	.	0	0	5	0.800641	1.15405	1.194206	0.246274
DENSITY	4	0	.	.	.	.	.	.
GENDER	2	0	.	.	.	.	.	.
HOMEKIDS	.	0	0	5	0.720567	1.11634	1.336299	0.628346
HOME_VAL	.	5.590605	0	885282.3	154523	129188.4	0.491969	-0.0354
ID	.	.	.	.	.	.	.	.
INCOME	.	5.532369	0	367030.3	61575.56	47456.01	1.161455	1.98847
INITDATE	513	0	.	.	.	.	.	.
JOBCLASS	8	6.454431	.	.	.	.	.	.
KIDSDRIV	.	0	0	4	0.169271	0.50649	3.343069	11.67806
MARRIED	2	0	.	.	.	.	.	.
MAX_EDUC	5	0	.	.	.	.	.	.
MVR_PTS	.	0	0	13	1.709987	2.158976	1.340631	1.33618
NPOLICY	.	0	1	9	1.695429	0.935207	1.750292	4.660369
OLDCLAIM	.	0	0	57037	4033.586	8732.81	3.119935	9.89925
PARENT1	2	0	.	.	.	.	.	.
PLCYDATE	513	0	.	.	.	.	.	.
POLICYNO	.	0	160	99992405	50042269	28939671	0.002339	-1.20411
RED_CAR	2	0	.	.	.	.	.	.
RETAINED	.	0	1	25	5.329224	4.110601	0.899408	0.479971
REVOLKED	2	0	.	.	.	.	.	.
SAMEHOME	.	6.202077	-3	28	8.298738	5.714449	0.280225	-0.76457
TRAVTIME	.	0	5	142.1206	33.41806	15.86392	0.436509	0.59543
YEARQTR	26	0	.	.	.	.	.	.
YOJ	.	5.318839	0	23	10.47391	4.10876	-1.20076	1.144943

14.3 SAS Model Manager

SAS Model Manager is a tool that manages analytic models. These models can be SAS Enterprise Miner models, SAS/STAT models, or Base SAS models. SAS Model Manager can define a model lifecycle that describes the required steps and milestones in the lifetime of a model. It supports a champion-challenger model approach and the monitoring of model performance over time.

From a data quality point of view, SAS Model Manager covers the scoring and model monitoring process. The following features are of particular interest:

- SAS Model Manager validates the score code with respect to the available score data. This validation ensures high-quality output from the model through the assessment of the completeness and correct mapping of available and required scoring input data.
- For score data, SAS Model Manager also monitors the statistical characteristics of the scoring input data for early detection of deviations that could impact the output quality of the model. Shifts in the distribution of the scoring data may also indicate changes in data quality that, in turn, could impact the model performance. Figure 14.5 shows a deviation index for a variable, where the deviation between the training and the scoring data increases over time.

Figure 14.5: Characteristic report for variable MONTHS_SINCE_LAST_GIFT showing the deviation index

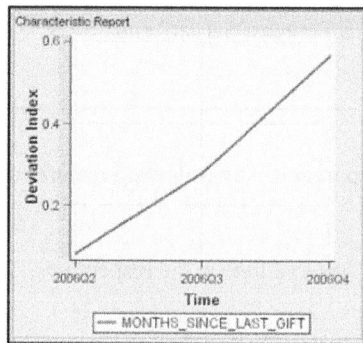

- SAS Model Manager provides reports to check the model quality over time. Because model quality also depends on data quality and stability, this report not only indicates whether a model will be retrained but also whether the quality of the score data and their correspondence to the assumptions in the model building phase will be rechecked.

14.4 SAS/STAT Software

SAS/STAT performs general statistical analyses. It offers a wide range of analytic procedures, from regression analysis procedures to survival analysis procedures to multivariate analysis procedures.

Many of the SAS/STAT capabilities for data quality are similar to the SAS Enterprise Miner capabilities and are thus only briefly discussed here.

- Procedures like PROC REG and PROC LOGISTIC provide multivariate methods to select the most important predictor variables.
- PROC GLMSELECT provides methods for variable selection. This includes the LARS and LASSO methods (compare [16]).
- PROC PLS (partial least squares) provides a method for selecting the best potential predictor variables from a highly correlated input variable space.
- PROC VARCLUS performs variable clustering to identify relationships between individual variables.
- PROC MDS displays the relationship and closeness of variables graphically.

- Procedures like PROC GLM and PROC REG calculate reference values with predictive models (see also section 13.4).
- PROC SURVEYSELECT provides different sampling methods that are important for bootstrapping analyses.
- PROC MI performs multiple imputation of missing values and PROC MIANALYZE combines the results of the analyses on the different imputations (see also section 10.6).
- PROC POWER and the SAS Power and Sample Size Application facilitate the calculation of the minimum number of observations for the planning of controlled experiments. See also appendix B for details.

14.5 SAS Forecast Server and SAS Forecast Studio

Short description of SAS Forecast Server

SAS Forecast Server is a powerful offering for automated time series forecasting. It can process a large number of high-dimensional time series. SAS Forecast Server automatically evaluates different time series models for each time series and selects the model with the best fit for producing forecasts for future periods. The analyst can predefine catalogs of the most appropriate models, build models manually, and override automatic model selection. SAS Forecast Studio is the graphical user interface for SAS Forecast Server. As many operations can be performed in both ways, coding and point-and-click, the two terms "SAS Forecast Server" and "SAS Forecast Studio" are used interchangeably in this section.

Data preprocessing

SAS Forecast Server includes time series forecasting methods for data preparation. The following features are especially relevant from a data quality point of view:

- **Alignment of data**: SAS Forecast Studio automatically aligns the observations in the respective intervals and ensures the time period matches for all records.
- **Treatment of missing and zero values**: The need for a treatment of missing and zero values in time series has already been discussed in chapter 11, which introduced PROC TIMESERIES. SAS Forecast Studio provides a UI for interactive parameter control. An example is presented in Figure 14.6.

Figure 14.6: SAS Forecast Studio with options for handling missing and zero values

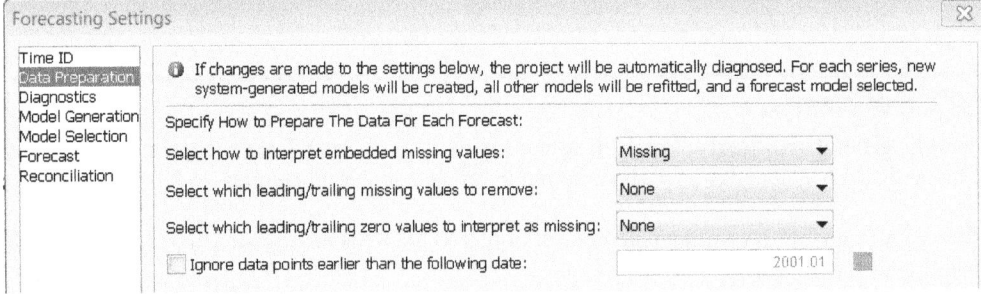

These options are available in the FORECAST SETTINGS window in the DATA PREPARATION tag:

- Missing values that occur in the time series (embedded missing values) can be treated as missing values or they can be replaced, for example, by a mean or by the previous or next value.
- Leading or trailing missing values can be removed from the analysis data.
- Leading or trailing zero values can be interpreted as missing values. This is relevant, for example, for market introduction or market retirement of products where the product is available in the system, but it has not had any sales so far. Alternatively, the zero value interval can be removed from the analysis to avoid bias.

Outlier detection

As already described in section 13.5, SAS Forecast Server facilitates the detection of outliers in time series data. Each detected outlier is flagged automatically for consideration in the time series model building process.

The option can be set in the DIAGNOSTICS tab of the FORECAST SETTINGS menu. Figure 14.7 shows an example.

Figure 14.7: Outlier detection setting in SAS Forecast Studio

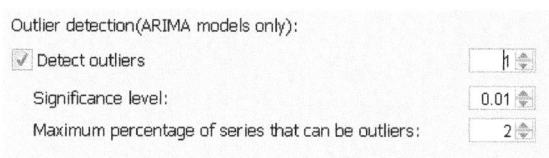

Note that this form of outlier detection can only be performed for ARIMA(X) models.

SAS Forecast Server also allows the manual definition of events. These events are used to flag certain intervals in the time series data, where the time series may be expected to show a different behavior.

- These events can be specified for recurrent events like Easter or Christmas.
- They can, however, also be used to manually flag intervals, where for special reasons the time series shows a specific behavior, such as product promotions and out-of-stock events. Figure 14.8 shows an example of the definition of events.

Figure 14.8: Definition of events in SAS Forecast Studio

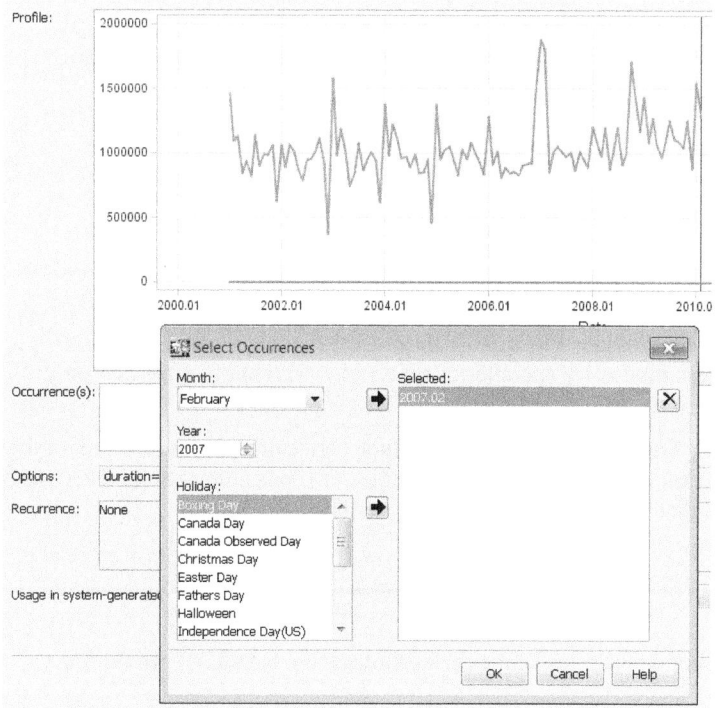

Model output data quality

- To control model output data quality, SAS Forecast Studio allows you to set an option to determine whether negative forecasts are allowed.

190 *Data Quality for Analytics Using SAS*

- SAS Forecast Studio also provides a user interface to manually override the automatically created forecasts. This is usually done for planning and budgeting reasons or to adapt the statistical forecast to expectations resulting from the underlying business process. Figure 14.9 shows an example of overriding.

Figure 14.9: Override calculator in SAS Forecast Studio

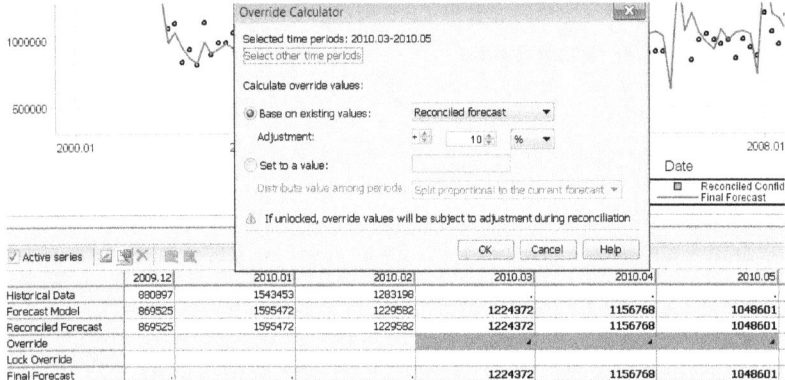

Data quantity

- SAS Forecast Studio has default settings for the minimum number of historic intervals required for model training. For example, for a seasonal model, by default at least two seasonal cycles have to be available in the data. These settings ensure that the model is based on sufficient data. The default settings can be changed if needed. An example of the DIAGNOSTICS tab in the FORECAST SETTINGS menu is shown in Figure 14.10.

Figure 14.10: Options to define the minimum number of historic periods in SAS Forecast Studio

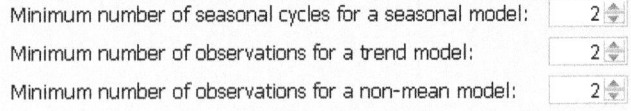

- The results view of SAS Forecast Studio also includes a list of failed forecasts that shows which models could not be built for a particular time series.

- SAS High-Performance Forecasting provides PROC HPFIDMSPEC that allows you to specify intermittent demand models. These modes are suitable for time series with many zeros and only occasional values.

- For time series with hierarchies, SAS Forecast Studio fits the time series model on each level of the hierarchy. Thus, in the case of a time series with many zero values, it is possible to build forecasts on a higher level of the hierarchy and distribute the values to the lower level hierarchies.

14.6 SAS/ETS Software

SAS/ETS is the SAS software used for econometrics and time series forecasting. SAS/ETS provides the following features for data quality:

- PROC EXPAND can be used to impute missing values in a time series using spline interpolation. An example appears in section 11.5.

- PROC EXPAND also helps convert data from one time interval to another. This includes the capability to disaggregate data from lower to higher granularity (for example, from quarters to months).

- PROC TIMESERIES provides many features for time series preparation. Examples appear in sections 11.3 and 11.4 and include the following:
    o inserting missing records in time series data
    o replacing leading and trailing zero values with missing values
    o replacing leading, trailing, and embedded missing values with zero values
- PROC ARIMA and PROC UCM allow you to include explanatory variables for the closer definition of the time series model. These input variables can, for example, be dummy variables that flag periods with outliers and, thus, filter the effect of outliers from the model.

14.7 Base SAS

Base SAS provides many features for general data quality control. The SAS language has many features to check data correctness and completeness. The following list only shows an extract of the main features:

- PROC MEANS lets you quickly profile the number of observations and the number of missing values with the N and the NMISS options. For examples, refer to section 10.2.
- PROC COMPARE compares two data sets and outputs the differences in the metadata of the table like column names and formats. Additionally, it also compares the data values.
- The FORMAT procedure defines format catalogs. These format catalogs can be used to validate the data content through lookup tables for acceptable values for categorical variables or for acceptable ranges for interval variables.
- The SAS language also enables the definition of a hash, which improves performance when you are using large lookup tables for data quality control.
- PROC CONTENTS gives a quick overview of the metadata of a table and lists variable names and formats.
- To compare data quality across different tables, methods like the SAS DATA step and SQL joins are available. Creating multiple output data sets from a DATA step merge allows you to route the records to separate tables according to their match.
- The SAS language provides a rich set of character functions and PERL regular expressions. They facilitate the extraction of information from unstructured text fields and allow pattern matching to verify that the values in data fields conform to the definition.

14.8 JMP

General

JMP is a desktop product for statistical and graphical analysis that has a very strong connection to SAS software. JMP can read and write SAS data and access SAS on a server.

Using user interface, the analyst can perform a wide range of statistical and analytical methods. The product provides a wide range of graphical visualization capabilities. More complex multistep tasks in JMP can be scripted using the JMP script language. JMP offers many ways to check and improve data quality. One example of visually detecting complex relationships and two specific types of JMP analyses, the "missing data pattern" and the "sample size and power calculation," are illustrated here.

Detecting complex relationships and data quality problems with JMP

This example is based on the data from a sailboat race, which is similar to the case study that was described in chapter 1. A GPS device has been used to record GPS trackpoint data during a sailboat race. These data include longitude, latitude, compass heading, and speed.

The points-plot for the compass heading in the right part of Figure 14.11 shows that there are different clusters of compass heading measurements.

- The two clouds around 120 and 210 degrees belong to the upwind part of the race on the backboard and the starboard course.
- The clouds between 300 and 10 degrees belong to the downwind course.

Figure 14.11: Screenshot of the visual data inspection in JMP using a points-plot and a scatterplot

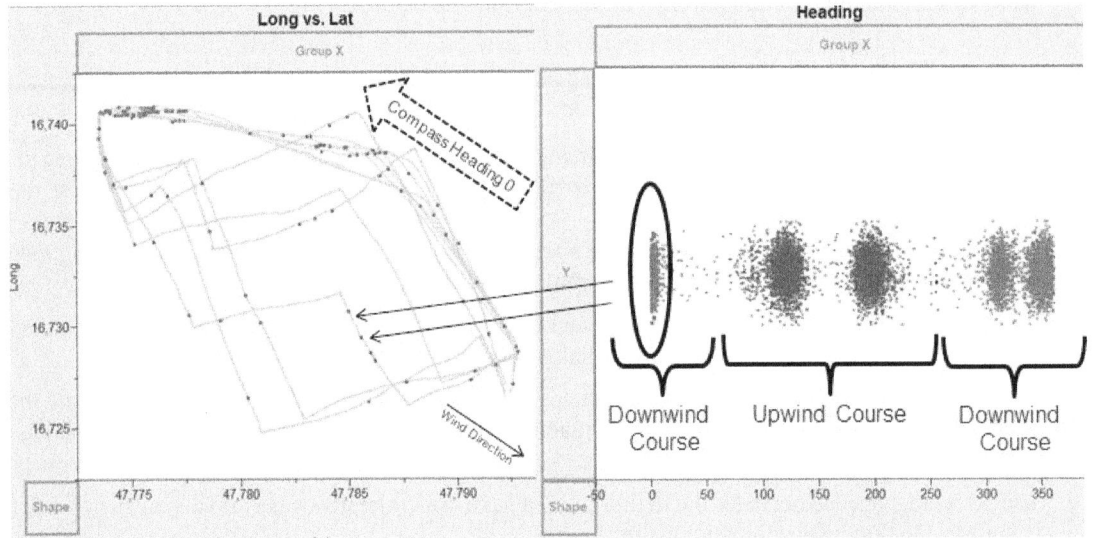

Combining this view with the geo-coordinates of the trackpoints, as shown on the left of Figure 14.11, the following can be seen:

- There are data points that have a compass heading around 0, which means that the boat should be on the downwind course and sails towards the upper left corner of the geo-coordinate graphs.
- However, some of these data points lie on the upwind section of the course (see the two arrows in the graph point to two example cases).
- It is very unlikely that on the upwind courses with a usual compass heading of around 120 and 210 the boat has turned around for just a moment and sailed toward 0 degrees compass heading for a second.

In order to perform closer inspection of the data, the respective points can easily be extracted to a subset table in JMP as shown in Figure 14.12.

Closer data inspection shows that the GPS device recorded the data correctly but did not output values right of the decimals if the heading value was integer (for example, in the case of heading = 198).

The data integration interface, however, was expecting the heading data with 2 digits after the decimal and shifted values without decimals by 2 digits to the right, and thus divided values without decimals by 100.

This led to compass heading values close to zero even for the upwind courses.

Figure 14.12: Screenshot of the data subset of suspicious data points in JMP together with the source data

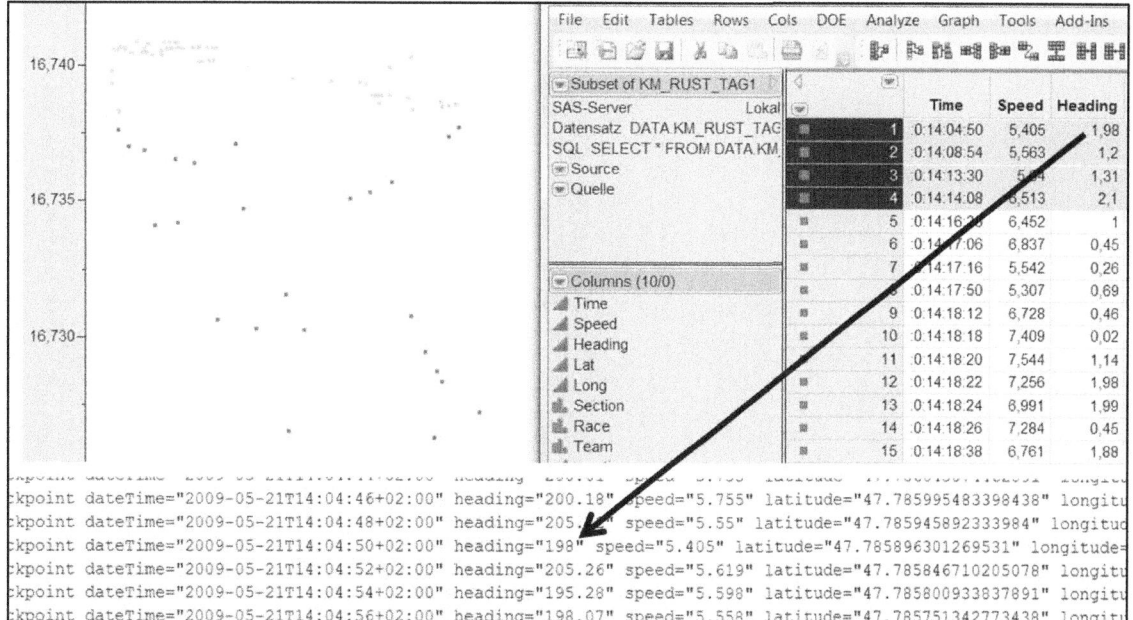

These complex relations could only be detected in a graphical way as shown above. All compass heading values fulfill the formal criteria of lying between 0 and 360.

With the ability of JMP to graphically display the geo-coordinate plot and interactively highlight the values in different sections of the race, however, the data inconsistency could be highlighted, which otherwise would have probably remained undetected.

Missing data pattern

The missing data pattern task profiles the structure of missing values in the analysis data. This task triggered the creation of the MV_PROFILE macro that is discussed in section 10.3. Note that the MV_PROFILE macro is not just a copy of the missing data pattern task in JMP; it expands on the idea with variable clustering and principal component analysis.

Output 14.1 shows the TREE MAP DIAGRAM for sample data of insurance customers.

Output 14.1: Tree map diagram as output of the missing data pattern task in JMP

As you can see, approximately 70% of the observations have no missing value (rectangle at the left) and a large portion of the data contains records with just one missing value.

The output also shows the frequency of each pattern in a table. An example is shown in Output 14.2.

Output 14.2: Frequency list of missing data pattern

	Count	Number of columns missing	Patterns	ID	KIDSDRIV	PLCYDATE	TRAVTIME	CAR_USE
1	7657	0	000000000000000000000000000000	0	0	0	0	0
2	497	1	000000000000000000000000000100	0	0	0	0	0
3	435	1	000000000000000000000000001000	0	0	0	0	0
4	29	2	000000000000000000000000001100	0	0	0	0	0
5	506	1	000000000000000000000000100000	0	0	0	0	0
6	45	2	000000000000000000000000100100	0	0	0	0	0
7	42	2	000000000000000000000000101000	0	0	0	0	0
8	3	3	000000000000000000000000101100	0	0	0	0	0
9	437	1	000000000000000000001000000000	0	0	0	0	0
10	32	2	000000000000000000001000000100	0	0	0	0	0
11	26	2	000000000000000000001000001000	0	0	0	0	0
12	6	3	000000000000000000001000001100	0	0	0	0	0
13	31	2	000000000000000000001000100000	0	0	0	0	0
14	2	3	000000000000000000001000101000	0	0	0	0	0
15	435	1	000000000000000000010000000000	0	0	0	0	0
16	20	2	000000000000000000010000000100	0	0	0	0	0
17	21	2	000000000000000000010000001000	0	0	0	0	0
18	2	3	000000000000000000010000001100	0	0	0	0	0
19	32	2	000000000000000000010000100000	0	0	0	0	0
20	1	3	000000000000000000010000100100	0	0	0	0	0
21	2	3	000000000000000000010000101000	0	0	0	0	0
22	26	2	000000000000000000011000000000	0	0	0	0	0
23	2	3	000000000000000000011000000100	0	0	0	0	0
24	5	3	000000000000000000011000001000	0	0	0	0	0
25	1	4	000000000000000000011000001100	0	0	0	0	0
26	1	4	000000000000000000011000100100	0	0	0	0	0
27	4	1	000000000000000001000000000000	0	0	0	0	0
28	2	2	000000000000000001000000001000	0	0	0	0	0
29	1	2	000000000000000001001000000000	0	0	0	0	0

For each pattern, this table shows the frequency, the number of missing values in the pattern, and whether a variable is missing in the respective pattern.

Note that the table and the graph are interactively linked to each other. Selecting a field in the tree map automatically selects the corresponding row in the table.

Sample size and power calculation

JMP also provides the functionality to calculate required sample sizes and powers for controlled experiments. From a data quality perspective, this is relevant to determine the minimum number of observations needed to run a particular analysis. Figure 14.11 shows the available methods for sample size planning in JMP.

Figure 14.13: Available methods for sample size planning in JMP

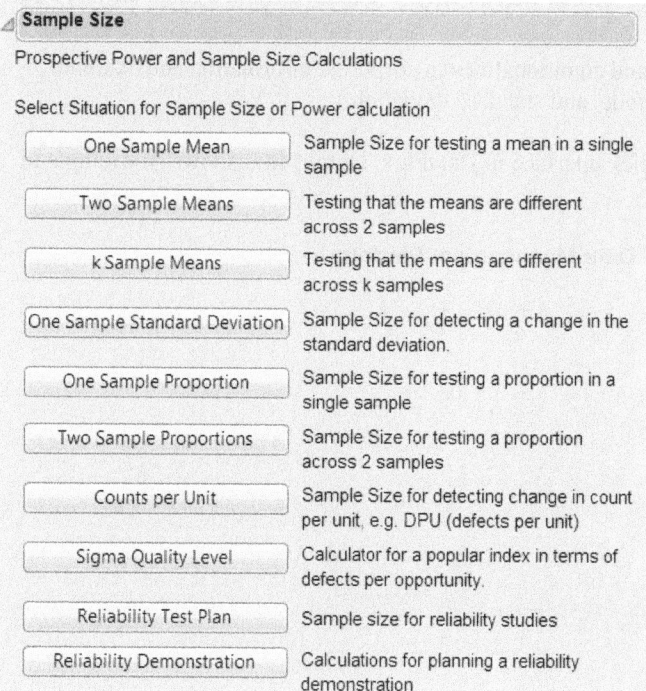

14.9 DataFlux Data Management Platform

Short description of DataFlux Data Management Platform

The DataFlux Data Management Platform allows organizations to cleanse, correct, and enhance any type of enterprise data and to create an accurate view of the organization's business environment.

DataFlux technology allows you to analyze, improve, and control enterprise data and to successfully address data quality issues. This includes the ability to

- profile data to discover errors, inconsistencies, redundancies, and incomplete information
- correct, standardize, and verify information across the enterprise from a single platform
- match, merge, or link data from a variety of disparate sources
- enrich data using information from internal and external data sources
- check and control data integrity over time with real-time data monitoring, dashboards, and scorecards

Data profiling

DataFlux focuses on data profiling by supporting the following capabilities:

- Business rule validation ensures that data meet organizational standards for data quality and business processes by validating them against standard statistical measures as well as customized business rules.
- Relationship discovery uncovers relationships across tables and databases and across different source applications.
- Outlier detection finds data that fall outside of predetermined limits and gains insight into source data integrity.
- Data validation verifies that data in the tables match the appropriate description.
- Pattern analysis ensures that data follow standardized patterns to analyze underlying data and build validation rules.
- Statistical analysis establishes trends and commonalities in corporate information and examines numerical trends via mean, median, mode, and standard deviation.

These features are available via a point-and-click interface in DataFlux. Output 14.3 shows an example of profiling output.

Output 14.3: Output of profiling in DataFlux Data Management Platform

Data standardization and record matching

DataFlux provides extensive functionality to standardize and correct data.

- Names, addresses, phone numbers, and customer IDs can be automatically identified and standardized.
- Product data like product names, product IDs, manufacturer names, quantities, and packaging information can be parsed and standardized.

This leads to a standardized picture of the data and reduces the number of entry variations in table columns. The benefit is that identical information is now represented with unique values in the data.

Based on this standardization, a single table or related tables can be checked for duplicates. For this procedure, sophisticated fuzzy matching technology and clustering methodologies are available. Records that have a high probability of similarity are allocated to the same cluster, and if they are dispersed over different tables, they receive a match key.

Figure 14.14 shows the results of record-matching. Three similar records have been assigned to a cluster. The record with the ID 9054 has been selected as the surviving record.

Figure 14.14: Match Report in DataFlux Data Management Platform

	ID	COMPANY	CONTACT	ADDRESS	CITY	STATE	PHONE
Surviving record:							
☑	9054	SAS	Robert Brauer	6512 Six Frks Road Ste 404B	Raleigh	NC	323-198-3282
Cluster records:							
☐	4267	SAS	Robert Brauer	6512 Six Forks #404B	Raleigh	NC	
☐	10	DataFlux Corporation	Bob Brauer	6512 Six Forks Road - 404B	Raleigh	North Carolina	323-198-3282
☑	9054	SAS	Robert Brauer	6512 Six Frks Road Ste 404B	Raleigh	N.C.	

Defining a data quality process flow

Dataflux also allows you to define a data quality process flow as shown in Figure 14.13. This responds to the need for continual data quality improvement. See also section 9.5.

Figure 14.15: Process flow for data quality in DataFlux Data Management Platform

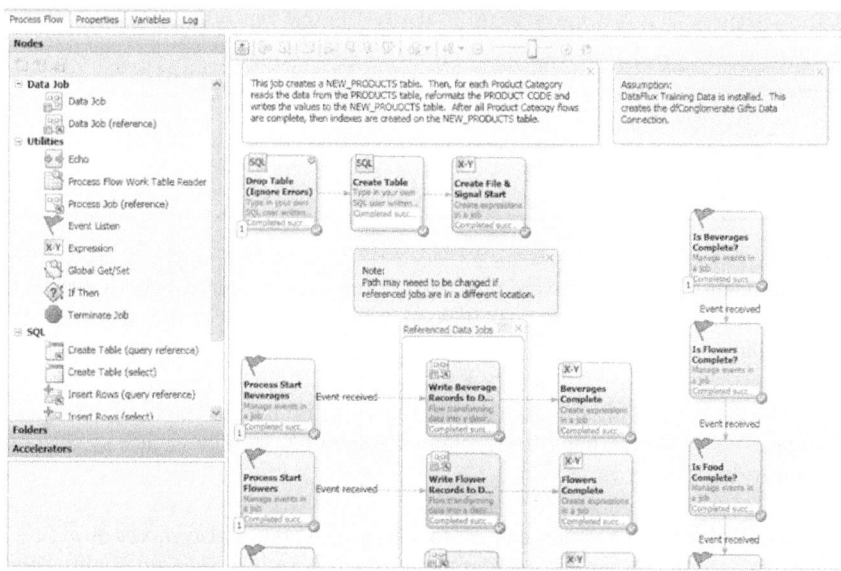

Section 9.6 shows that a data quality process can be monitored on a regular basis. For this task, DataFlux offers a way to customize a project-specific data quality dashboard that shows key performance indicators for data quality over time. See the example in Figure 14.16.

Figure 14.16: Data quality dashboard in DataFlux Data Management Platform

14.10 Conclusion

This chapter summarizes the functionality of SAS software products for data quality control and improvement. It has been shown that the ability to analyze data quality goes beyond defining rules for outlier detection and reporting the number of missing values.

The powerful set of features in SAS contributes significantly to the profiling and improvement of data quality (for example, replacing individual missing values, calculating segment-specific reference values, or selecting and evaluating the predictive power of different variables).

This chapter not only describes the features of SAS with respect to data quality, but it also serves as a benchmark for the essential features organizations must have in their arsenal for advanced data quality control and improvement.

Part III: Consequences of Poor Data Quality— Simulation Studies

Chapter 15: Introduction to Simulation Studies

15.1 Rationale for Simulation Studies for Data Quality 201
    Closing the loop 201
    Investment to improve data quality 202
    Further rationale for the simulation studies 203
15.2 Results Based on Simulation Studies 203
    Analytical domains in the focus of the simulation studies 203
    Data quality criteria that are simulated 203
15.3 Interpretability and Generalizability 203
    Simulations studies versus hard fact formulas 203
    Illustrate the potential effect 204
15.4 Random Numbers: A Core Ingredient for Simulation Studies 204
    The simulation environment 204
    Random number generators 204
    Creation of random numbers in SAS 205
    Random numbers with changing start values 205
    Code example 205
15.5 Downloads 206

15.1 Rationale for Simulation Studies for Data Quality

Closing the loop

The cyclic data quality process shown in Figure 15.1 has already been presented in the introduction to this book. The goal of this book is to define the criteria for data quality for analytics, which appears in part I. Part II shows how the data quality status can be profiled, and it illustrates analytical methods to improve the data quality.

Figure 15.1: Cyclic data quality process

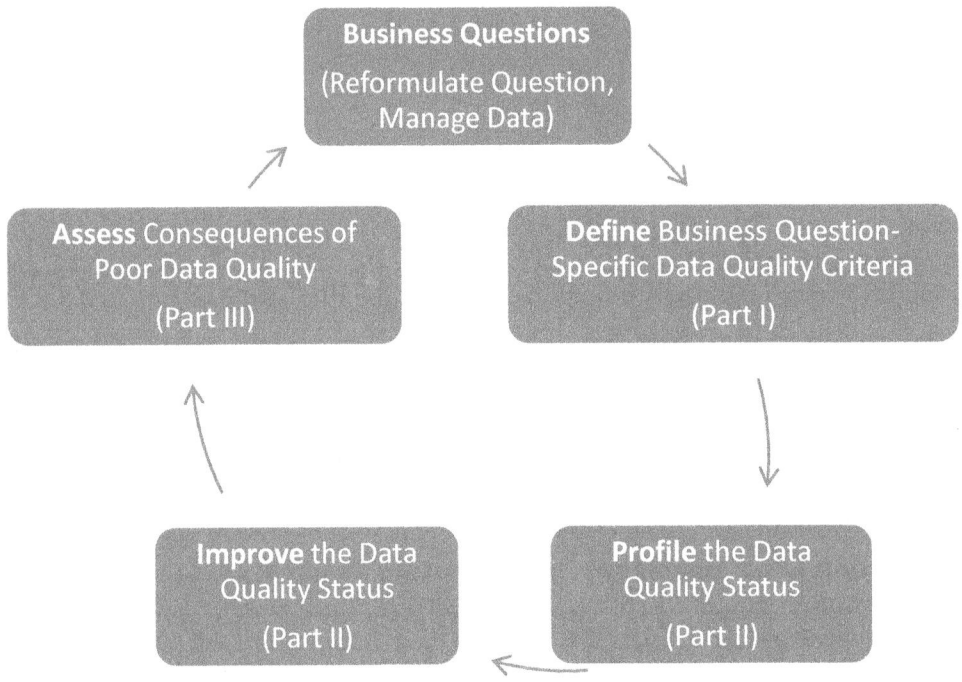

Part III now closes this loop by showing possible consequences of poor data quality. This is demonstrated for two reasons:

- First, it illustrates the consequences poor data quality can have on model performance. This information can support the decision to perform an analysis of specific data.
- Second, the presentation of the consequences in the following chapters also highlights their effect from an accuracy and monetary point of view using a business case calculation. This information can then help you decide whether to invest additional efforts on data quality improvement.

Investment to improve data quality

The second point is important because, along with the analytical methods to improve data quality shown in part II, additional effort may be needed to manually check data correctness and improve data completeness.

While analytical methods can help to improve the data quality (for example, by suggesting most likely replacement values for missing values), inserting the true values by accessing additional data sources or by manually checking data records usually provides a more complete and correct picture of the actual situation.

These steps, however, are more cost- and time intensive. Such investments need to be justified compared to the expected outcomes. Therefore, considerations of the monetary benefit are important. They are discussed in this part.

This part of the book deals with the **consequences of poor data quality in an analysis** and uncovers the effects on the accuracy of the analysis results.

Further rationale for the simulation studies

In additional to showing the consequences of poor data quality, the simulation studies also

- allow you to view the data quality characteristics presented in part I of this book in the context of the respective analytical method
- describe the effect and consequences of data quality problems on analytical models and the accuracy of the predictive and forecasted outcome
- analyze the effects of data quality problems from different perspectives

15.2 Results Based on Simulation Studies

Analytical domains in the focus of the simulation studies

Part III provides a practical approach to assessing potential consequences of poor data quality by evaluating results from simulation studies. The simulations studies concentrate on two topics: predictive modeling and time series forecasting.

- For predictive modeling, the case of a prediction of a binary event variable with a logistic regression model is analyzed (refer to chapters 16 through 19).
- In the case of time series forecasting, a time series on a monthly aggregated basis is analyzed. The forecasting models that are used in this context only depend on the forecast variable itself. No additional explanatory variables are used as co-variables in these models (refer to chapters 20 through 22).

Of course, there are many other analytical methods that could be included in the simulation studies presented here to show the consequences of bad data quality. However, to stay within the scope of this book, the two most prominent analytical methods in business decision-making have been selected.

Data quality criteria that are simulated

For the two analytical domains, predictive modeling and time series forecasting, the following data quality criteria are analyzed:

- random and systematic missing values in the input data.
- random and systematic errors in the input data AND the target variables.
- reduction of the available data quantity. Quantity in this regard refers to the available number of observations and the number of events in predictive modeling and the available length of the history of the time series in time series forecasting.

15.3 Interpretability and Generalizability

Simulations studies versus hard fact formulas

The simulations studies in the following chapters give an indication about the size of the effect of data that do not fully meet the completeness, correctness, availability, and quantity criteria. The following questions are answered by the simulations studies:

- Does it make sense to run analyses on data that do not fulfill the data quality criteria as shown in part I of this book?
- How much trust should be placed on results that are produced from these data?

- What is the expected loss in accuracy when dealing with such data?
- Which criteria have a strong effect on forecast accuracy? Where should effort be placed in data collection and data quality improvement to enhance forecast quality?

The data sources have been carefully selected so that they generalize well across industries and analysis domains. Thus, the conclusions that are drawn from the simulations can also be assumed to generalize well.

Note, however, that these results are not hard facts that are calculated by a formula and should not be interpreted as such.

Illustrate the potential effect

The purpose of the simulation studies is to highlight what a potential outcome might be in a certain situation and how a gradual change in a data quality characteristic may affect the model quality.

15.4 Random Numbers: A Core Ingredient for Simulation Studies

The simulation environment

A very important ingredient for simulation studies is definitely a good simulation environment. SAS is an excellent integrated simulation environment that provides

- programming structures for simulations
- data preparation and simulation-specific data manipulation
- analytical methods
- evaluation of results

SAS Enterprise Miner has been used here for the predictive analysis simulation studies. See appendix C for more information.

For the time series forecasting simulation studies, SAS High-Performance Forecasting, which is part of SAS Forecast Server, has been used. See appendix D for more information.

Note that downloads of programs, macros, and data sets are available from
http://www.sascommunity.org/wiki/Data_Quality_for_Analytics.

Random number generators

Random numbers are a core ingredient for simulation studies to ensure that the different simulations differ by a random factor. The variations in the simulation runs can be achieved in the following ways:

- Random numbers are used to draw samples from the entire base of available records.
- In predictive analysis, random numbers control the split of the available analysis records into training, validation, and test data.
- For the generation of artificial random missing values and random bias in the data, random numbers are used to decide for each record whether a missing value or bias will be introduced.
- In the case of random errors, the magnitude and the sign of the error are also defined randomly.
- The reduction of the number of observations and the number of events in the analysis data is performed on the basis of random numbers.

Uniform random numbers are in the interval from 0 to 1. Two examples of the application of random numbers are the following:

- If the data will change for 10% of the observations (for example, a delta of 0.2 will be added), a random number is generated for each observation. If the random number for a particular observation is lower than the threshold of 0.1, the value is changed.
- If the delta of 0.2 in this example will be added for a random selection of 50% observations and subtracted for the other half, a random number is generated for each observation. This random number is used to control the sign. If the value of the random number is larger or smaller than 0.5, the delta is added or subtracted.

Creation of random numbers in SAS

SAS offers a wide range of random number generators. SAS functions and CALL routines are available to generate random numbers from different distributions (for example, the uniform, the normal, the Poisson, and the Beta distribution, to name a few). For more details, refer to the SAS online help and the SAS documentation.

Random number generation for the scenarios is based on the uniform random number generator RANUNI. Note that it can also be called with the synonym UNIFORM.

The following syntax generates a random number for each observation in the data set with the RANUNI function:

```
RandNumber = RANUNI(1234);
```

Note that the number 1234 initializes the random number generated and is called SEED. If the seed value that is specified is positive, the starting value of a random number list is always the same and an identical list of random numbers is produced.

While you might want to retrieve an identical list of random numbers in order to reproduce the same results with each run, this is not the case for simulations because variation in sample selection, sample size, and other criteria is desired to study its effects.

Random numbers with changing start values

In order to generate lists of changing random numbers, a non-positive seed value (for example, 0) can be used. Then the computer clock is used to initialize the stream.

The following code can be used in a SAS DATA step to create random numbers with different starting values. The seed is set to 0. The CALL function generates a random number of each observation in the data set and stores the value in the RND variable.

```
retain seed 0;
call ranuni(seed, rnd);
```

The random number can now be used to set 50% of the values of the AGE variable to missing.

```
if RND < 0.5 then call missing(AGE);
```

Code example

The following code example shows the numbers generated with a positive and a non-positive seed. In a SAS DATA step, the first five observations of the SASHELP.CLASS data set are used. Two random numbers are generated, RND_FIXED and RND_FLEXIBLE, one with a positive seed value and one with a 0 seed value.

```
data class5obs;
 set sashelp.class(obs=5);
 retain seed1 123;
 call ranuni(seed1,RND_Fixed);
 retain seed2 0;
 call ranuni(seed2,RND_Flexible);
 keep name rnd_fixed rnd_flexible;
run;
```

The results of the three runs are shown here. The RND_FIXED is the same for all runs, while RND_FLEXIBLE changes with each run.

Run 1

```
              RND_    RND_
Obs  Name    Fixed   Flexible
 1   Alfred  0.75040 0.31942
 2   Alice   0.32091 0.34762
 3   Barbara 0.17839 0.87860
 4   Carol   0.90603 0.59829
 5   Henry   0.35712 0.05846
```

Run 2

```
              RND_    RND_
Obs  Name    Fixed   Flexible
 1   Alfred  0.75040 0.41262
 2   Alice   0.32091 0.85679
 3   Barbara 0.17839 0.42090
 4   Carol   0.90603 0.79547
 5   Henry   0.35712 0.70354
```

Run 3

```
              RND_    RND_
Obs  Name    Fixed   Flexible
 1   Alfred  0.75040 0.50015
 2   Alice   0.32091 0.47748
 3   Barbara 0.17839 0.54997
 4   Carol   0.90603 0.00870
 5   Henry   0.35712 0.59494
```

15.5 Downloads

The results presented in the following chapters are also available at
http://www.sascommunity.org/wiki/Data_Quality_for_Analytics.

Make sure to check this site for updates on the results and additional simulation scenarios that are presented here.

You can also download SAS programs, macros, and data sets from the site.

Chapter 16: Simulating the Consequences of Poor Data Quality for Predictive Modeling

16.1 Introduction 208
    Importance of predictive modeling 208
    Scope and generalizability of simulations for predictive modeling....................................208
    Overview of the functional questions of the simulations...208

16.2 Base for the Business Case Calculation 209
    Introduction..209
    The reference company Quality DataCom 210

16.3 Definition of the Reference Models for the Simulations 210
    Available data..210
    Data preparation..210
    Building the reference model 211
    Process of building the reference model 211
    Optimistic bias in the models in the simulation scenarios...212
    Detailed definition of the data and the reference model results 212

16.4 Description of the Simulation Environment..213
    General 213
    Input data source node CHURN 214
    START GROUPS and END GROUPS node..214
    DATA PARTITION node..214
    INSERT MISSING VALUES node...214
    IMPUTE MISSING VALUES node 214
    REGRESSION node 214
    MODEL COMPARISON node...215
    STORE ASSESSM. STATISTICS node 215

16.5 Details of the Simulation Procedure..215
    Validation method..215
    Process of building the scenario models 215
    Validation statistic..216
    Data quality treatment in training data and scoring data..216
    Box-and-whisker plots...217

16.6 Downloads...217

16.7 Conclusion 217

16.1 Introduction

Importance of predictive modeling

Predictive modeling is the most prominent discipline in data mining to answer business questions. Across industries, data mining models play an important role in predicting the most likely behavior of analysis subjects. In customer analytics, these include predicting the probability of churn or the propensity to buy, predicting the claim risk or the failure risk to pay back loans, and assessing the probability of fraudulent activities or events.

Predicting events plays an important role but so too does predicting the values themselves, like the claim amount, the loss caused by a loan default, or the expected revenue a customer will create on a new product or service. Analyses like the prediction of survival time (for instance, the time to a cancellation event or the time to a purchase event) fall into the category of predictive analytics.

Because this is a very important topic in data mining, the goal of this chapter is to study the effect of different data quality features on both the model results and the model quality. The effect of poor data quality on the performance of predictive models is the focus of chapters 16–19.

Scope and generalizability of simulations for predictive modeling

The simulations that are presented in the next three chapters are based on real-life data sets from different industries. These real-life data illustrate the effect of different quality characteristics and their impact on model results.

The results that are presented here serve as illustrations of how different data quality deficiencies can influence model performance.

The simulation results can provide an anchor point to systematically assess the direction and approximate size of poor data quality. They also allow you to compare the effect of different simulation studies. Finally, the simulation studies illustrate in more detail the different aspects of data quality you need to consider.

Thus, the simulation studies allow you to draw important conclusions for planning and prioritizing data quality actions.

Section 16.2 introduces a fictional reference company that is used to quantify the influence of the improvement or decline in data quality in financial numbers. The numbers that are presented here are illustrative only; however, they are an indication of the leverage effect for quantified levels of data quality improvement.

The simulation studies are based on the prediction of binary event variables, like customer churn, product purchase, or failure to repay a loan. The analytical algorithm of choice is logistic regression. The reason for the focus on that method is that event prediction is a very important discipline in predictive modeling and data mining, and it may also be assumed that the results generalize on other methods.

Overview of the functional questions of the simulations

Table 16.1 gives an overview of the main topics that have been investigated with the simulation studies.

Table 16.1: Overview of the functional questions for the simulations

	Topic	Functional Question
Chapter 17	Number of Observations	Influence of the available number of observations on model quality
	Number of Events	Influence on the available number of events on model quality
	Availability of Data	Influence of the availability/nonavailability on model quality
Chapter 18	Missing Values	Influence of the percentage of random missing values in the input variables on model quality
		Influence of the percentage of systematic missing values in the input variables on model quality
Chapter 19	Data Bias	Influence of random errors in the input variables on model quality
		Influence of systematic errors in the input variables on model quality
		Influence of systematic and random errors in the target variable on model quality

Chapters 17 through 19 are structured in the following way:

- definition of the simulation and specification of the simulation settings
- presentation of the simulation results in graphical and tabular form
- interpretation of the results and relationship to a business case

16.2 Base for the Business Case Calculation

Introduction

As mentioned earlier, a fictional reference company is used to calculate a business case based on the outcome of the different simulation scenarios. The change in model quality is transferred into a response rate and the respective profit is expressed in US dollars.

This is done in order to illustrate the effect of different data quality changes from a business perspective. As a note of caution, the numbers in US dollars should only be considered as rough indicators based on the assumption of the simulation scenarios and on the business case as described here. In individual cases, these values and relationships are somewhat different.

The reference company Quality DataCom

The reference company, Quality DataCom, in the following chapters operates in the communications industry. The company has 2 million customers and runs campaigns to promote the cross- and up-sell of its products and services. In a typical campaign, usually about 100,000 customers (5% of the customer base) with the highest response probability are contacted. The average response of a customer to an offer (offer on take or product upgrade) represents a profit of $25.

Assume that you use an analytic model that predicts 19% of the positive responses in the top 5% of the customer base correctly. Response here means that the campaign contact results in a product purchase or upgrade. This leads in total to 19,000 responding customers in this campaign, which generate a profit of $475,000 (19,000 x $25).

These numbers are used in the next three chapters to assess the effect of different degrees of data quality, not only on model quality but also on the return of investment for a campaign. This also allows the analyst to quantify the value of measures to improve the data quality.

16.3 Definition of the Reference Models for the Simulations

Available data
The data used for the simulations are taken from four real-life cases:

- CHURN: customer data from the telecommunications industry, where the cancellation event on a customer level is predicted
- AFFINITY: customer data from the communications industry, where the affinity for buying a new product is predicted
- CLAIM: customer data from the insurance industry, where the claim event in car insurance is predicted
- RESPONSE: data from the retail industry, where the response probability on a marketing mailing is predicted

Data preparation
In order to create a stable reference model with a good fit, before running the simulation studies, you need to create a predictive model for each data set using SAS Enterprise Miner. For each of these models, the data are prepared as part of the data mining process, and a final data table with quality checked, non-missing values is used for modeling.

- The number of missing values for each variable has been calculated. Based on this result, observations or variables have been deleted from the data using the following rules:
    - If a variable has less than 5% of missing values, the observations with missing values are deleted from the data.
    - If a variable has more than 5% of missing values the variable is deleted from the data.
    - This results in data without missing values. Following is a detailed description of each data set with the number of observations and variables that have been deleted.
- Variables with a large number of categories have been removed from the data or categories with a small number of observations have been grouped into an Others group variable.

This provides a data set with good data quality as a reference level. This procedure is a tradeoff between altering the input data from their original value and keeping only variables with a high degree of completeness. Only when the data in the base table are complete and of good formal data quality can a stable reference model be trained. This model can then be used to perform manual alterations of the input data interventions and to test their effect on the model performance outcome.

If data for the reference model itself already contain data quality features that need to be corrected, the correction adds an additional layer of fuzziness. For example, if the data contain missing values, the imputation for these missing values would already introduce an error that also affects the models.

Building the reference model

There are two reasons for building the reference models:

- A subset of variables is defined to be used in the simulation scenarios as the standard set of variables.
- A benchmark value for model performance is produced, which is used in the scenarios to study the relationship between degrees in data quality and model accuracy.

Different data modification and data modeling nodes of SAS Enterprise Miner were used to build a list of candidate models. The candidate models have then been compared in model assessment to define the model with the best performance.

The following methods were used to find the best models and to define the list of variables:

- Variable Clustering Node
- Variable Selection Node
- Decision Tree Node with Variable Selection
- (Logistic) Regression Node with Stepwise Selection
- Partial Least Squares Node

In order to retrieve a stable definition of the reference model, this procedure has been repeated for 10 different data partitions in training and validation data. For each repetition, the model performance, the best model, and the list of variables in the best model were documented. This set of variables was then used to define the final set of variables for the reference model and for the simulations.

The analytical model that is finally used in the scenarios is the logistic regression model without variable selection. Thus, all variables that are in the set of final variables in the reference model are used in the subsequent models in the scenarios.

Process of building the reference model

Figure 16.1 explains the process of building the reference model on perfect data:

- The available data are split into training, validation, and test data.
- The training and validation data are used to build the model. In the case of a stepwise model, the subset of variables is selected. In the case of a full model, only the coefficients are estimated.
- The model logic (score code) is then applied to the test data to calculate the assessment statistics per data partition and to assess the model quality.
    - Note that the assessment statistics are also calculated on the training and validation data partitions to allow comparison model assessment statistics across data partitions.
    - Also note that SAS Enterprise Miner automatically applies the model logic that is defined on the training data to all data partitions to calculate predictions.

Figure 16.1: Creation of the baseline model

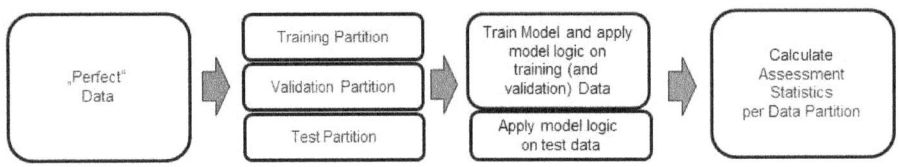

Optimistic bias in the models in the simulation scenarios

Note that providing a predefined set of optimal variables for the predictive models in the different simulation scenarios generates a bias toward too optimistic model performance. The model does not need to find the optimal set of predictor variables. For data with data quality problems, such as not enough observations, high numbers of missing values, bias in the input data, and selection of the optimal predictor variables may not result in the best data set.

Data quality can impact the selection of the optimal input. Thus, the simulation scenarios are influenced by this a priori variable selection (and, consequently, on a priori knowledge). Nevertheless, it has been decided to provide a predefined set of variables in order to compare apples to apples and to remove possible bias in the scenario results caused by differences in the variable selection.

Detailed definition of the data and the reference model results

The following subsections provide an overview of the different business scenarios that have been defined for the simulations. For each model, the input data are described as well as the model results.

Churn

- The final data set for the analysis has 147,534 observations, of which 50.93% are cancellation events (n_{events} = 75,135). Note that in this data set the events were already oversampled when it was provided.
- Before removing the observations with missing values for important variables, the original data set had 215,235 observations. Some 67,701 observations (31.5%) were removed.
- From 117 candidate predictors in the data set, 12 variables were selected for the final model: 7 interval variables and 5 categorical variables. The categorical variables have 26 categories in total.
- The reference model predicts 18.56% of the responses correctly in the top 5% of the data, which gives a lift of 3.71 (after applying a 5% prior event probability). The average squared error (ASE) is 0.1913.

Affinity

- The final data set for the analysis has 266,185 observations, of which 8.28% are product purchase events (n_{events} = 22,039).
- Before removing the observations with missing values for important variables, the original data set had 409,462 observations. Some 143,277 observations (35.0%) were removed.
- From 117 candidate predictors in the data set, 5 variables were selected for the final model: 4 interval variables and 1 categorical variable.
- The categorical variable has 9 categories.
- The reference model predicts 12.23% of the responses correctly in the top 5% of the data, which gives a lift of 2.44 (after applying a 5% prior event probability). The average squared error (ASE) is 0.0736.

Claim

- The final data set for the analysis has 7,657 observations, of which 26.89% are claim events (n_{events} = 2,059).

- Before removing the observations with missing values for important variables, the original data set had 10,303 observations. Some 2,646 observations (25.7%) were removed.
- From 32 candidate predictors in the data set, 12 variables were selected for the final model: 5 interval variables and 7 categorical variables. The categorical variables have 24 categories in total.
- The reference model predicts 33.55% of the responses correctly in the top 5% of the data, which gives a lift of 6.71 (after applying a 5% prior event probability). The average squared error (ASE) is 0.1386.

Response

- The final data set for the analysis has 14,423 observations, of which 13.71% are response events (n_{events} = 1,977).
- Before removing the observations with missing values for important variables, the original data set had 14,841 observations. Some 418 observations (2.8%) were removed.
- From the 13 candidate predictors in the data set, 4 variables were selected for the final model: 3 interval variables and 1 categorical variable. The categorical variable has 9 categories.
- The reference model predicts 10.77% of the responses correctly in the top 5% of the data, which gives a lift of 2.15 (after applying a 5% prior event probability). The average squared error (ASE) is 0.1142.

16.4 Description of the Simulation Environment

General

For each of the main functional questions, as described in Table 16.1, simulations have been performed with SAS Enterprise Miner. In these main simulations, crucial parameters are varied systematically (for example, the percentage of missing values that is introduced into the data). These simulations are called "specific simulations" in the context of this chapter.

A specific simulation, for example, is the scenario with 10% random missing values. For each specific simulation, a number of iterations are applied. In each iteration, the data partitioning, as well as the sampling, is redone. This results in different analysis data sets for each iteration of a specific simulation.

The simulations have been performed with SAS Enterprise Miner 6.2 on a SAS 9.2 M03 platform. The analysis and the preparation of the results have been performed with Base SAS and SAS/GRAPH software.

Figure 16.2 shows an example of a process flow in SAS Enterprise Miner 6.2.

Figure 16.2: Example process flow chart for the simulations in SAS Enterprise Miner

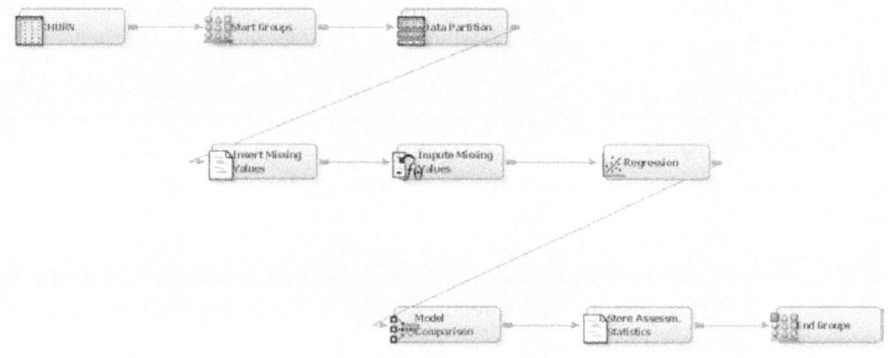

As an example, the nodes in the flow chart describe the process for simulating the impact of missing value imputation on the performance of a logistic regression model for the prediction of a churn event.

Input data source node CHURN

In this node, the simulation data are defined (for this example, the CHURN data). The simulation was run for the AFFNITY, INSURANCE, and RESPONSE data sets as well. Note that in this node only those variables are set as input variables that have been selected for the respective reference model.

Figure 16.3 shows variable settings for the CHURN data.

Figure 16.3: Variable settings for the CHURN data in SAS Enterprise Miner

Name	Role	Level	Report	Order	Drop
ID	ID	Nominal	No		No
Age	Input	Interval	No		No
TrafficCAT	Input	Nominal	No		No
Bindung5CAT	Input	Nominal	No		No
Traffic6	Input	Interval	No		No
Region	Input	Nominal	No		No
NL18	Input	Interval	No		No
Duration5	Input	Interval	No		No
Demo12	Input	Interval	No		No
NL_CAT	Input	Nominal	No		No
Demo5_CAT	Input	Nominal	No		No
NL60	Input	Interval	No		No
NL37	Input	Interval	No		No

START GROUPS and END GROUPS node

The start groups node, together with the end groups node, defines a loop in the process flow. The nodes between the two nodes are executed a specified number of times. The loop number has been set to 10. Thus, for each data set, 10 iterations are performed, which results in 40 iterations in total. For each iteration, the data partition and the treatment of the data in a SAS code node that varies the data quality status are randomly differentiated.

DATA PARTITION node

The data partition node has been used to partition the available data into training, validation, and test data with a ratio of 40:30:30. Note that the XML file of this node in the SAS Enterprise Miner installation has been edited in order to allow a zero seed value for the partitioning. The zero seed value is important to ensure a different initialization value for the generation of the random values with every run (see also section 15.4 and appendix C).

INSERT MISSING VALUES node

The insert missing values node is a SAS code node that randomly inserts a certain proportion of missing values for each input variable. Again, the program uses a negative seed value in order to allow varying random values with every simulation run.

IMPUTE MISSING VALUES node

The impute missing values node uses standard SAS Enterprise Miner functionality to impute the previously generated missing values.

Note that the insert missing values node and the impute node are examples of how to introduce a certain change in data quality. In some scenarios, perturbed data are used directly for the model training; in other scenarios, like this one, the data are treated to improve data quality.

REGRESSION node

The regression node trains a full regression model without variable selection on the input data.

MODEL COMPARISON node

The model comparison node calculates validation statistics on the training, validation, and test data. This node automatically stores the respective data in a data set.

STORE ASSESSM. STATISTICS node

The Store Assessm. Statistics node is a SAS code node that stores the table with the assessment statistics for the respective run in a separate data set. This data set holds the loop indicator of the respective loop that is defined by the START and END group nodes. The content of this data set is used in the evaluation phase of the simulation runs, where it is appended to a results repository.

Appendix C describes the simulation environment in more detail.

16.5 Details of the Simulation Procedure

Validation method

The validation method compares the results of different simulation scenarios. The available data for the scenarios are split into training, validation, and test data partitions.

- **The training and validation partition is used for model training.** Note that when only a full model is created, the validation partition remains untouched by the regression node. The validation partition is, however, separated from the available data for training to mirror the real life situation where the available data would need to be split into training and validation during model training.

- **The test partition is used for model evaluation** and calculation of the model quality statistics (%Response). Note that, in this case, the test data set takes over the role of the scoring data to which the model usually is applied in production.

Process of building the scenario models

The process flow of the predictive models for the simulation scenarios differs from the previous illustration by the following two points:

- The model building only consists of the estimation of the regression coefficients and does not involve the selection of the optimal set of predictor variables.
- The data used for model training are being perturbed with data quality problems for the simulations.

The different simulation scenarios impose a certain treatment of the data before they are entered into the modeling procedure. The number of events is varied in the data or missing values are inserted into the data. This has been shown, for example, in section 16.4 (Figure 16.2 with the SAS code node that inserts missing values into the data).

Figure 16.4 shows this situation in an example similar to Figure 16.1. Here, however, an additional step (the third rectangle), "Perform Scenario Specific Treatment of Data," is included.

Figure 16.4: Creation of the simulation models with data treatment

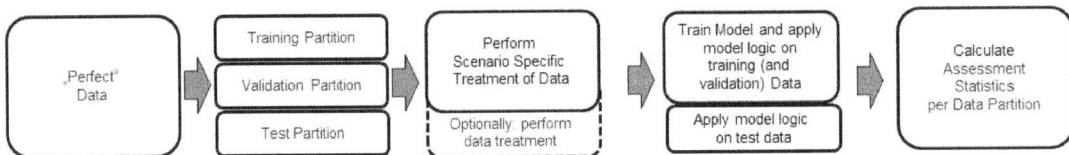

Validation statistic

Validation of the results for the various scenarios is based on the %Response in the 5% of the cases with the highest event probability. The percentage of correctly predicted event cases compared to all cases in this group is used to compare the model accuracy between scenarios.

For each scenario run, the percent response rate is calculated with the Model Comparison node in SAS Enterprise Miner. The output data are then stored in a data set that is appended to the results repository. For each specific simulation, the results repository contains validation statistics for 40 simulation runs (10 iterations for each of the four data sources).

Using previous event probabilities of 5%, the different baseline event probabilities for each data source have been standardized. Thus, the results for different data sources can be evaluated together. In the box plots that are presented in chapters 17 through 19, showing the %Response for the different scenarios, each box (and whisker) is based on 40 simulation results.

Data quality treatment in training data and scoring data

An important simulation parameter in defining the scenarios is whether this treatment of data is applied only to the training and validation data partition or also to the test data partition. From a business perspective, you need to know if the bad data quality only occurs on the historic data used for model training or whether it also occurs in the scoring data when the model is applied.

For example, it could be that for model building the training data from historic periods have poor data quality like missing values or bias. But for future periods, these data quality issues are resolved and model quality is good in the scoring data.

- Figure 16.4 expresses this situation that in the third rectangle the lower section is shown with a dashed line. This means that data quality treatment is not performed in all cases on the test (scoring) data.

Thus, for each functional question in the simulation scenarios, the analyst needs to decide whether in practice the data quality feature is only present in the actual training data or whether it is also present in the data that is used for model scoring in later periods. For example:

- The scenario evaluating the impact of reducing the available observations or the proportion of events applies the data treatment only on the training and validation data. In practice, the problem of insufficient observations or events is only present in the model phase. As soon as the logic is built, it is irrelevant how many observations it will be applied to. Thus, removing observations only takes place on the training and validation data in this context.

- The scenario evaluating the impact of changing proportions of missing values in the data has to be seen from different viewpoints:
  o If missing values **only occur in the training data** (for example, for historic data), but the data quality is good in the future, the scenario mirrors this situation by only inserting missing values for the training and validation partition of the data and assessing the model results on untreated test data. This can happen, for example, if a model is built for historic data with missing values, but the

business process requires that all variables be available at the time of model application (for example, a loan is only granted when all required questions are answered). If the model will be applied in real-time scoring in the future, where all variables are available, no missing values are present in the application of the model.

- o If missing values, however, **not only occur for historic periods**, but it is expected that they will **also occur in future periods**, the scenario will also insert the missing values for future periods in order to assess the model outcome.

The simulation scenarios in the next chapter that deal with data quantity and availability do not involve changing the test data. Only the scenarios in chapters 18 and 19 that deal with input variables (missing values and bias) involve changing the test data as well.

Box-and-whisker plots

The box-and-whisker plots that are shown in the results graphics in chapters 17–19 are defined as follows:

- The box ranges from the first to the third quartile.
- The median is shown as a horizontal line in the box.
- The mean is shown as a diamond.
- In a graph, a number of boxplots is shown, one boxplot per setting of the simulation parameters. In order to show a trend over different scenarios, the mean is connected with a straight line across the boxplots.
- Observations falling outside the 1.5-fold interquartile distance added to the third quartile and subtracted from the first quartile are plotted as outliers.

16.6 Downloads

The results that are presented in the following chapters are also made available at http://www.sascommunity.org/wiki/Data_Quality_for_Analytics.

Make sure to check this site for updates on the results and additional simulation scenarios that are presented there.

SAS programs, macros, and data sets can also be downloaded from this site.

16.7 Conclusion

This chapter has introduced the necessary prerequisites for the simulation studies shown in chapters 17 through 19.

- Chapter 17 shows the results of scenarios with different data quantity and different data availability.
- Chapter 18 shows the results of scenarios when random or systematic missing values are introduced into the data.
- Chapter 19 shows the results of scenarios where random and systematic biases are generated in the input and target variables.

Chapter 17: Influence of Data Quantity and Data Availability on Model Quality in Predictive Modeling

17.1 Introduction..219
    General 219
    Data quantity..220
    Data availability...220

17.2 Influence of the Number of Observations 220
    Detailed functional question 220
    Data preparation..220
    Simulation settings..221
    Results...221
    Business case 223

17.3 Influence of the Number of Events 223
    Detailed functional question 223
    Data preparation..223
    Simulation settings..224
    Results...224
    Business case 226

17.4 Comparison of the Reduction of Events and the Reduction of Observations...226

17.5 Effect of the Availability of Variables 227
    Alternate predictors 227
    Availability scenarios 227
    Results...228
    Interpretation 229
    Business case 229

17.6 Conclusion 229

17.1 Introduction

General

This chapter examines the influence of data quantity and data availability on the predictive power of logistic regression models. Results of simulation studies are shown that simulate variations of data quantity and that study the respective effect on the predictive power. The "Correct Response rate" (denoted by "%Response" in the results) in the Top -5% is used to quantify the predictive power.

The following subsections show how data quantity and data availability are defined for the scenarios.

Data quantity

For the simulations, the variation of the data quantity has been performed in two different ways:

- **The variation of the number of events**
    - Here events are gradually deleted from the data in order to study the change in predictive power if fewer events are available for modeling. Non-events are not deleted from the data.
    - This reflects the situation that is frequently encountered where there is a sufficient amount of non-events in the data, but only a limited amount of observations with events have been observed.
    - The variation of the number of events only, while keeping the number of non-events constant, implicitly causes a variation of the sampling ratio between events and non-events. This, however, is not the focus of the study that is performed here.
- **The variation of the number of observations in general**
    - Here the number of observations is systematically reduced in order to study the impact on the predictive power. Both events and non-events are deleted from the data so that the sampling ratio between events and non-events stays (approximately) constant.
    - This reflects the situation where a limited amount of observations is available for the analysis. This scenario illustrates the effect of additional observations on the predictive power (for example, by providing more customer records or interviewing more respondents).

The simulations were performed based on the data and in the environment described in chapter 16. The results are presented in the following sections.

Data availability

In addition to varying the number of observations, the effect of the availability of data (more precisely, the availability of certain variables) has also been studied here.

In order to study this effect, the variables that have been included in the best performing model are removed in order to train a model on data with reduced variable availability. This has been iterated three times.

These results allow you to study the effect on model quality if the top predictors are not available in the data. In many cases, the benefit of a certain variable is compared to the cost of providing this variable in order to judge whether the effort for data retrieval and preparation should be invested. Section 17.5 also shows the effect on model quality if the variable Age is not available.

17.2 Influence of the Number of Observations

Detailed functional question

The functional question of this simulation code block is to assess the influence of the total number of observations on the accuracy of the predictions.

Data preparation

For the different simulation runs, a certain percentage of the data is systematically deleted without distinguishing between events and non-events. The following SAS statements have been used in a SAS Enterprise Miner code node to reduce the number of observations according to the simulation settings. Note that the following code uses macros and macro variables that are provided by default in SAS Enterprise Miner. Refer to appendix C for more details.

The number of events, %EM_TARGET=1, for the training and validation data are selected into respective macro variables, &NE_TRN and &NE_VLD:

```
proc sql noprint;
 select count(*) into :ne_trn
 from &em_import_data where %em_target = 1;
 select count(*) into :ne_vld
 from &em_import_validate where %em_target = 1;
quit;
```

The total number of event observations that remain in the data are set with macro variable &LIMIT. Note that this number represents the total amount of event observations that is then split between training and validation data. This mirrors the practical situation. For example, if 300 observations are available for model building, they usually have to be split between training and validation data.

The following code shows the use of a random number to remove observations from the training and validation data so that the sum of event observations in both data sets equals on average &LIMIT:

```
data &em_export_train;
 set &em_import_data;
 retain seed -1; ** use flexible seed;
  call ranuni(seed, rnd);
  if rnd < &limit/(&ne_trn+&ne_vld) then output;
 drop rnd seed;
run;

data &em_export_validate;
 set &em_import_validate;
 retain seed -1;
 call ranuni(seed, rnd);
 if rnd < &limit/(&ne_trn+&ne_vld) then output;
 drop rnd seed;
run;
```

Note that the treatment of the data is only performed on the training and validation data, not on the test data. For the test data, the unrestricted data are used in order to allow a proper estimation of the model quality.

Simulation settings

Ten simulation scenarios are run for each data source (Churn, Affinity, and so on) for the following number of available events in the data (settings of the LIMIT variable): 50, 100, 150, 200, 300, 400, 500, 750, 1,000, 2,000, and 5,000.

Results

Figure 17.1 shows a boxplot for the %Response for the different scenarios. The line connects the means, which is represented by a diamond, for each scenario. As expected, the %Response increases with the increasing number of events. There is a strong relationship between an increase of available event cases and model performance gains in the range from 50 to 100 observations. Above 400 events, the improvements get smaller. Above 1,000 (2,000 and 5,000), only a slight increase in model quality can be seen.

222 *Data Quality for Analytics Using SAS*

Figure 17.1: Boxplot with %Response for the different scenarios for the reduction of events and non-events

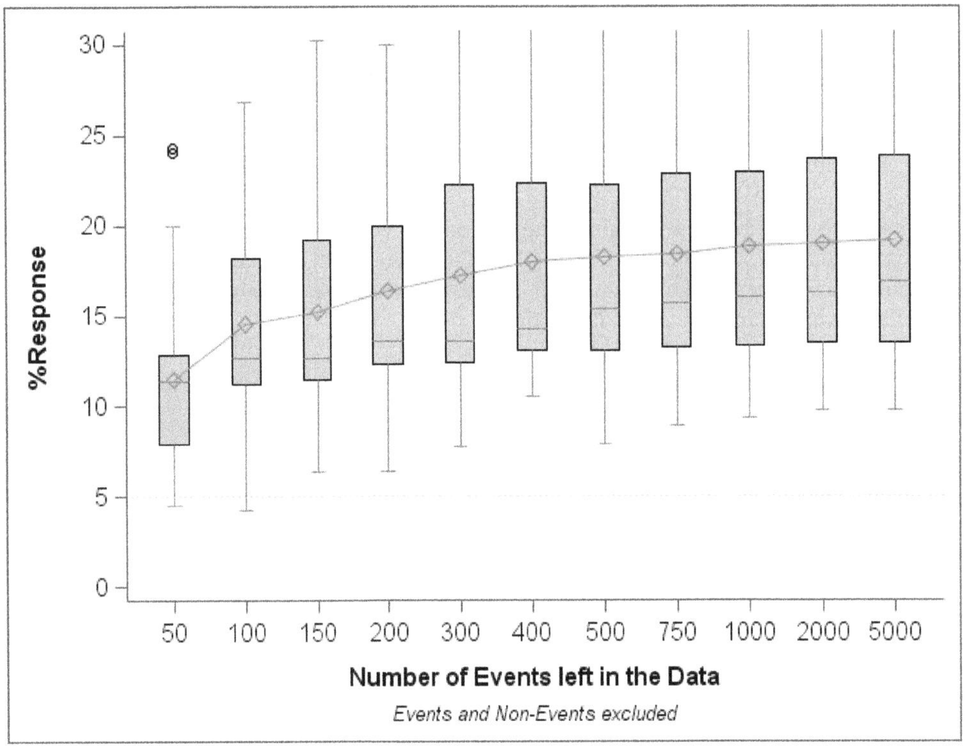

Table 17.1 shows the detailed numbers for the %Response statistic. Additional columns have been created with relative response rates in relation to the scenarios with 50 and 500 target events.

Table 17.1: %Response with figures for the different scenarios for the reduction of events and non-events

Scenario	%Response	Relative %Response (Scenario'500'=100)	Relative %Response (Scenario'50'=100)
50	11.46	62.7	100.0
100	14.56	79.6	127.0
150	15.21	83.2	132.8
200	16.40	89.7	143.1
300	17.30	94.6	151.0
400	17.99	98.4	157.0
500	18.29	100.0	159.6
750	18.46	100.9	161.1
1,000	18.84	103.0	164.4
2,000	19.02	104.0	166.0
5,000	19.21	105.1	167.7

From this table, the following facts can be seen:

- As expected, there is a strong increase in model quality when moving from 50 events to 100 events (+27%) and from 100 to 150 events (+5%).
- Above 750 events, there is only a slight increase in model quality.

Business case

For the reference company described in section 16.2, these results mean the following:

- Running a campaign that has been trained on 50 events has an expected cumulative profit of $268,500, whereas 100 events in the training data have an expected cumulative profit of $364,000.
- Thus, an additional profit of $77,500 can be created per campaign, if the training data contain 100 events instead of 50 events. This is an additional profit of $1,550 for each additional event in the training data.
- Increases in total event numbers on a higher base (for example, from 500 to 1,000) still generate an additional total profit of $13,750 per campaign. Broken down to additional profit for each additional event in the training data, the result is $27.50.

17.3 Influence of the Number of Events

Detailed functional question

The functional question of this simulation code block is to assess the influence of the number of events on the accuracy of the predictions. This scenario differs from the scenario described in section 17.2 because now only the numbers of events have been reduced in the data and the non-events have remained unchanged. As already mentioned in the introduction, this changes the event rate in the various scenarios, whereas in the previous section the event rate stayed constant.

Data preparation

For the different simulation runs, a certain percentage of the data has been deleted. Only events have been deleted from the data. The following SAS statements have been used in a SAS Enterprise Miner code node to reduce the number of events according to the simulation settings. Note that the following code uses macros and macro variables that are provided in a SAS code node. Refer to appendix C for more details.

The number of events, %EM_TARGET=1, for the training and validation data are selected into respective macro variables, &NE_TRN and &NE_VLD:

```
proc sql noprint;
  select count(*) into :ne_trn
  from &em_import_data where %em_target = 1;
  select count(*) into :ne_vld
  from &em_import_validate where %em_target = 1;
quit;
```

The total number of event observations that remain in the data are set with macro variable &LIMIT. Note that this number represents the total amount of event observations that is then spilt between training and validation data. This mirrors the practical situation. For example, if 300 observations are available for model building, they usually have to be split between training and validation data.

The following code shows a random number is used to remove observations from the training and validation data so that the sum of event observations in both data sets equals on average &LIMIT:

```
data &em_export_train;
 set &em_import_data;
 retain seed -1;
  call ranuni(seed, rnd);
  if  %em_target = 0 or
      (%em_target = 1 and rnd < &limit/(&ne_trn+&ne_vld))
                                                    then output;
  drop rnd seed;
run;

data &em_export_validate;
 set &em_import_validate;
 retain seed -1;
 call ranuni(seed, rnd);
  if  %em_target = 0 or
      (%em_target = 1 and rnd < &limit/(&ne_trn+&ne_vld))
                                                    then output;
  drop rnd seed;
run;
```

Note that the treatment of the data is only performed on the training and validation data, not on the test data. For the test data, the unrestricted data are used in order to allow a proper estimation of the model quality.

Simulation settings

The simulation scenario has been iterated 10 times for each data source (Churn, Affinity, and so on) in SAS Enterprise Miner. The simulation scenarios have been run for the following number of available events in the data (settings of the LIMIT variable): 50, 100, 150, 200, 300, 400, 500, 750, 1,000, 2,000, and 5,000.

Results

Figure 17.2 shows a boxplot for the %Response for the different scenarios. The line connects the means, which is represented by a diamond, for each scenario. As expected, the %Response increases with the increasing number of events. There is a somewhat stronger increase when moving from 50 to 100 observations. Above 400 observations with a target event, the line starts to flatten out. Above 1,000 (2,000 and 5,000), only a slight increase in model quality can be seen.

Figure 17.2: Boxplot with %Response for the different scenarios for the reduction of events

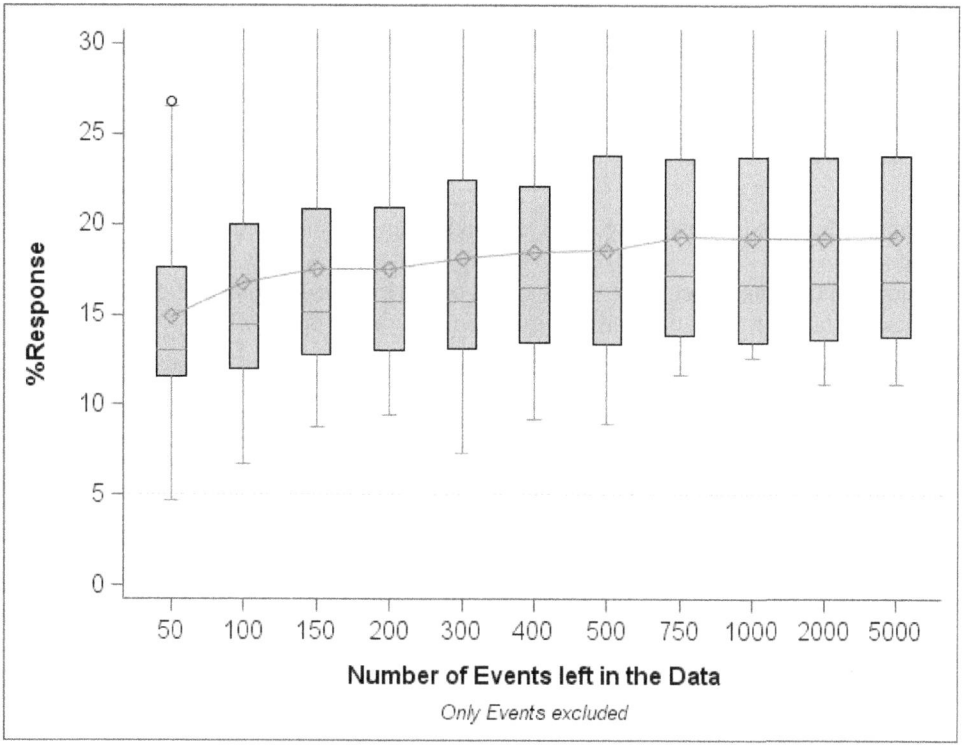

Table 17.2 shows the %Response in numbers. Additional columns have been created with relative numbers where the scenarios with 50 and 500 target events are the baseline.

Table 17.2: %Response with figures for the different scenarios for the reduction of events and non-events

Scenario	%Response	Relative %Response (Scenario '500'=100)	Relative %Response (Scenario '50'=100)
50	14.91	80.5	100
100	16.75	90.4	112.3
150	17.49	94.4	117.3
200	17.51	94.5	117.4
300	18.14	97.9	121.7
400	18.42	99.4	123.5
500	18.52	100.0	124.2
750	19.26	103.9	129.2
1,000	19.17	103.4	128.6
2,000	19.19	103.6	128.7
5,000	19.26	103.9	128.2

From this table, the following facts can be seen:

- As expected, there is a strong increase in model quality when moving from 50 events to 100 events (+12%) and from 100 to 150 events (+5%).
- Above 750 events, there is almost no increase in model quality.

Business case

For the reference company described in section 16.2, these results mean the following:

- Running a campaign that has been trained on 50 events has an expected cumulative profit of $372,750, whereas 100 events in the training data have an expected cumulative profit of $418,750.
- This leads to an additional profit of $46,000 per campaign, if the training data contain 100 events instead of 50 events. This is an additional profit of $920 for each additional event in the training data.
- In the scenarios with higher event numbers (for example, 500 and 750) the additional profit is still $18,500 per campaign. And the additional profit for each additional event in the training data is still $74.

17.4 Comparison of the Reduction of Events and the Reduction of Observations

These results implicitly show that there is a difference between the two cases, where either only events or events and non-events are removed from the data. The model quality is better if just events are removed from the data because the non-events still contribute information to the model.

Figure 17.3 makes this clearer by showing the mean response rate per scenario.

Figure 17.3: Comparison of the %Response between "only events excluded" and "events and non-events excluded"

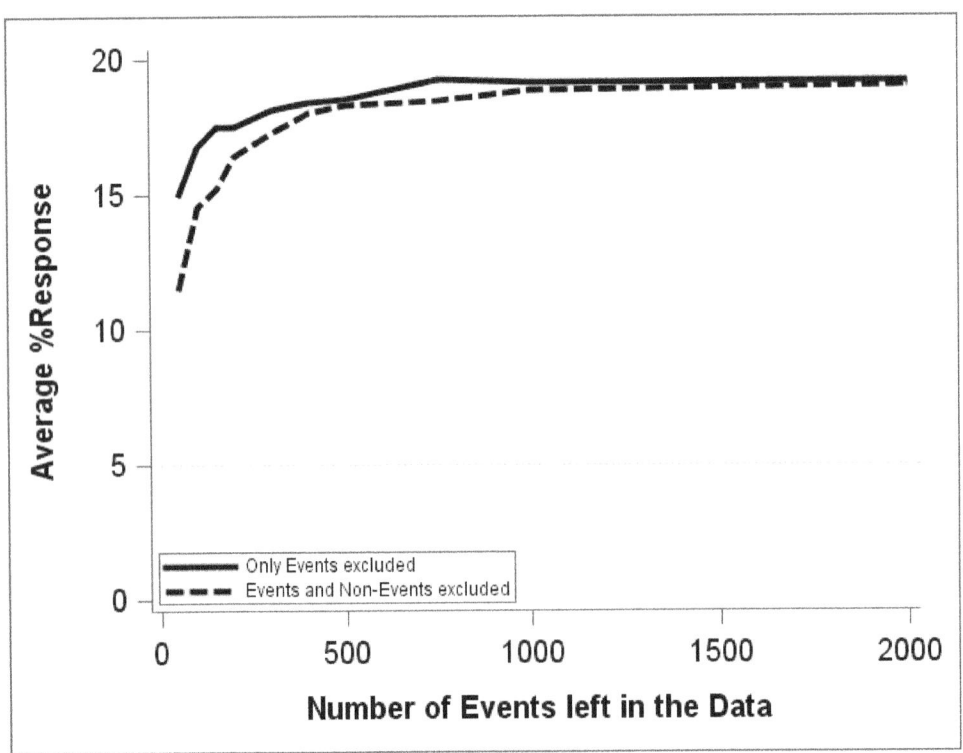

For scenarios up to 1,000 events, it makes a difference whether just events or both events and non-events are removed from the data. Thus, as soon as the total sample size does not get sufficiently large, **also additional non-event records in the data contribute to the predictive power of the model**, as this increases the total "n."

Note that for the proportional sampling of events and non-events in the simulation scenarios, the limited available number of observations in the Response and Claims data significantly decreases the sample size.

- In the 500 events scenario, for example, the average number of observations is 84,147 in the case where only the events are sampled to 500 (represented by the solid line).

- Compare this to an average number of only 3,132 observations, where events and also proportionally non-events are sampled.

Especially in the two scenarios compared here, the reduction of events and non-events substantially decreases the number of total observations.

This shows that in samples with an event count below 1,000, a large number of non-events contributes to the predictive power of the model, even if the proportion of events in the data falls under 1%.

17.5 Effect of the Availability of Variables

Alternate predictors

The data that were used to build the CHURN and AFFINITY models contain more than 200 candidate input variables. As described in section 16.3, the baseline models that were used for the simulations in this chapter contain a subset of these variables. This subset was defined by a modeling exercise to build the model with the highest predictive power and good stability.

The analysis that is described in this section is based on the availability of variables. If a certain subset allows you to build a good model, study how much the predictive power of this model decreases if this set of variables is removed from the training data. In this case, the analytical model has to use other variables for model building.

Section 8.2 noted that the correlation between variables is important to some extent to determine a representative replacement value for missing values and for alternate predictor variables. If, for example, a certain variable is deemed to be important for the model but cannot be made available, other variables will probably take over its predictive content and power as they are correlated.

Availability scenarios

In order to roughly illustrate the effect of the non-availability of certain data, the following scenarios are assessed for the CHURN and AFFINITY data sets:

- The unconstrained model that can use all available variables (**Model 1**). "All available variables" here (and in the following paragraphs) means that for variable selection all variables available in the final data set could be used. For the CHURN and the AFFINITY data sets, 117 candidate input variables were available. See also section 16.3.

- The unconstrained model that can use all available variables except the Customer Age variable (**Model 1 without Age**).
    - Comment: Customer age has intentionally been chosen here because this is, in many cases, a very important variable (also all four baseline models presented in chapter 16 use the customer age variable), but often it cannot be made available or it can only be made available with (systematic) missing values.

- The unconstrained model that can use all available variables except the AGE variable and those variables that are in the same variable cluster as age (**Model 1 without Age Cluster**). The variable cluster has been defined with the Variable Cluster node in SAS Enterprise Miner.
  - Comment: This scenario reflects the situation where the non-availability of a certain variable often triggers the non-availability of correlated variables. This is especially the case if different variables are derived from the same variable in transactional data. If this variable is not available, all derived variables that are based on it will not be available, too.
- The model that can use all variables, except those that are in the unconstrained model (**Model 2 [without Model 1 Variables]**).
- The model that can use all variables, except those that are in Model 1 and 2 (**Model 3 [without Model 1 and 2 Variables]**).
- The model that can use all variables, except those that are in Model 1, 2, and 3 (**Model 4 [without Model 1, 2, and 3 Variables]**).

Results

The average %Response over all simulations for each of these categories is shown in Figure 17.4.

Figure 17.4: Bar chart for the average %Response per availability of variables

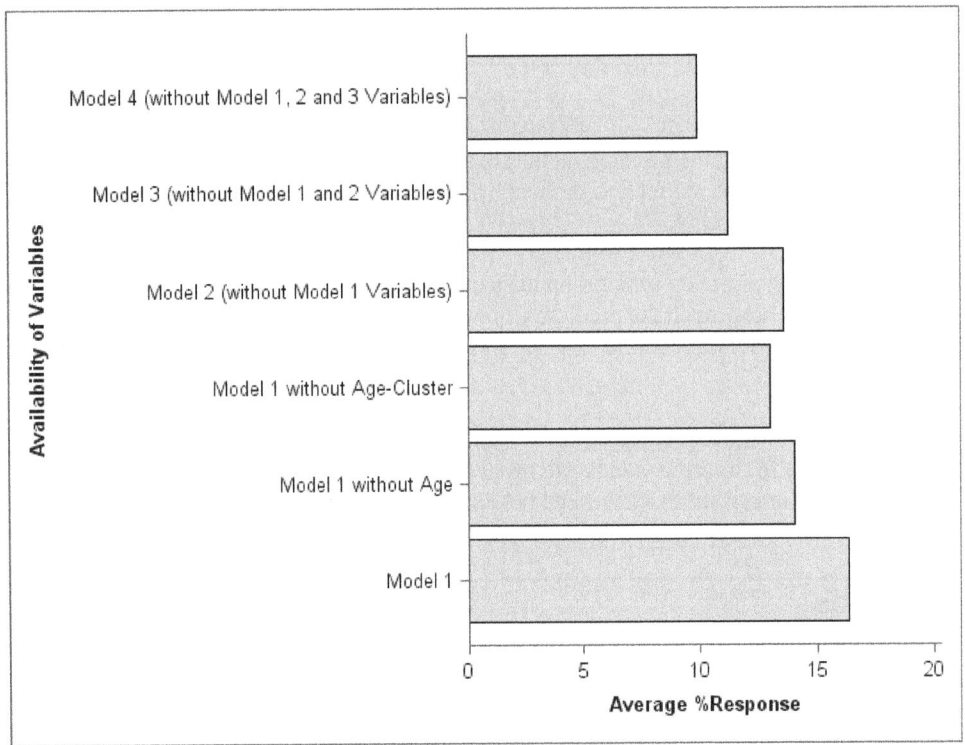

Table 17.3 shows the results in tabular form. The order to the availability scenarios has been changed so that there is an increasing average %Response with an increasing availability of variables. And there is a decreasing average %Response with the elimination of age and the age cluster variables.

Table 17.3: Table with the average %Response per availability of variables

Availability	Average %Response	StdDev
Model 4 (without Model 1, 2, and 3 Variables)	9.95	2.04
Model 3 (without Model 1 and 2 Variables)	11.23	1.32
Model 2 (without Model 1 Variables)	13.59	1.92
Model 1	**16.37**	**3.14**
Model 1 without Age	14.00	1.73
Model 1 without Age Cluster	13.03	1.50

Interpretation

For the simulation data, the results show a monotonic drop in the average %Response from availability in model 1 to model 4.

It is interesting to see that the removal of the age variable causes a drop in the average %Response from 16.4% to 14%, which is almost as strong as the decrease from model 1 to model 2. This means that the non-availability of the age variable solely accounts for almost all of the decrease in the %Response when excluding the model 1 variables.

You can also see that the Model 1 without Age Cluster variables falls below the Model 2 %Response. This means that Model 2, even with the Age variables removed, performs better because this model can include correlated substitute variables for age, which are not available in the model where the variables that are correlated with age are removed.

Business case

For the reference company described in section 16.3, the results for the removal of the age variable means that there is an expected cumulative profit of $410,000 with the age variable and an expected cumulative profit of $350,000 without the age variable. The absolute difference of $60,000 can be a trigger to pay particular attention to the completeness and correctness of the age variable in a customer database.

17.6 Conclusion

This chapter has quantified the effect on model quality of the reduction of the number of observations and the reduction of the number of events in the training data. Although these results are only based on simulations on selected data and may differ from the outcome in particular cases, they provide insight into the effect of data quantity for modeling.

It has been shown that in the low event number ranges (up to 500 events) additional non-event cases provide important additional information to improve model quality.

As long as the number of events does not exceed 500, additional events contribute greatly to model improvement and quality.

There is a business case to invest in higher data quantity because additional events contribute directly to a higher response rate. For the reference company described in section 16.2, this can mean an additional profit per campaign from $100 up to $1,000 for each additional event in the training data.

The availability of variables has also been studied in this chapter. It has been shown that for all the reference data used to run the simulations, it is important to include the age variable or at least variables that are highly correlated with age. Scenarios have also shown how model quality decreases if the top predictor set of variables is not available.

Chapter 18: Influence of Data Completeness on Model Quality in Predictive Modeling

18.1 Introduction	**231**
General	231
Random and systematic missing values	232
Missing values in the scoring data partition	232
18.2 Simulation Methodology and Data Preparation	**233**
Inserting random missing values	233
Inserting systematic missing values	233
Replacing missing values	233
Process flow	234
Simulation scenarios	234
18.3 Results for Random Missing Values	**234**
Random missing values only in the training data	234
Random missing values in the training and scoring data	236
Business case	237
18.4 Results for Systematic Missing Values	**237**
Systematic missing values only in the training data	237
Systematic missing values in the training and scoring data	238
18.5 Comparison of Results between Different Scenarios	**239**
Introduction	239
Graphical comparison	239
Differentiating between types of missing values	240
Multivariate quantification	241
18.6 Conclusion	**241**

18.1 Introduction

General

This chapter outlines the influence of missing values (data completeness) on the predictive power of the logistic regression models. Results of simulation studies with different percentages of random and systematic missing values in the input variables are shown. The correct response rate (**denoted by %Response in the results**) in the top 5% is used as comparison criterion.

Chapter 5 covered the scope and background of missing values. In many analytic and data mining projects, analysts encounter available data that contain missing values for certain characteristics for their analysis subjects.

Observations with missing values cannot be used by many modeling techniques like regression methods or neural network methods. Decision trees can handle missing values in the input data because they consider them as a separate category in the data and assign them to one of the branches.

The analyst faces the challenge of deciding whether to skip a certain record for the analysis or whether to impute the missing value with a replacement value. Here, however, the analyst must decide which percentage of missing values for a given variable imputation will be performed. Imputing a variable with 10% missing values is not considered a problem. But what about the situation with 30%, 50%, or more missing values?

The simulation studies that are presented in this chapter provide insight and rough guidance on how different proportions of missing values that are imputed affect model quality. The simulation studies also show whether there is a difference between random or systematic missing values and if just the training data or both the training and the scoring data contain missing values.

Random and systematic missing values

As shown in chapter 5, it is important to differentiate between random and systematic missing values. Random missing values are assumed to occur for each analysis subject with the same probability irrespective of the other variables. Systematic missing values are not assumed to occur for each analysis subject with the same probability.

For the simulations, for simplicity systematic missing values have been created in the data by defining 10 equal-sized clusters of analysis subjects. Specific to the definition of the respective simulation scenario, the variables for all observations in one or more of these clusters are set to missing.

Missing values in the scoring data partition

The simulations that are performed to assess the effect of missing values in the input variables must be differentiated by whether the data partition that is used to evaluate the model quality also has missing values inserted.

Note that for the context of performing the simulations in SAS Enterprise Miner, scoring Data refers to the test partition that is only used to calculate the model performance statistics like the %Response. For simplicity, the training partition and the validation partition are both denoted as training data here.

The simulations have been run for the following two cases:

- **No insertion of missing values into the scoring data.**
    - This scenario mirrors the perfect world in the application of the model, which means that data quality problems only occur in model training. In the application of the models, no data quality problems occur.
    - This occurs frequently in situations where models are built on historic data that are extracted from different operational or analytical systems from the current one or data are extracted from time periods where data quality measures were not yet in place.
- **Insertion of missing values into the training data and the scoring data.**
    - This scenario refers to situations where data quality problems like missing values occur not only in the model training phase but also during the application of the model.
    - This case reflects the situation where data quality problems could not (yet) be fixed and are present in the training and scoring data.
    - The missing values in the scoring data are replaced by the imputation logic that has been defined in the model training phase with the Impute node. The respective imputation logic is part of the SAS Enterprise Miner score code.

18.2 Simulation Methodology and Data Preparation

Inserting random missing values

For simulations with random missing values, the following code has been used to artificially insert missing values into the input data:

```
retain seed -1;
  %do i = 1 %to &em_num_input;
    call ranuni(seed, rnd);
    if rnd le &pct_missing then call missing(%scan(%em_input,&i));
  %end;
```

Within a DO loop for all input variables, a random number RND is created that controls whether for the respective input variable a missing value will be inserted or not. The macro variable &PCT_MISSING references to the simulation setting, which is the percentage of the observations that will get a missing value. The values of &PCT_MISSING range from 0 (no missing value) to 0.9 (90% of missing values).

Note that the &EM_NUM_INPUT variable is used to hold the number of input variables and %EM_INPUT represents the list of variable names of the input variables. The %SCAN function is used to receive the ith element of the variable list.

Depending on the simulation settings, this has been applied only for the training and validation data or it has also been applied for the test data.

Inserting systematic missing values

The following code has been used to insert systematic missing values for the simulations:

```
%do i = 1 %to &em_num_input;
  if (&pct_missing*10)-1 >= cluster then
                           call missing(%scan(%em_input,&i));
%end;
```

Again, a DO loop is used to loop over all input variables. Here, however, a random number is not used to control the insertion of missing values as in the case of random missing values. Instead, variable CLUSTER is used. Variable CLUSTER has previously been created and groups the observations into the 10 equal-sized groups. Variable CLUSTER is based on a simple cluster analysis on the age values of the analysis subject in the respective input data.

Depending on the desired proportion of missing values, the values of analysis subjects of one or more clusters are set to missing. Note that the calculation (&PCT_MISSING*10)-1 is used to relate the cluster numbers that range from 0 to 9 to the appropriate value between 0 and 1.

Replacing missing values

After the missing values have been inserted, the missing values for both interval and categorical input variables are replaced with the Impute node in SAS Enterprise Miner. Here the methodology of the TREE SURROGATE missing value imputation is used.

With the TREE SURROGATE method, replacement values are estimated by analyzing each input as a target in a decision tree, and the remaining input and rejected variables are used as predictors. Additionally, surrogate splitting rules are created. A surrogate rule is a backup to the main splitting rule. When the main splitting rule relies on an input whose value is missing, the next surrogate is invoked. If missing values prevent the main rule and all the surrogates from applying to an observation, the main rule assigns the observation to the branch that is assigned to receive missing values. (See also the help in SAS Enterprise Miner.)

Process flow

Figure 18.1 shows a partial process flow in SAS Enterprise Miner where, for example, 10% of random missing values are inserted into the data and consecutively replaced with the Impute node. Note that the replacement of missing values has been split for interval and categorical variables, mainly for didactical purposes. Replacing both variable types can also be accomplished in a single Impute node.

Figure 18.1: Partial SAS Enterprise Miner process flow for the scenario with missing values

Simulation scenarios

The simulations have been run for the following scenarios:

- Random Missing Values (0%, 10%, 20%, ... 90%)
  - Inserting missing values in the training data only
  - Inserting missing values in the training and scoring data
- Systematic Missing Values (0%, 10%, 20%, ... 90%)
  - Inserting missing values in the training data only
  - Inserting missing values in the training and scoring data

The results of these simulation scenarios are presented in sections 18.3 and 18.4. Section 18.5 contains a comparison between the different scenarios.

18.3 Results for Random Missing Values

Random missing values only in the training data

The first simulation scenario shows the effect of random missing values in the training data. Here, however, the scoring data are not affected by missing values. As already mentioned in the introduction, this reflects the situation where data already have good quality (for example, due to a system change or previous data quality efforts); however, the historic data that need to be used for model development (model training) still contain missing values that cannot be fixed retrospectively.

It can be seen from Figure 18.2 that there is surprisingly little decrease in model quality with the increasing percentage of missing values. This, however, shows very clearly that true random missing values indeed create some noise in the training data. The model logic that is trained on the basis of these data, however, still comes very close to the optimal model. If the data on which the model will be applied do not contain missing values, the prediction will still be quite accurate.

Figure 18.2: Box plot for %Response for different percentages of random missing values in the training data

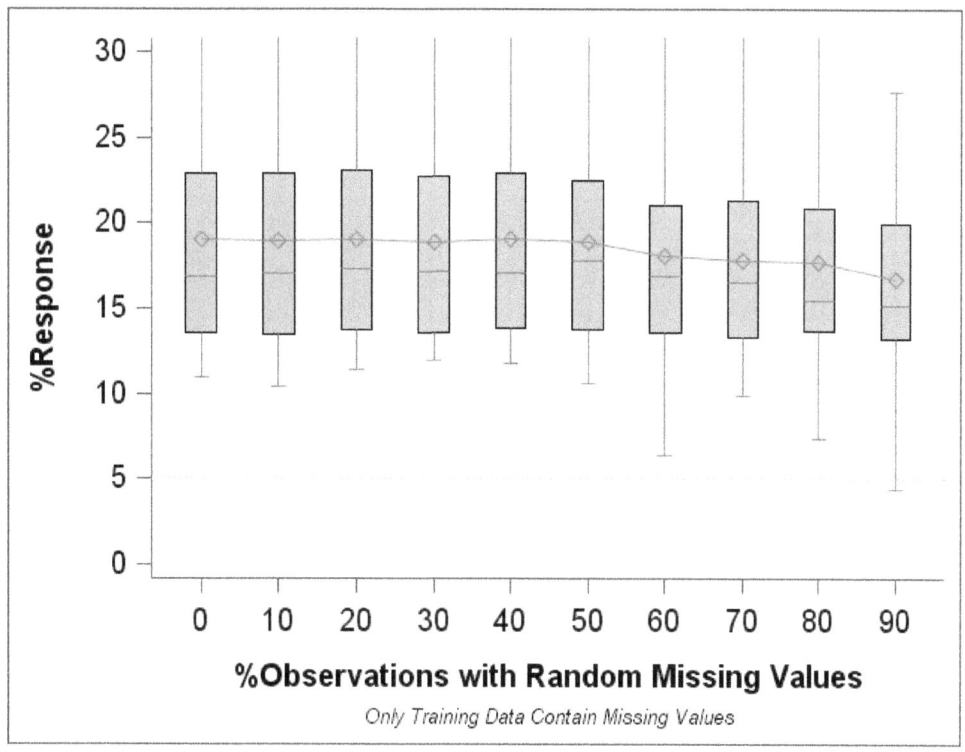

Table 18.1 illustrates the scenario with numbers. For the simulation data that have been used here, random missing values in the training data (up to 50%) do not have a major effect. Starting with 60% and 70% of missing values, a relative decline in model quality of 5% to 6% can be seen.

Caution: Do not take the problem of missing values lightly. Here we are only studying random missing values that only occur in the training. And, as mentioned in chapter 16, the baseline model with the predefined set of explanatory variables benefits from the fact that the variables for the optimal model have been preselected and do not need to be found based on the impaired data.

Table 18.1: Absolute and relative (no missing values = 100) %Response values for different percentages of random missing values in the training data

%Random Missing Values	%Response	%Response relative (no missing values = 100)
0	19.0	100.0
10	18.9	99.7
20	19.1	100.3
30	18.9	99.4
40	19.0	100.0
50	18.9	99.3
60	18.1	95.2
70	17.8	93.8
80	17.7	93.2
90	16.7	87.9

Random missing values in the training and scoring data

Now let's discuss the scenario where the scoring data, on which the model will be applied in production, has missing values. Here, it can be seen that the picture changes a lot.

Figure 18.3: Box plot for %Response for different percentages of random missing values in the training and scoring data

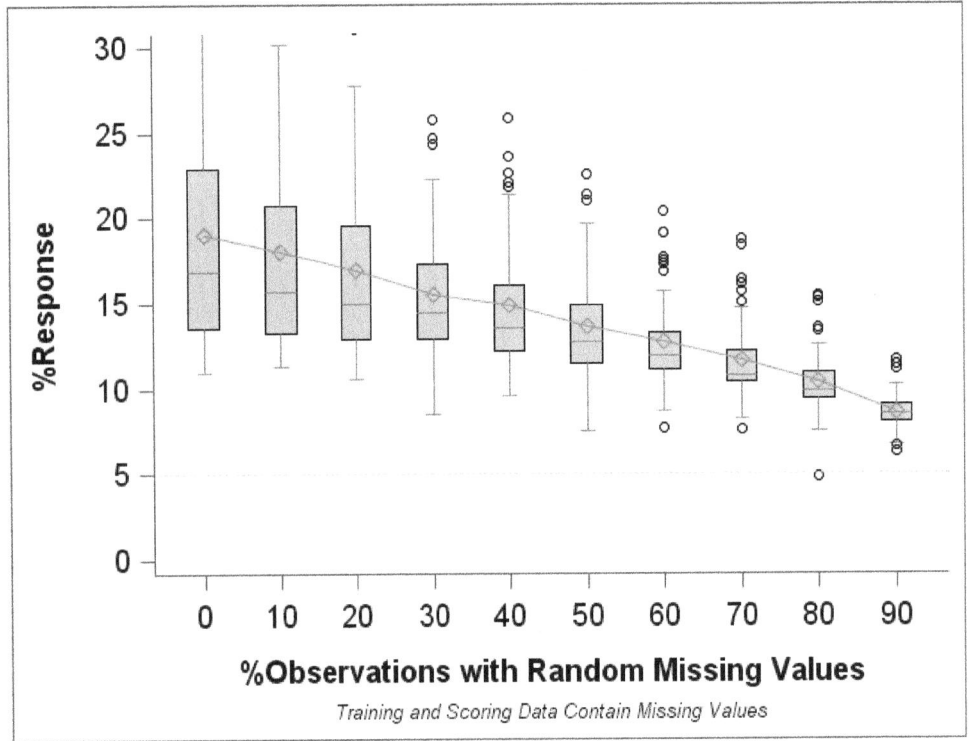

While there is almost no drop in model quality in the training data only case for 30% of missing values, there is already a substantial drop in the %Response, from 19.0 to 15.6, when both training and scoring data contain missing values. Table 18.2 also shows that in relative numbers model quality drops to 81.9% compared to the case where there are no missing values.

Table 18.2: Absolute and relative (no missing values = 100) %Response values for different percentages of random missing values in the training and scoring data

%Random Missing Values	%Response	%Response relative (no missing values = 100)
0	19.0	100.0
10	18.1	95.2
20	17.0	89.3
30	15.6	81.9
40	14.9	78.6
50	13.7	71.8
60	12.8	67.3
70	11.7	61.5
80	10.4	54.8
90	8.7	45.6

Business case

For the reference company, as described in section 16.2, this means the following:

- A reduction of missing values from 50% to 30% means an additional profit of $47,500 ($2,375 per percentage point).
- A reduction of missing values from 30% to 10% provides an additional profit of $62,500. This represents $3,125 per percentage point.

These numbers show that for the reference company a reduction of missing values by 10 percentage points equals an additional profit of $20,000 to $30,000 per campaign.

18.4 Results for Systematic Missing Values

Systematic missing values only in the training data

The first simulation scenario in this section shows the effect of systematic missing values that occur only in the training data. As explained previously, this reflects the situation where data already have good quality (for example, due to a system change or previous data quality efforts); however, the historic data that need to be used for model development (model training) still contain missing values that cannot be fixed retrospectively.

Figure 18.4 shows that unlike the case of random missing values (see Figure 18.2), systematic missing values affect the model quality even if they only occur in the training data.

Whereas random missing values introduce fuzziness into the data, which can still be treated effectively by missing value imputation, systematic missing values introduce an effect into the data that cannot be replaced by missing value imputation because these methods rely on the fact that the values are missing at random.

Figure 18.4: Box plot for %Response for different percentages of systematic missing values in the training data

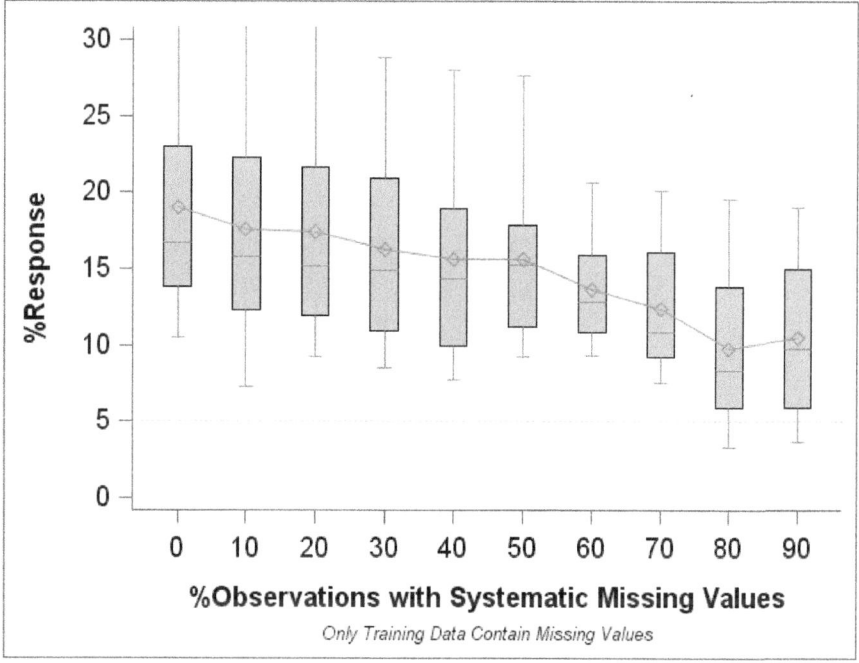

Table 18.3 shows the detailed numbers for these simulations. Note that the comparatively low value at 80% should be considered as a random outlier rather than as a meaningful change in patterns at 80%.

Table 18.3: Absolute and relative (no missing values = 100) %Response values for different percentages of systematic missing values in the training data

%Systematic Missing Values	%Response	%Response relative (no missing values = 100)
0	19.0	100.0
10	17.6	92.7
20	17.4	91.6
30	16.3	85.5
40	15.7	82.3
50	15.7	82.2
60	13.7	71.8
70	12.4	65.1
80	9.8	51.3
90	10.5	55.1

Systematic missing values in the training and scoring data

The picture changes substantially when the systematic error also occurs in the scoring data. Unlike the previous case, where only the model logic was biased by the missing values, here the scoring data also contain missing values.

Figure 18.5 shows a strong decrease in model quality with an increasing percentage of missing values. It can also be seen that with 70% and more missing values, the %Response rate even falls under the 5% (dotted) line. The 5% line represents the baseline event rate in the training data, which should be achieved by a random model. The systematic missing values in that range influence the model so strongly that it does not even predict as well as a random model.

Figure 18.5: Box plot for %Response for different percentages of random missing values in the training and scoring data

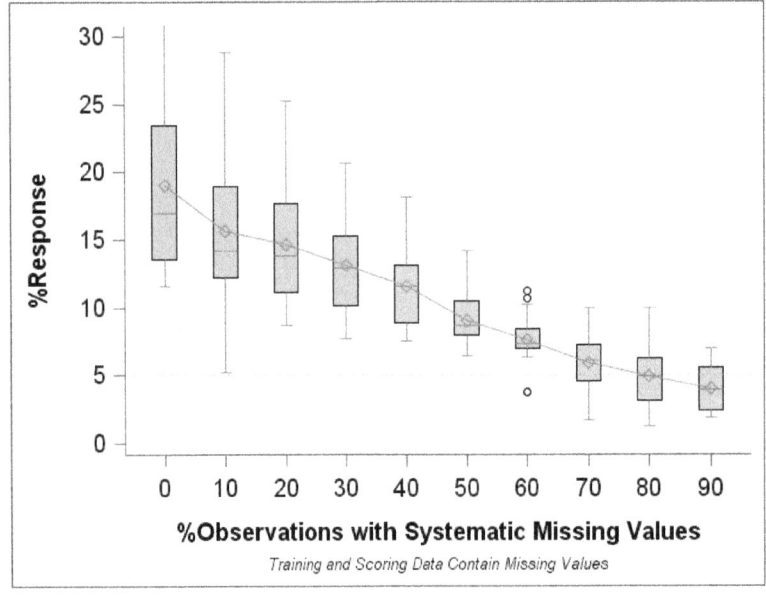

Table 18.4: Absolute and relative (no missing values = 100) %Response values for different percentages of systematic missing values in the training and scoring data

%Systematic Missing Values	%Response	%Response relative (no missing values = 100)
0	19.0	100.0
10	15.6	82.2
20	14.7	77.0
30	13.1	68.7
40	11.6	60.7
50	9.1	47.8
60	7.7	40.2
70	5.9	31.2
80	4.9	25.8
90	4.1	21.5

Table 18.4 shows the detailed numbers. It can be seen that even with only 10% of missing values, the %Response rate drops to 15.6%, which means that almost 18% of the predictive power of the perfect world model is lost.

18.5 Comparison of Results between Different Scenarios

Introduction

The previous sections have shown the results of the simulation scenarios for each of the different settings:

- random and systematic missing values in the training data only.
- random and systematic missing values in both training and scoring data.

This final section compares the results of the different simulation scenarios graphically and provides additional interpretation. Also, a multivariate quantification of the various simulation results by a linear model is shown.

Graphical comparison

The line plot that is shown in Figure 18.6 combines the mean %Response value over different percentages of missing values for all the scenarios presented in this chapter. The plot visually illustrates the differences between the scenarios.

Table 18.5 explains the line types for each scenario and shows the average %Response rate for 30% of missing values.

Table 18.5: Different missing value scenarios with line type and %Response

Scenario	Line Type	%Response
Random missing values in the training data only	solid	18.9
Systematic missing values in the training data only	short-dash	16.3
Random missing values in the training and scoring data	medium-dash-dot	15.6
Systematic missing values in the training and scoring data	long-dash	13.1

The numbers show that there is a relative difference of 30% between the best scenario, "Random missing values in the training data only," and the worst scenario, "Systematic missing values in the training and scoring data." This result can also be seen in the line chart in Figure 18.6.

Figure 18.6: Line plot for the average %Response for different percentages for all the different scenarios

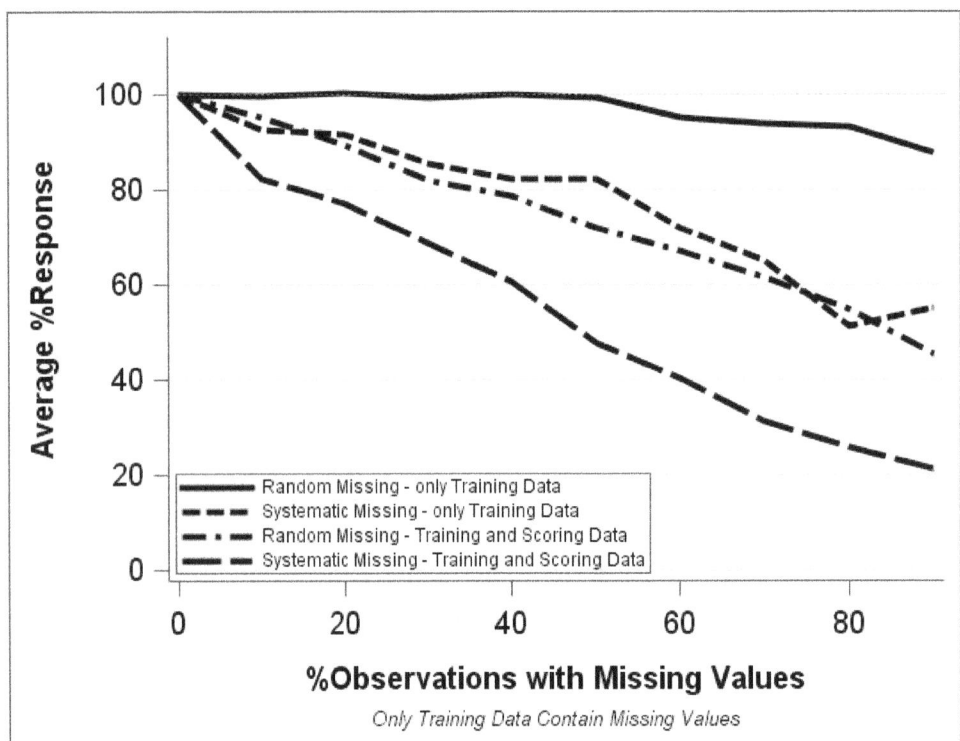

Differentiating between types of missing values

The broad range of the four lines over different missing value percentages illustrates how important it is to clearly differentiate between different types of missing values. Very often the topic of missing values is discussed in analytics as one monolithic subject. However, the numbers presented in the tables and figures here, and especially the line chart, make visually clear that it is wrong to consider all missing values as the same.

In every discussion about missing values in predictive analytics, you should bear in mind whether

- the missing values occur only in the training data or whether they also occur in the scoring data when the model is applied
- the missing values are random or systematic by nature

It is important to adequately address the problem of missing values because, otherwise, the influence of missing values may be over- or underestimated.

The first point may be easier to qualify than the second point. In many cases, the analyst knows in advance whether bad data quality in terms of missing values only affects the (historic) training data.

It is sometimes harder to decide whether missing values occur randomly or systematically. Here, business expertise as well as a more detailed analysis of both the profile of missing values and the correlation of the indication of missing values with other variables need to be performed to gain more insight.

Multivariate quantification

In order to quantify the effect of the following three potential influential factors on the model quality measured by %Response, a linear model has been run on these data:

- the percentage of missing values
- the random or systematic missing values
- the missing values only in the training data or in both the training and scoring data

The following code has been used to perform the analysis:

```
PROC GLM DATA=MissingValue_Scenarios;
 CLASS RandomSystematic;
 MODEL Response = Percentage ScoringData RandomSystematic
      /SS1 SS3 SOLUTION;
 RUN;
QUIT;
```

The outcome in terms of the coefficients for each of the variables is as follows:

- Intercept: 19.29
  - This represents the estimate for 0% of missing values when only the training data are biased.
- Percentage: -0.0996
  - %Response decreases on average by 0.1 percentage points for each additional percentage point of missing values.
- ScoringData: -4.23
  - %Response decreases on average by -4.23 percentage points if both the training data and the scoring data are biased.
- RandomSystematic=Random: 3.6
  - %Response is on average 3.6 percentage points higher if the missing values occur randomly.

18.6 Conclusion

This chapter has shown the effect of missing values in predictive modeling. The various parameters of missing values—the percentage of missing values, the random or systematic missing values, and existence of the missing values only in the training data or in both the training and scoring data—have been discussed, illustrated, and quantified.

The results are plausible from an interpretational point of view. The visual relationships in the graphs as well as the signs and the magnitudes of the coefficients of the linear model mirror what would have been expected from a business point of view.

Even if the results are based only on four different data sources that may not be exactly the same for all different analytical questions, industries, and business domains, they give a rough quantification of the effect of missing values.

The results also provide an impression about the key drivers of the missing values problem itself.

Chapter 19: Influence of Data Correctness on Model Quality in Predictive Modeling

19.1 Introduction 243
  General 243
  Non-visible data quality problem 244
  Random and systematic bias 244
  Biased values in the scoring data partition 244
19.2 Simulation Methodology and Data Preparation 245
  Standardization of numeric values 245
  Inserting random biases in the input variables 245
  Inserting systematic biases in the input variables 245
  Inserting a random bias in the target variable 246
  Inserting a systematic bias in the target variable 246
  Simulation scenarios 247
19.3 Results for Random and Systematic Bias in the Input Variables 247
  Scenario settings 247
  Bias in the input variables in the training data only 247
  Bias in the input variables in the training and scoring data 248
  Comparison of results 249
19.4 Results for Random and Systematic Bias in the Target Variables 250
  General 250
  Examples of biased target variables 250
  Detecting biased target variables 251
  Results for randomly biased target variables 251
  Results for systematically biased target variables 252
19.5 Conclusion 253
  General 253
  Treatment of biased or incorrect data 253
19.6 General Conclusion of the Simulations for Predictive Modeling 253
  Increasing the number of events and non-events matters 253
  Age variable is important and there are compensation effects between the variables ... 254
  It makes a difference whether data disturbances occur in the training data only
  or in both the training and scoring data 254
  Random disturbances affect model quality much less than systematic disturbances ... 254

19.1 Introduction

General

This chapter studies the influence of biases in the data on the predictive power of prediction models. Results of simulation studies are shown where a bias is introduced into the input variables in the model. The quality of the model is described with the correct response rate (denoted by "%Response" in the results) of the predictions in the top 5% of the analysis subjects.

Chapter 6 discussed the topic of data correctness from a conceptual point of view. This chapter shows the consequences of different types and graduations of bias in the data. The following three types of bias are studied here:

- random bias in the input variables
- systematic bias in the input variables
- random and systematic bias in the target variable

Non-visible data quality problem

Data correctness is, in many cases, a **non-visible data quality problem** because it often cannot be explicitly seen. Unlike missing values, where a missing value can explicitly be queried from the data and summarized for each variable, whether the available data are biased or not cannot just be detected by simply checking the value. Often business and validation rules are applied here to decide whether a value is correct or not.

Chapter 13 has shown different methods to profile the data quality status in terms of data correctness. In some cases, hard fact rules that base on value ranges, list of values, or integrity checks can be used to check whether a value is correct or not. In other cases, however, the outcome of checking for data correctness can only be done on a probabilistic basis (like "value is probably not correct"). For example, statistical methods can be used to flag observations that fall outside the upper and lower validation limits that are calculated based on the standard deviations.

Random and systematic bias

As stated in chapter 6, an analyst must differentiate between random and systematic bias in the data. Random biases occur for each analysis subject with the same probability, irrespective of the other variables. Also, the bias itself does not point systematically in one direction.

Systematic missing biases are assumed to occur for each analysis subject with a different probability. Also the direction of the bias can be systematically upward, downward, or toward the center. Consider, for example, the cases where all observations show the identical value or all observations with a large value show a small value.

For simplicity, systematic bias values have been created in the data by defining 10 equal-sized clusters of analysis subjects. Specific to the definition of the respective simulation scenario, the variables for all observations in one or more of these clusters are biased.

Biased values in the scoring data partition

The simulations that are performed to assess the effect of bias in the input variables have to be differentiated by whether the data partition that is used to evaluate the model quality also has biased values.

Note that when performing the simulations in SAS Enterprise Miner, scoring data refers to the test partition, which is used to calculate the model performance statistics like %Response. For simplicity, the training partition and the validation partition are both denoted as training data here.

The simulations have been run for the following two cases:

- **No biasing of data in the scoring data**
    o This scenario mirrors the perfect world in the application of the model, which means that data quality problems only occur in model training. In the application of the models, no data quality problems occur.
    o This case is frequent in situations where models are built on historic data that are extracted from different operational or analytical systems than the current one or from time periods where data quality measures were not yet in place.

- **Insertion of biased values in the training and scoring data**
  - This scenario refers to situations where data quality problems like biased values occur not only in the model training phase, but also in the application of the model.
  - This case refers to the situation where data quality problems have not yet been fixed and are present in the training and scoring data.

19.2 Simulation Methodology and Data Preparation

Standardization of numeric values

For the simulations, only interval variables have been biased. For this purpose, the values have been standardized to mean = 0 and standard deviation = 1. Based on these standardized values, a random bias of one standard deviation can be introduced simply by adding the constant 1 to the data.

Inserting random biases in the input variables

For simulations with random biases, the following code artificially inserts biases into the input variables:

```
retain seed -1;
%do i = 1 %to &em_num_interval_input;
call ranuni(seed, rnd);
call ranuni(seed, rnd2);
if rnd2 < &prop then
%scan(%em_interval_input,&i)=
%scan(%em_interval_input,&i)+(1-2*(rnd>0.5))*&factor;
%end;
```

- Within a DO loop for all interval input variables, a random number RND2 is created that controls whether a bias will be introduced for the respective input variable.
- The macro variable &PROP controls the percentage of records that are biased. For the simulations presented here, &PROP has been set to 30%.
- Macro variable &FACTOR controls the number of standard deviations that is added or subtracted from the data. For these simulations, &FACTOR has been set to 0.5, 1, 2, and 3.
- The expression (1-2*(rnd>0.5)) controls whether the bias is added or subtracted from the data. A random number RND is used to generate an expression that is 0 or 1 in 50% of the cases each. The calculation 1-2*expression transforms this value into the set -1 and 1.

Note that the &EM_NUM_INPUT variable is used to hold the number of input variables, and %EM_INPUT represents the list of variable names of the input variables. The %SCAN function is used to receive the ith element of the variable list.

Depending on the simulation settings, this has been applied only for the training and validation data or for the training, validation, and test data.

Inserting systematic biases in the input variables

The following code inserts systematic biases into the input variables:

```
if (&prop*10)-1 >= cluster then do;
%do i = 1 %to &em_num_interval_input;
%if &factor = -1 %then
%scan(%em_interval_input,&i)=
%scan(%em_interval_input,&i)*(-1);
%else %if &factor = 0 %then %scan(%em_interval_input,&i)=0;
%end;
end;
```

Again, a DO loop is used to loop over all interval input variables. Here, however, a random number is not used to control the insertion of missing values as in the case of random missing values. Instead, the variable CLUSTER is used. Variable CLUSTER has previously been created and clusters the observations into 10 equal-sized groups. Variable CLUSTER is based on a simple cluster analysis on the age values of the analysis subjects in the respective input data.

The macro variable &PROP controls the percentage of records that are biased. For the simulations presented here, &PROP has been set to 30%. Note that the calculation (&PROP*10)-1 is used to relate the cluster numbers that range from 0 to 9 to the appropriate value between 0 and 1.

Depending on the desired proportion of missing values, the analysis subjects of one or more clusters are set to missing.

For the simulations, two options for systematic biases in the data have been provided:

- Setting the input variable to zero. Note that because these are standardized values, this equals setting the value of the variable to its mean.
    - From a business point of view, this mirrors the situation where the data are systematically biased towards the mean value, for example if the respondent gives an average answer. In the previous code, this can be controlled by setting the factor to 0.
- Changing the sign of the input variable. For a standardized value, this means that for non-standardized values, a high value is converted to a low value and vice-versa. In the code, this can be accomplished by setting the factor to -1.
    - From a business point of view, this reflects the situation where the reported value is inverted from its true nature (for example, persons with a low salary report a high salary and vice-versa).

Inserting a random bias in the target variable

A random bias in the target variables has been introduced with the following code:

```
retain seed -1;
 call ranuni(seed, rnd);
 if rnd < &prop then %em_target = 1-%em_target ;
```

Macro variable &PROP is used to control the percentage of records with biased target variables. A random number is used to select the records that are biased.

Bias here means that the value of the binary target is reversed. This is performed by subtracting the value of the target variable from 1 (0 gets 1, 1 gets 0). From a business point of view, this reflects the situation where either the event is not observed accurately or the respondents intentionally reverse the true answer into the opposite.

Inserting a systematic bias in the target variable

A random bias in the target variables has been introduced with the following code:

```
if (&prop*10)-1 >= cluster then %em_target = 1-%em_target ;
```

Macro variable &PROP is used to control the percentage of records with biased target variables. The selection of observations is performed systematically based on the variable CLUSTER, which groups the observations into 10 equal-sized clusters.

Bias here means that the value of the binary target is reversed. This is performed by subtracting the value of the target variable from 1 (0 gets 1, 1 gets 0).

Simulation scenarios

The simulations have been run for the following scenarios:

- Random and systematic biases in the input variables for 30% of the observations.
  - Inserting missing values in the training data only
  - Inserting missing values in the training and scoring data

- Random and systematic biases in the target variables for 0%, 10%, 20%, 30%, and 40% of the observations.

The results of these simulation scenarios are presented in sections 19.3 and 19.4.

19.3 Results for Random and Systematic Bias in the Input Variables

Scenario settings

Table 19.1 shows which settings for random and systematic bias in the input variables have been made in the scenarios.

Table 19.1: Legend and settings for random and systematic bias scenarios

Legend	Explanation
R0	No bias is introduced into the data.
R0.5	Half a standard deviation is randomly added/subtracted from the standardized interval input variables.
R1	One standard deviation is randomly added/subtracted from the standardized interval input variables.
R2	Two standard deviations are randomly added/subtracted from the standardized interval input variables.
R3	Three standard deviations are randomly added/subtracted from the standardized interval input variables.
S0	The values of the input variables are set to 0.
S-1	The sign of the input variables is changed.

Bias in the input variables in the training data only

The first simulation shows the effect of bias in the input variables in the training data only. As already mentioned in the introduction, this reflects the situation where data already have good quality (for example, due to a system change or previous data quality efforts); however, the historic data that need to be used for model development (model training) still have biased values that cannot be fixed retrospectively.

Figure 19.1 shows that there is a slight decrease in model quality with an increasing bias in the data (scenarios R0–R3). As long as only the training data are biased, however, the model quality does not decrease dramatically (for example, only down to 16.9% when the bias is +/-2 standard deviations).

Figure 19.1: Box plot for %Response for different settings of random and systematic bias in the training data

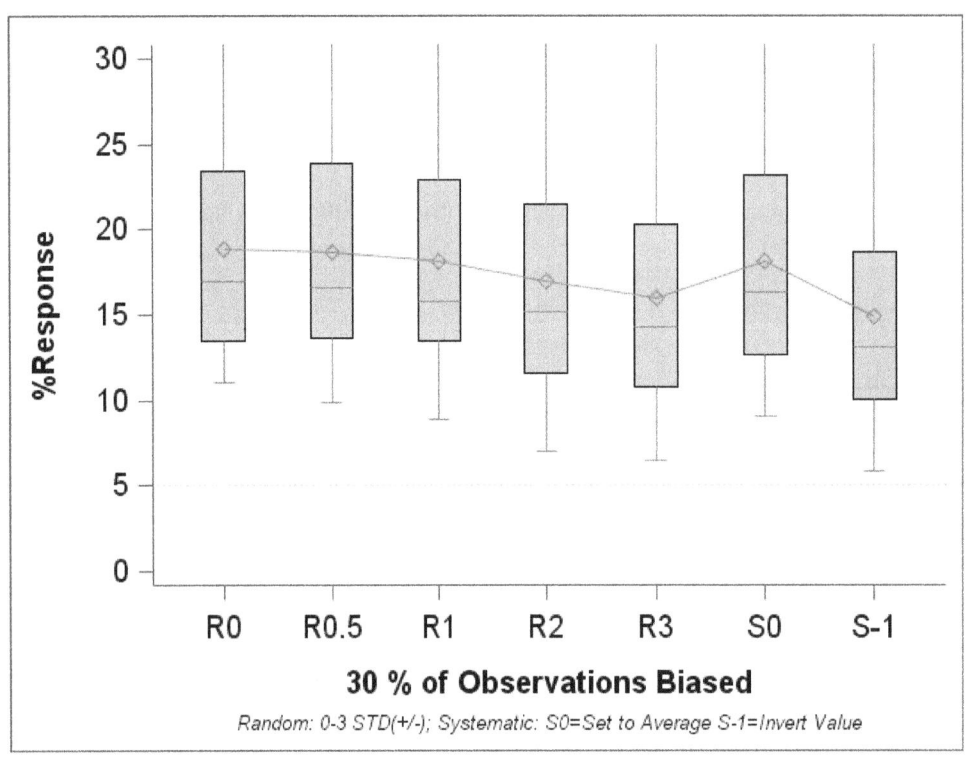

The results also show that scenario S0, where the interval input variables are set to 0, almost has the same effect as randomly adding/subtracting one standard deviation of bias.

Inverting the values obviously shows much worse results, decreasing the %Response for about 22% from 18.9 to 14.9 percentage points.

Table 19.2: Absolute and relative (no bias = 100) %Response for different settings of random and systematic bias in the training data

%Obs with Random Biased Target	%Response	%Response relative (no bias = 100)
No Bias	18.87	100.0
+/- 0.5 Std	18.73	99.3
+/- 1 Std	18.15	96.2
+/- 2 Std	16.94	89.8
+/- 3 Std	15.96	84.6
Set to average	18.16	96.2
Invert Value	14.88	78.8

Bias in the input variables in the training and scoring data

Biasing the input variables in the training and scoring data shows results similar to those in Figure 19.2 and Table 19.3. The scenarios with systematic errors especially show a decrease in model performance.

Figure 19.2: Box plot for %Response for different settings of random and systematic bias in the training and scoring data

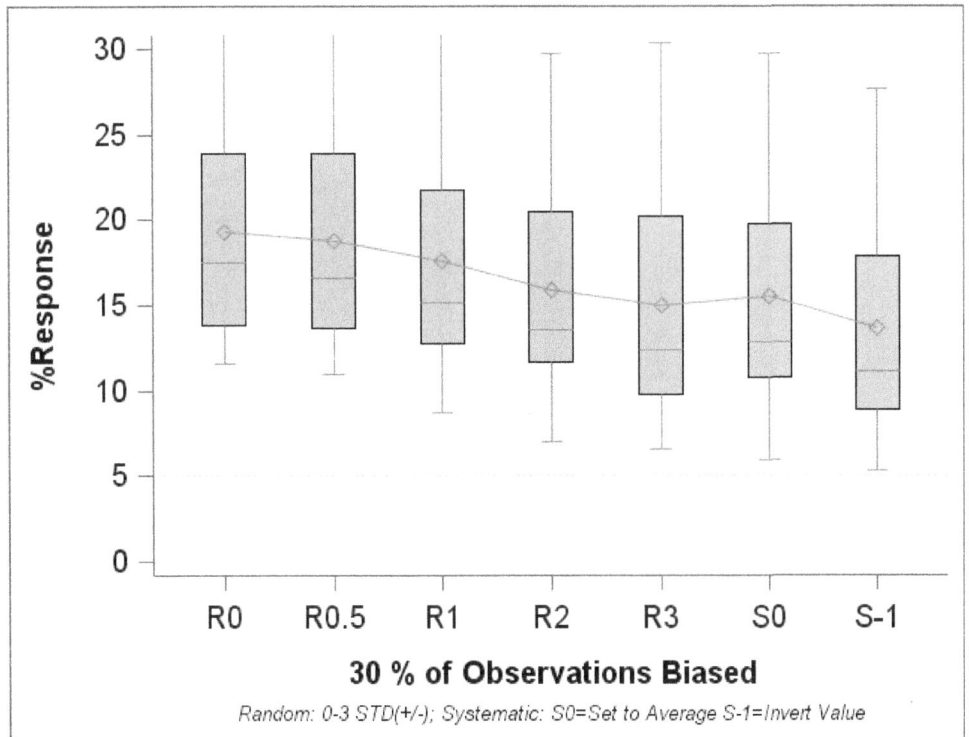

Table 19.3: Absolute and relative (no bias = 100) %Response for different settings of random and systematic bias in the training and scoring data

%Obs with Random Biased Target	%Response	%Response relative (no bias = 100)
No Bias	19.29	100.0
+/- 0.5 Std	18.81	97.5
+/- 1 Std	17.63	91.4
+/- 2 Std	15.88	82.3
+/- 3 Std	14.99	77.7
Set to average	15.55	80.6
Invert Value	13.63	70.7

Note that the %Response value for the no-bias scenario here is higher than those in the other tables presented in this and other chapters. However, this has only occurred randomly and should not be interpreted in any detail.

Comparison of results

Figure 19.3 compares the outcome of the scenarios between the settings for introducing bias only in the training data or in both the training and scoring data.

As previously discussed in the missing values scenarios and in the case of biased input variables, the decrease in model quality is stronger when the scoring data become biased as well. Figure 19.3 illustrates this situation with a line chart for the relative average %Response in the different scenarios.

Figure 19.3: Line plot for the relative average %Response (no bias = 100) for different settings of random and systematic bias in the training data only or in the training and scoring data

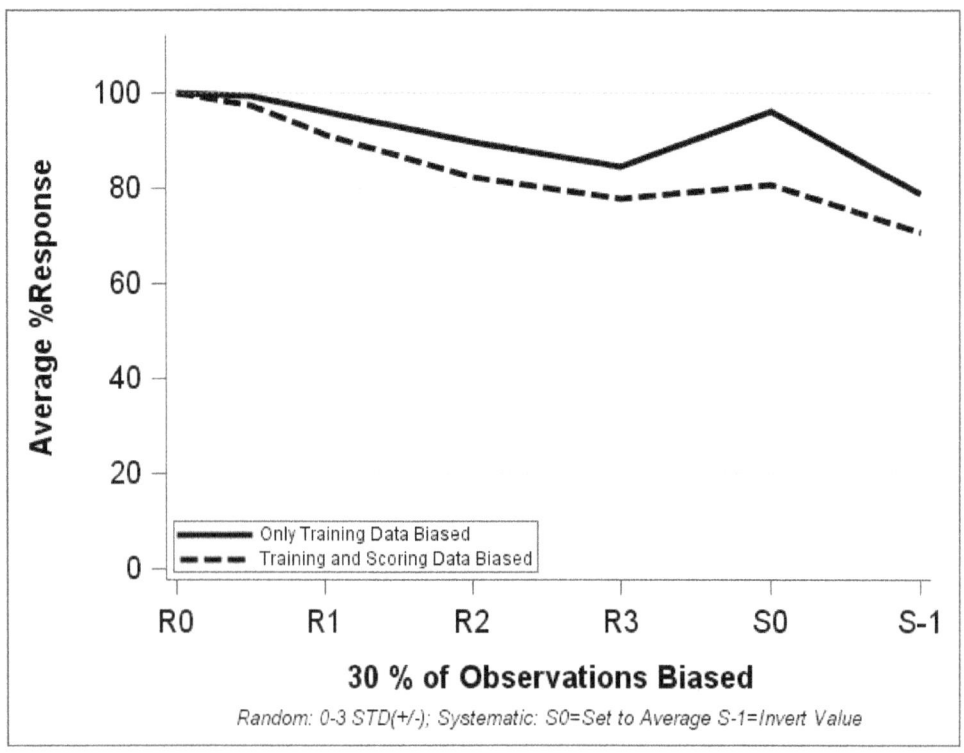

For the mean response of the random bias scenarios R0.5, R1, R2, and R3, an absolute difference in %Response of 0.6% (17.4% vs. 16.8%) can be seen between cases where only the training data or both the training and scoring data are biased,. The average decrease in %Response with the increasing random bias of the data is 1.31% per standard deviation that is added or subtracted from the interval variables.

The difference in mean response of systematic bias scenarios S0 and S-1 between the training data only scenarios and the training and scoring data scenarios is 1.9% (16.5% vs. 14.6%) larger than those in the random bias cases.

19.4 Results for Random and Systematic Bias in the Target Variables

General

This section shows the effect of random and systematic bias in the target variables. In the case of a binary target variable, the "bias" is reflected by the fact that the value is reversed when an event becomes a non-event and a non-event becomes an event.

These scenarios are only evaluated for bias in the training data because the scoring data usually do not contain a target variable. In the case of the simulation scenarios, the scoring data (= test data) contain a target variable that is needed for evaluation purposes.

Biasing a binary target variable and changing it to the other value weakens the signal in the data and makes it harder to train a good and predictive model.

Examples of biased target variables

Systematically biased target events can occur in practice (for example, if, for a certain customer group, the recording of the outcome is not done as precisely as in other groups). If data from all groups are being analyzed together, the relationship between input and target variables in some groups is much weaker or shows in the opposite direction. If relationships remain undetected, the model will generate biased estimates, and model quality will suffer.

Randomly biased target variables can occur if, across the set of analysis records, recording the value of the target variable is not performed precisely. Here, the bias weakens the signal in the data; however, it does not distort or invert the relationship between the input and target variables.

Detecting biased target variables

In general, it is difficult to profile and detect biased target variables. You could eventually do so by training and scoring a predictive model on the training data and identifying those observations that receive a prediction that deviates from the true value. In this case, however, it is hard to decide whether the target variable has been inaccurately recorded or whether the model itself fails to make predictions for these data.

Results for randomly biased target variables

A box plot for the change in %Response for scenarios with randomly biased target variables in 0%, 10%, 20%, 30%, and 40% of the cases is shown in Figure 19.4. As expected, a downward trend can be seen, which, however, only starts after a percentage of 20% biased target events.

Figure 19.4: Box plot for %Response for different percentages of randomly biased target variables

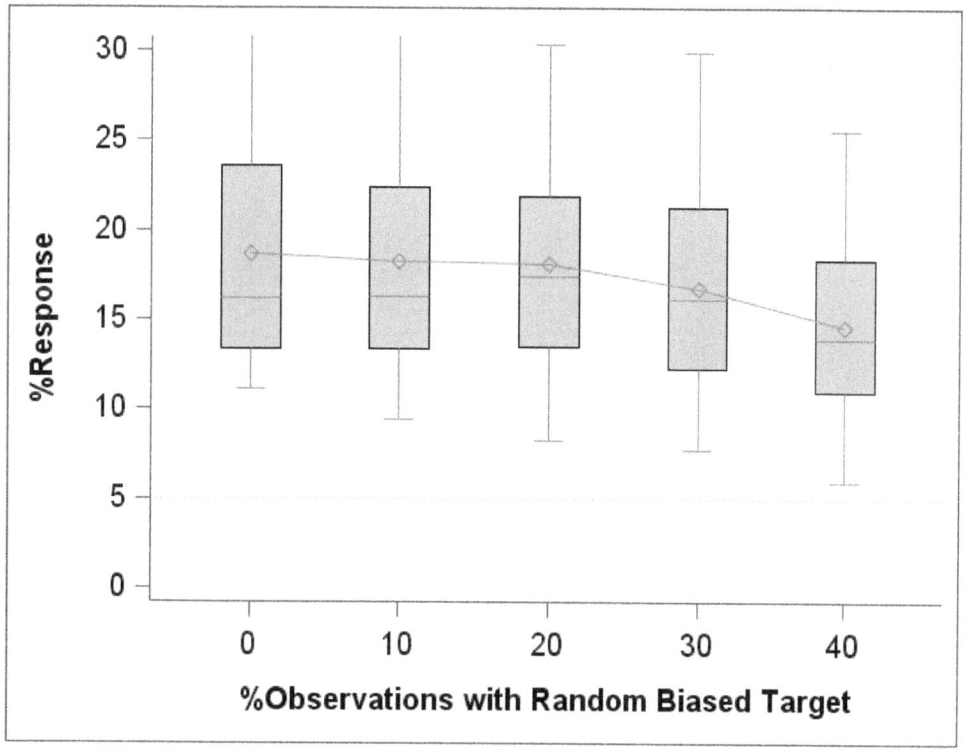

Table 19.4 shows the results in numbers. It can be seen that the %Response relatively decreases by 10.3% if 30% of the target values are randomly biased.

Table 19.4: Absolute and relative (no bias = 100) %Response for different settings of random bias in the target variable

%Obs with Random Biased Target	%Response	%Response relative (no bias = 100)
0	18.7	100
10	18.3	97.8
20	18.1	96.7
30	16.7	89.7
40	14.5	77.8

Comparing the results in Tables 19.2 and 19.4, it can be seen that a random error of +/-2 standard deviations in the input data in 30% of the observations has approximately the same effect as randomly exchanged target values in 30% of the observations.

Results for systematically biased target variables

Figure 19.5 shows a box plot with the decrease in the %Response in the case of systematically biased target events in 0%, 10%, 20%, 30%, and 40% of the observations.

Figure 19.5: Box plot for %Response for different percentages of systematically biased target values

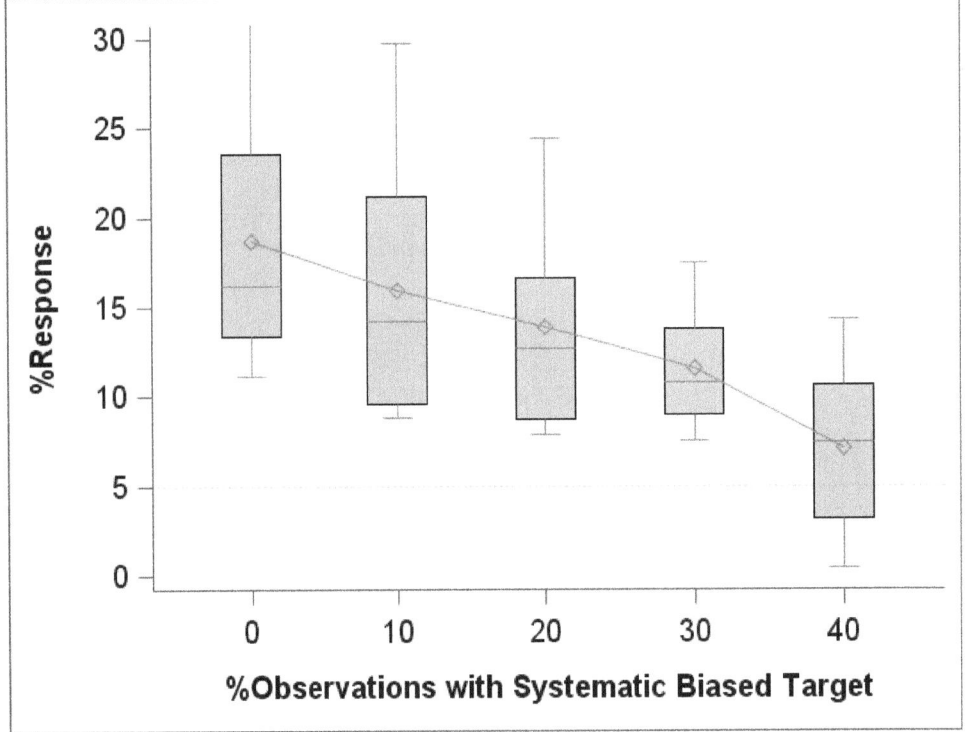

As expected, the downward trend is now much stronger, as in the random case. Table 19.5 shows that in the case of 30% biased target events, the relative %Response goes down to 62.1% compared to the no-bias scenario.

Also note that with 40% of systematically biased target events, the %Response goes down to 7.1% compared to the 5% baseline probability (which would be achieved with a random model). It can also be seen that the first quartile limit of the box even falls below the 5% level, showing that some simulation runs have even worse results than the random model.

Table 19.5: Absolute and relative (no bias = 100) %Response for different settings of systematic bias in the target variable

%Obs with Systematically Biased Target	%Response	%Response relative (no bias = 100)
0	18,68	100
10	15.98	85.5
20	13.90	74.4
30	11.59	62.1
40	7.14	38.2

19.5 Conclusion

General

This chapter has shown the effect of random and systematic missing values on the input and target variables. As expected, a systematic bias has a stronger effect on the decrease of the %Response than a random bias. Also, there is a decrease in the %Response with the increasing size of the random error in terms of standard deviations that are added or subtracted from the data.

It has been shown that in the case of systematic errors, changing the standardized input values to 0, which equals the usage of the mean value instead of the true value, does not have such a severe effect as other systematic biases or as higher level random biases.

Treatment of biased or incorrect data

This also leads to a possible answer to the question of how to treat detected biased values in the data. Data values that do not meet the validation rules and are apparently incorrect should undergo data improvement.

In this case, data values that are flagged as incorrect or potentially incorrect can be exchanged with a replacement value. Methods like calculating an individual replacement value can be performed, as shown in chapter 13.

Another method is to simply replace this value with the most frequent category, in the case of categorical variables, or with the mean, in the case of interval variables.

Usually it is better to replace this value with a most likely value or at least with a mean value instead of filtering the entire observation set because then the information is lost.

19.6 General Conclusion of the Simulations for Predictive Modeling

From a data quality point of view, the following conclusions can be drawn from the results presented in chapters 17 through 19.

Increasing the number of events and non-events matters

The simulation results have shown that with the increasing number of events and non-events, the model quality improves. This is especially true for cases with up to 250 events, but it is also true for cases with more events. Additional data still improve model quality to some extent.

Age variable is important and there are compensation effects between the variables

In all the simulation data sets, an age variable was present. Removing the age variable from the list of available input variables showed a decrease in variable quality. It has, however, also been shown that the non-availability of the input variables in the best model can be compensated, to some extent, by other input variables.

It makes a difference whether data disturbances occur in the training data only or in both the training and scoring data

The simulations have shown that it makes a difference whether only the training data are biased by missing or incorrect data or whether both the training and scoring data are biased. Biasing the scoring data decreases model quality.

Random disturbances affect model quality much less than systematic disturbances

It has been shown that many models can cope with random disturbances to a certain extent. The introduction of systematic biases, however, causes a larger decrease in model quality.

Chapter 20: Simulating the Consequences of Poor Data Quality in Time Series Forecasting

20.1 Introduction 255
    General 255
    Purpose and application of time series forecasting 255
    Methods to forecast time series 256
    Scope and generalizability of simulations for time series forecasting 256

20.2 Overview of the Functional Questions of the Simulations 257

20.3 Base for the Business Case Calculation 258
    Introduction 258
    The reference company Quality DataCom 258

20.4 Simulation Environment 258
    General 258
    Available data for the simulation environment 259
    Time series methods 259

20.5 Simulation Procedure 260
    Basic simulation procedure 260
    Insertion of disturbances for the data in the simulation procedure 261
    Loop over TIME HISTORIES 262
    Loop over shifts 262
    Qualification of time series for the simulations 262
    Assessment of forecast accuracy 263

20.6 Downloads 263

20.7 Conclusion 264

20.1 Introduction

General

This chapter deals with how time series forecasting, and especially the resulting forecasts, are affected by different levels of data quality and quantity. Using a simulation study based on real life data, the effect of the quality and quantity of the historic data on the accuracy of the forecasts for future periods is studied.

Purpose and application of time series forecasting

Forecasting the values of a time series in future periods is important to companies and organizations across many industries:

- In retail, companies want to forecast the number of purchases that customers make. These forecasts are often detailed (for example, by product group, geographic region, and sales channel). The influence of marketing campaigns, promotions, and competitor prices on the number of sold quantities is studied here.

- In the manufacturing industry, the demand forecasts for finished products, semi-finished products, and raw materials are necessary to better plan the optimal set up of the supply chain.
- For financial planning, companies want to forecast their sales numbers for future periods. Insurance companies, for example, want to forecast their premium streams for the next year, as well as the expected costs and the claim cases and amounts.

Methods to forecast time series

Time series forecasting methods can use the historic values of the dependent variable to forecast its values in future periods. Here, patterns in the course of the dependent variable over time (like trends, seasonal patterns, or cycles) are analyzed and used to define a forecast model for the values in future periods. These methods include, for example, exponential smoothing models, seasonal exponential smoothing models, Winters Additive and Multiplicative methods, ARIMA models, and UCM (unobserved components models). This book does not deal with the specific properties of these methods and refers the reader to forecasting books in SAS Press. See also [19], [20].

Additionally, input variables can be used in the model equation, which provide additional input forecasts for the values in future periods. These input variables can, for example, include product prices and competitor's prices, events like promotions or product changes, or econometrical data like indices, exchange rates, or economic cycle data. The applicable methods here include autoregressive methods, ARIMAX methods, and UCM models, for example.

Forecasting problems can also roughly be divided into two groups, based on the resources that can be allocated from a business point of view to optimally fit the model to the data:

- Modeling a minimum of four different econometric figures should allow the analyst to analyze and model each time series in detail. Here it is possible to individually study the autocorrelations between different lags in the time series, to study the influence and correlation of different input variables, and to create an individually fitted model.
- A retail company that, for example, has 25,000 different stock keeping units (SKUs) that need to be forecast on a weekly basis is most likely not in a position to model each time series individually, but it could model the different time series automatically.

These two (extreme) examples also describe the two most frequent ways time series forecasting is applied in practice. The first case is often referred to as econometric analysis and forecasting, whereas the second case is referred to as automatic forecasting.

SAS Forecast Server allows you to perform large-scale forecasting. Here, thousands of time series can be automatically forecasted. For each time series, the optimal model that minimizes forecast error either is selected from a predefined list of typical forecast models or is individually fit to the respective data of the time series.

Scope and generalizability of simulations for time series forecasting

The simulation studies that are presented here are based on large-scale automatic forecasting and performed using SAS High-Performance Forecasting procedures, which are part of SAS Forecast Server.

The simulations are based on real-life data sets from different industries. The data are used to illustrate the influence of different quality characteristics and their impact on model results. The results presented here do not offer ultimate hard-and-fast rules. Instead, they serve as illustrations of the types of effects different data quality deficiencies can have on forecast quality.

The simulation results provide an anchor point to assess the approximate size and influence of different degrees of poor data quality and available data quantity. The simulation studies allow you to study how data quality changes if a certain data quality characteristic is changed. They also allow you to compare the effect of different simulation studies against each other and the size of the outcome. Finally, the simulation studies illustrate in a detailed way the different aspects of data quality that have to be considered.

Thus, the simulation studies allow you to draw important conclusions for the planning and prioritization of data quality actions.

Section 20.3 shows a fictional reference company that is used to quantify the effect of the improvement or decline in data quality in financial numbers. The numbers that are presented are illustrative only; however, they are an indication about the leverage effect for quantified levels of data quality improvement.

The simulation studies shown here are all based on the forecasting of quantities in different industries and business domains. Seasonal exponential smoothing is used as the method to produce the forecasts. Focusing on this single method provides one stable reference method that remains constant across the different scenarios.

20.2 Overview of the Functional Questions of the Simulations

Table 20.1 gives an overview of the main topics that have been investigated in the simulation studies.

Table 20.1: Overview of the functional questions for the simulations

	Topic	Functional Question
Chapter 21	Available length of data history	Influence of the available number of historic periods on forecast quality.
	Best length of data history	Identification of the optimal length of data history.
Chapter 22	Missing values	Influence of the percentage of random missing values in the input variables on forecast quality.
		Influence of the percentage of systematic missing values in the input variables on forecast quality.
Chapter 23	Data bias	Influence of random errors in the input variables on forecast quality.
		Influence of systematic errors in the input variables on forecast quality.
		Influence of systematic and random errors in the target variable on forecast quality.

Chapters 21 through 23 show the simulations in the following structure:

- definition of the simulation and specification of the simulation settings
- presentation of the simulation results in graphical and tabular form
- interpretation of the results and relation to a business case

20.3 Base for the Business Case Calculation

Introduction

A fictional reference company is used to calculate a business case for the outcome of the different simulation scenarios. The change in forecast quality is transferred into a quantification of the avoidance of over- and underforecasting and the respective profit is expressed in US dollars.

This is done to illustrate the effect of different data quality changes. The numbers in US dollars should be considered only as rough indications based on the assumption of the simulation scenarios and on the business case as described. In individual cases, these values and relationships are somewhat different. The business case allows you to compare the outcomes in different simulation scenarios and make the results and the findings much more visible.

The reference company Quality DataCom

The reference company discussed in the following chapters is the same company that was used in the predictive modeling case studies. Quality DataCom operates in the communications business.

The company has a business line that produces and sells electronic accessories for end user devices for both wholesale and retail sales. The company makes 10,000 products that are sold to customers. For each of these products, 1,000 units on average are sold per month. This leads to a total number of 10,000,000 units sold per month.

Quality DataCom currently has a mean absolute percentage error (MAPE) rate of 15%. This means that the forecast on average deviates by 1,500,000 units in total.

The benefit of improving forecast quality is to avoid out-of-stock situations, where a profit of $1 is lost per unit. Also overforecasting causes a loss of $1 due to excessive stock keeping and transportation as well as producing non-sellable goods.

Thus, the decrease in MAPE by one percentage point from 15% to 14% generates an additional profit for this company of 10,000,000 units multiplied by the difference in forecast errors of 1% equals a reduction of the forecast error by 100,000 units. Multiplying this by $1 results in a monthly profit of $100,000. Thus, a decrease in MAPE by 1 percentage point results in a profit of $1,200,000 per year.

20.4 Simulation Environment

General

In order to answer questions about the relationship between forecast accuracy and the features of the available data, a simulation environment has been set up. This simulation environment is used to perform time series forecasts based on different scenarios of the data and to compare the forecast for future periods with real out-of-sample data.

The simulations have been performed with SAS 9.2 TSLEVEL 2M3 on X64_VSPRO platform. The simulations have been performed with SAS High-Performance Forecasting 2.3. Base SAS and SAS/GRAPH software have been used for data preparation, simulation definition, and results analysis. SAS Enterprise Guide has been used as the simulation user interface.

Available data for the simulation environment

For the simulation, time series from real-world situations with monthly aggregated data have been used. The time series IDs have been made anonymously. In order to qualify for usage in the simulation environment, a particular time series had to fulfill the following conditions:

- History of at least 72 months:
  - The simulations were run on time series with an available history of up to 48 months for time series forecasting.
  - An additional 12 months of real data were used to validate the performance of the results.
  - A further 12 months were needed to shift the time series by up to 12 months in order to start in the simulations the time series at different points in time and, thus, provide a larger sampling base. See section 20.5 for more details.
- No leading, trailing, or embedded missing values or zero values:
  - Time series with missing values or zeros were excluded in order to make sure that only proper and complete time series that have no known data quality defect were used. Thus, the effect of a controlled modification of the input data can be measured.
- No visual or other obvious data specifics like periods with zero quantities:
  - Note that the macro %PROFILE_TS_MV, which was discussed in chapter 11, has been used for these data screening purposes.

A total of 788 time series qualified based on these criteria. In order to get a representative sample of the time series, data have been taken from the following industries: leisure, retail, steel manufacturing, paper production, and oil and gas production. Depending on the business context of the time series, the quantity variable represents different units like tons, units, sales amounts, or persons.

While the time series were taken from different time periods, ranging from March 1997 to July 2009, each time series has been shifted to an artificial end date of January 2010 in order to make data preparation easier for the simulation environment.

Time series methods

Automatic time series forecasting with no manual intervention has been performed. The HPFENGINE procedure, which is part of SAS High-Performance Forecasting, has been used to forecast the time series.

The following were used to parameterize the HPFENGINE procedure:

- A model selection list SEASONAL_SL has been created, which contains only the model SEASONAL (seasonal exponential smoothing).
- To avoid the failure of fitting the seasonal model to the data and defaulting to exponential smoothing, the following parameters have been set:
  - The number of necessary seasonal cycles for the usage of a seasonal model has been set to one: MINOBS = (SEASON = 1).
  - The significance test for seasonality in the data has been turned off: SEASONTEST = NONE.

Note that the selection list has been limited to the SEASONAL model for a reason. Different outcomes in the simulations scenarios as a result of different forecasting models can be avoided. Because the available model has been fixed to a specific model, the differences between the scenarios can be assessed more directly.

Also note that SAS Forecast Server and SAS High-Performance Forecasting provide a much wider set of available methods for time series forecasting:

- The TSFSSELECT selection list of the HPFDFLT catalog, for example, provides a rich set of default forecasting models, which can be used to pick the best possible forecast model for the particular time series:
    - Exponential smoothing models (for example, Seasonal Exponential Smoothing, Multiplicative, and Additive Winters models)
    - Linear Trend models with or without seasonal dummies
    - ARIMA models and log-ARIMA models with different parameterization of the p, q, and r parameters with and without intercepts
- PROC HPFDIAGNOSE can be used to create model selection lists that are specific to the individual data that will be forecast. Here, the data are analyzed and the optimal forecasting model for each time series is added to a catalog. The models in this catalog can then be used to forecast the data. It is also possible to define models by hand and manually add them to the model selection list.

20.5 Simulation Procedure

Basic simulation procedure

The core of the simulation procedure involves the following four main steps:

1. For each time series, preserve the last 12 periods from the forecasting process as out-of-sample data in order to use them to assess accuracy.
2. Perform data treatments, such as inserting random or systematic disturbances or missing values into the data.
    - If missing values have been inserted into the data, use PROC EXPAND or PROC TIMESERIES to impute those values.
3. Run PROC HPFENGINE to fit the forecast model and produce forecasts for the next 12 periods.
4. Calculate the forecast error MAPE (mean absolute percentage error) by comparing the preserved actual values in the out-of-sample data and the forecast values for the 12 future periods for each time series.

These four steps are then iterated over different TIME HISTORIES and different SHIFTS:

- The iterations over TIME HISTORIES are performed to generate results for different lengths of available data history. Different TIME HISTORIES here means that the number of available historic months is continuously increased.
- The iterations over SHIFTS are performed to get results that do not only depend on a single start point of the forecast period but on different start points. SHIFTS here means that a number of months from the beginning of the time series are deleted in order to start the time series at a later point in time. This, for example, causes the forecast period to start in different calendar months and not just depend on a single starting month. Thus, the data are more representative and generalize better.

Figure 20.1 visualizes the simulation process graphically.

Figure 20.1: Simulation process for the time series forecasting simulations

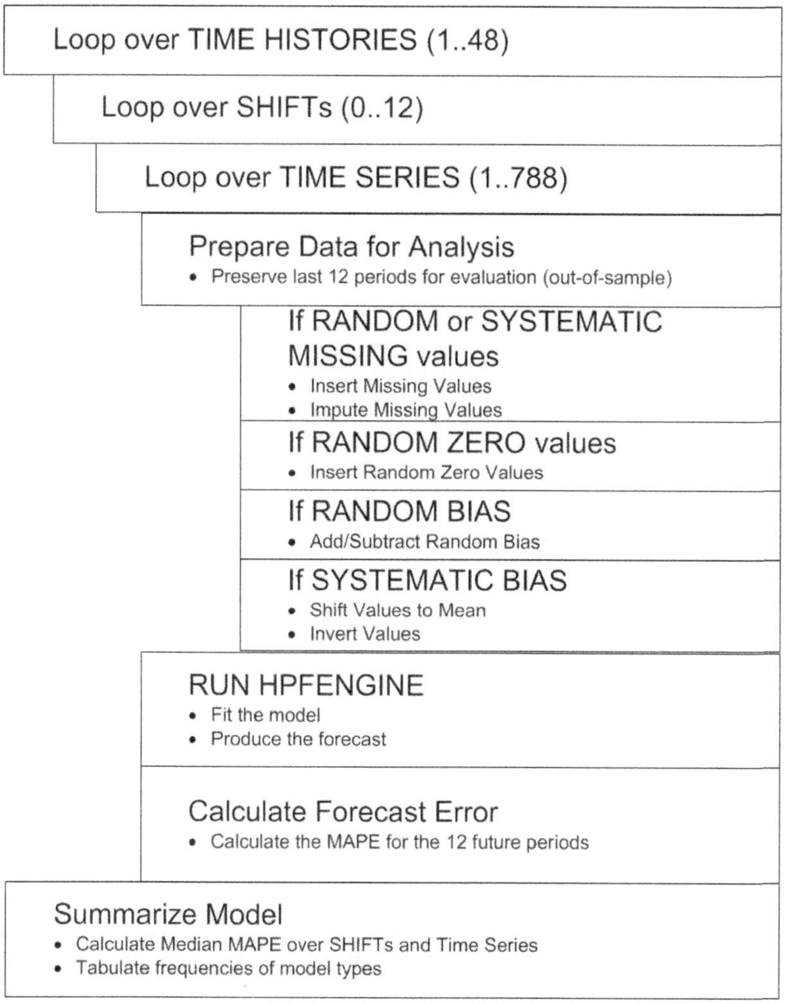

```
┌─────────────────────────────────────────────────────────────────────┐
│ Loop over TIME HISTORIES (1..48)                                    │
│ ┌─────────────────────────────────────────────────────────────────┐ │
│ │ Loop over SHIFTs (0..12)                                        │ │
│ │ ┌─────────────────────────────────────────────────────────────┐ │ │
│ │ │ Loop over TIME SERIES (1..788)                              │ │ │
│ │ │                                                             │ │ │
│ │ │  Prepare Data for Analysis                                  │ │ │
│ │ │   • Preserve last 12 periods for evaluation (out-of-sample) │ │ │
│ │ │                                                             │ │ │
│ │ │     If RANDOM or SYSTEMATIC                                 │ │ │
│ │ │     MISSING values                                          │ │ │
│ │ │      • Insert Missing Values                                │ │ │
│ │ │      • Impute Missing Values                                │ │ │
│ │ │     If RANDOM ZERO values                                   │ │ │
│ │ │      • Insert Random Zero Values                            │ │ │
│ │ │     If RANDOM BIAS                                          │ │ │
│ │ │      • Add/Subtract Random Bias                             │ │ │
│ │ │     If SYSTEMATIC BIAS                                      │ │ │
│ │ │      • Shift Values to Mean                                 │ │ │
│ │ │      • Invert Values                                        │ │ │
│ │ │                                                             │ │ │
│ │ │  RUN HPFENGINE                                              │ │ │
│ │ │   • Fit the model                                           │ │ │
│ │ │   • Produce the forecast                                    │ │ │
│ │ │                                                             │ │ │
│ │ │  Calculate Forecast Error                                   │ │ │
│ │ │   • Calculate the MAPE for the 12 future periods            │ │ │
│ │ └─────────────────────────────────────────────────────────────┘ │ │
│ │                                                                 │ │
│ │  Summarize Model                                                │ │
│ │   • Calculate Median MAPE over SHIFTs and Time Series           │ │
│ │   • Tabulate frequencies of model types                         │ │
│ └─────────────────────────────────────────────────────────────────┘ │
└─────────────────────────────────────────────────────────────────────┘
```

Insertion of disturbances for the data in the simulation procedure

Depending on whether the scenario also involves the insertion of disturbances into the data, one or more of the following steps are also applied directly in the step to prepare the data for analysis:

- If a proportion of the data will be randomly set to 0, then
  - set the quantity variable for those observations to 0
  - enter the data directly into the forecasting process as described in step 3
- If a proportion of the data will be randomly set to missing, then
  - set the quantity variable for those observations to missing
  - run PROC EXPAND or PROC TIMESERIES to replace the missing values with an imputation value
- If random or systematic biases will be applied to the data, then
  - apply the bias to the data (random bias, set to mean, invert value)
  - enter the data directly into the forecasting process as described in step 3

The detailed procedure for the insertion of systematic or random disturbances is explained in the respective chapters where the results of the simulations are shown.

Loop over TIME HISTORIES

The iteration over time histories is used to continuously increase the number of available months with historic data and to assess their effect on the forecast accuracy. Time histories have been looped from 1 to 48. The purpose of this iteration is to study the effect of an increase of the time history on the forecast accuracy.

Figure 20.2: Illustration of the TIME HISTORIES for the simulation environment

History	02/2006	...	06/2009	07/2009	08/2009	09/2009	10/2009	11/2009	...	10/2010
1							Use	Evaluation period		
2						Use	Use	Evaluation period		
3					Use	Use	Use	Evaluation period		
4				Use	Use	Use	Use	Evaluation period		
5			Use	Use	Use	Use	Use	Evaluation period		
...										
48								Evaluation period		

Loop over shifts

The iteration over shifts has been used to shift each time series back by m months. By shifting for one month, for example, the last (most recent) observation from the available data has been deleted, and only the observations until period t-1 are used for the analysis. These settings offer a more representative sampling over the time line and avoid having the evaluation results always depend on the same periods.

Figure 20.3: Illustration of the SHIFTS and TIME HISTORIES for the simulation environment

Shift	History	04/2009	05/2009	06/2009	07/2009	08/2009	09/2009	10/2009	11/2009	...	10/2010
0	1							Use	Evaluate 11/2009-10/2010		
0	2						Use	Use	Evaluate 11/2009-10/2010		
0	3					Use	Use	Use	Evaluate 11/2009-10/2010		
0	4				Use	Use	Use	Use	Evaluate 11/2009-10/2010		
0	5			Use	Use	Use	Use	Use	Evaluate 11/2009-10/2010		
1	1						Use	Evaluate 10/2009-09/2010			
1	2					Use	Use	Evaluate 10/2009-09/2010			
1	3				Use	Use	Use	Evaluate 10/2009-09/2010			
1	4			Use	Use	Use	Use	Evaluate 10/2009-09/2010			
1	5		Use	Use	Use	Use	Use	Evaluate 10/2009-09/2010			
2	1					Use	Evaluate 09/2009-08/2010				
2	2				Use	Use	Evaluate 09/2009-08/2010				
2	3			Use	Use	Use	Evaluate 09/2009-08/2010				
2	4		Use	Use	Use	Use	Evaluate 09/2009-08/2010				
2	5	Use	Use	Use	Use	Use	Evaluate 09/2009-08/2010				

Qualification of time series for the simulations

As mentioned earlier, the scenarios are run with different TIME HISTORIES and SHIFTS. The simulation procedure makes sure that a particular time series is only used for the simulation if it can be used for the simulations with all history lengths and with all SHIFTS.

For example, in SHIFT=0, a time series with an available history of 42 months would qualify for the run of the simulations with time histories from 1 to 30 (+12 month validation period). However, it would not be possible to run the simulations with time histories 31 or above.

Thus, this time series is excluded from all the simulations in order to have a stable and comparable set over all different lengths of time histories. Otherwise, the results may be biased more for the shorter time histories; therefore, a different subset of time series is used.

Additionally, a time series must have enough history to allow the simulation scenarios with a shift of up to 12 months. Thus, for each time series in the simulations, a minimum history length of 72 (TIME HISTORY: 48 months, + VALIDATION: 12 months + SHIFT: 12 months) is needed.

Assessment of forecast accuracy

Different validation statistics exist for time series forecasting methods, which measure the closeness of the forecast values and the actual values. Literature and research papers that discuss the advantages and disadvantages of various methods are available. This chapter does not enter this discussion.

Using the MAPE as a validation statistic is not undisputed; it has some weaknesses, for instance it is not defined for ACTUAL=0 cases and MAPE tends to favor underprediction. However, MAPE is often chosen as the default accuracy measure because it is the most widely used measure in business forecasting. MAPE is also easy to interpret. Compare, for example, [19]. For simplicity, the MAPE (mean absolute percentage error) is used here.

Calculating the validation statistic to compare the outcome of the different scenarios is performed in the following steps:

- For each time series and different length of time histories and shift, the MAPE for 12 future months is calculated with the following formula (n is the number of observations in the series, which is 12 for all cases; Y_t is the actual value; $\hat{Y}_t$ is the forecast value):

$$\text{MAPE} = \frac{1}{n}\sum_{t=1}^{n} |Y_t - \hat{Y}_t|/Y_t$$

- This results for each scenario in a MAPE value per time series, length of history, and shift.
- In the next step, the median is used to aggregate the MAPE values per time series and shift, resulting in one value for each length of history.
- This median value is then used to analyze the results, for example:
    - to plot the course of median MAPE on the Y axis over increasing numbers of time histories on the X axis
    - to compare different scenarios

20.6 Downloads

The results that are presented in the following chapters are also available at http://www.sascommunity.org/wiki/Data_Quality_for_Analytics.

Make sure to check this address for updates on the results and additional simulation scenarios that are presented here.

SAS programs, macros, and data sets can also be downloaded from this site.

20.7 Conclusion

This chapter has introduced the necessary prerequisites for the simulation studies that are shown in chapters 21 through 23:

- Chapter 21 shows the results of scenarios with different lengths of the data history and presents results for the optimal data history length.
- Chapter 22 shows the results of scenarios when random or systematic missing values are introduced in the data.
- Chapter 23 shows the results of scenarios where random and systematic biases are generated in the time series.

Chapter 21: Consequences of Data Quantity and Data Completeness in Time Series Forecasting

21.1 Introduction	265
21.2 Effect of the Length of the Available Time History	265
General	265
Simulation procedure	266
Graph results	266
Interpretation	268
Results in numbers	268
Business case calculation	269
21.3 Optimal Length of the Available Time History	269
General	269
Results	269
Interpretation	271
21.4 Conclusion	271
General	271
Data relevancy	271
Self-assessment of time series data	272

21.1 Introduction

This chapter examines the influence of the length of the available data history of a time series on the quality of the forecast for future periods. The results of simulation studies that are presented in this chapter address the importance of long time histories to performing good time series forecasting.

The results shown in this chapter are based on a simulation study where the available time history is continuously increased and the respective model quality is analyzed. Thus, the relationship between length of time history and model quality can be studied in a visual way.

Additionally, for each of the time series and the different simulation repetitions per time series, the length of the time series that delivers the best forecast quality is selected. Based on these results, the distribution of the optimal length of the time history over all-time series examples is shown and discussed.

21.2 Effect of the Length of the Available Time History

General

The simulation environment described in chapter 20 is used to simulate the different availability of the length of the history in time series data. Results on how the quality of the forecast model changes over the available history length are shown in a line chart.

The general assumption is that with increasing time history, the quality of the forecast model, which is trained on these data, improves (expressed by a decrease in MAPE). In general, we expect to see a relationship between increasing data quantity (length of available time history) and forecast quality.

Simulation procedure

The simulation procedure is as follows:

- For all time series that are available for analysis, the time history is truncated to the length of 1. Based on this 1 value data, a forecast for the next 12 periods is performed and evaluated against the true (untouched) data from the out-of-sample data.
- In the next step, the available time history is increased backwards to the last 2 months. Again, the forecast quality is assessed on the next 12 months.
- This procedure is iterated by adding an additional historic month to each iteration until a maximum data history of 48 months is reached.

For illustration purposes, Figure 20.2 from the previous chapter is repeated here.

Figure 21.1: Illustration of the TIME HISTORIES for the simulation environment

History	02/2006	...	06/2009	07/2009	08/2009	09/2009	10/2009	11/2009	...	10/2010
1							Use	Evaluation period		
2						Use	Use	Evaluation period		
3					Use	Use	Use	Evaluation period		
4				Use	Use	Use	Use	Evaluation period		
5			Use	Use	Use	Use	Use	Evaluation period		
...										
48								Evaluation period		

The results of each iteration are stored and plotted in a line plot.

Graph results

Figure 21.2 shows the results of the continuous increase of the available time history and the creation of a seasonal exponential smoothing model (Modelname=SEASONAL) with PROC HPFENGINE. The MAPE over all-time series and SHIFTS is aggregated using the median.

Figure 21.2: Median MAPE by available length of history

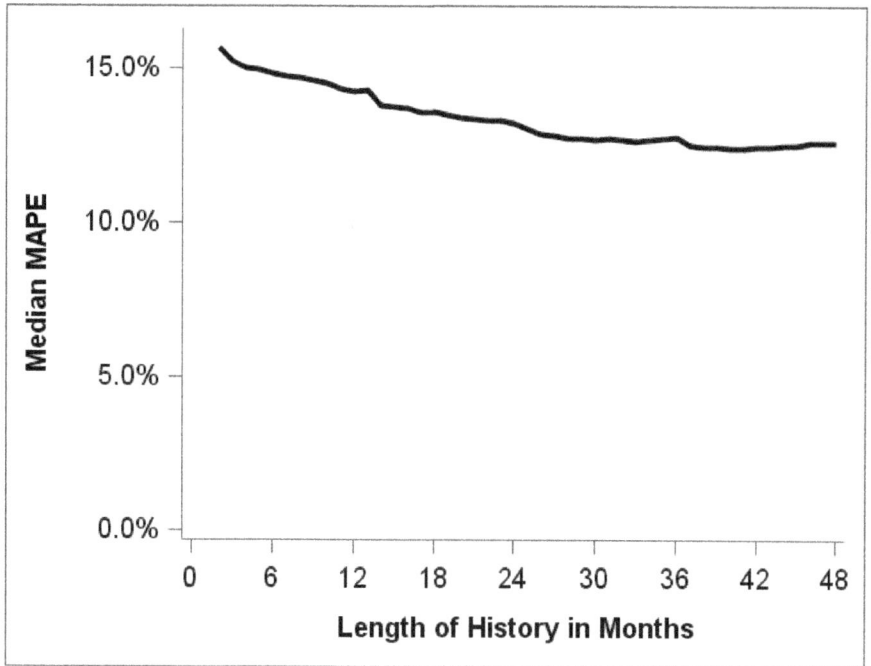

The X-axis represents the respective number of available history months. The Y-axis represents the median MAPE value over all-time series and all shift scenarios.

Figure 21.3 shows the same line plot as in Figure 21.2; however, it has a different Y-axis scale that allows a more detailed insight into the characteristics of the line. Figure 21.2 is still important because it more truly visualizes the relative change in MAPE with the increasing time history.

Figure 21.3: Median MAPE by available length of history (different Y-axis scaling)

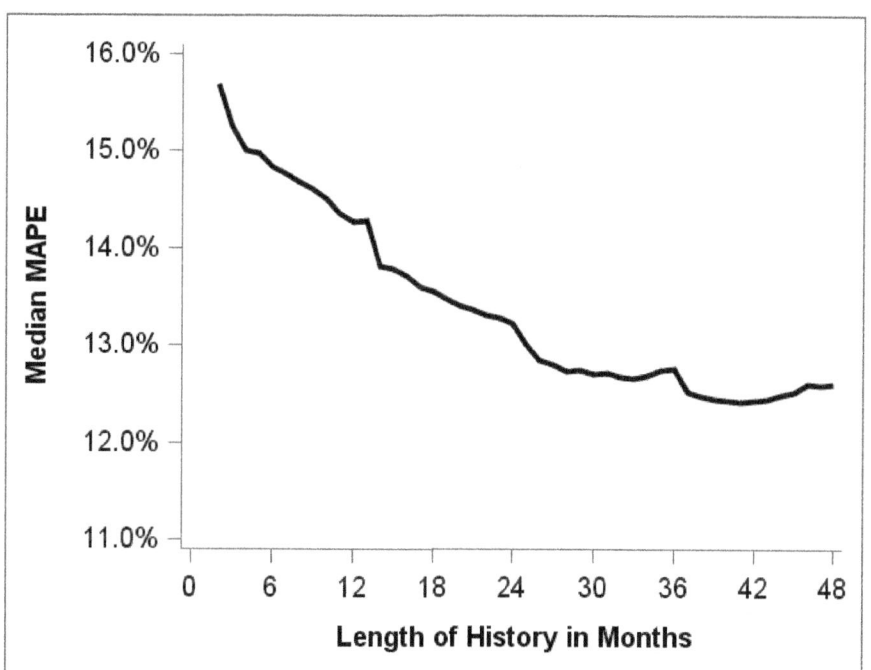

From the graph, you can see that the MAPE decreases with more available months in the time history. Studying the course of the line in more detail, you can see the following features:

- There is a steep descent from month 2 to 4, which shows that at least a couple of months are necessary to be able to forecast the mean level of the series to some extent.
- The line decreases further until month 12, showing that the relative contribution of each additional month plays an important role here because only a short time history is available.
- From month 12 to month 13, again a steep descent can be seen, when the first repetition of the seasonal cycle starts. Starting from here, the seasonal specifics of the time series can be integrated into the model.
- The MAPE line decreases further, showing again a decrease after month 24 where a full second cycle is available in the data.
- This repeats after the third full season at month 36. The potential increase at the very edge of the X-axis might also be due to a data artifact and should not be important.

The analysis of the **interquartile range** of the MAPE statistics per length of time history also shows a decrease of the variability (6 months: 14.7%, 18 months: 12.9%, 30 months: 12.5%, 42 months: 12.2%). This indicates that with increasing length of the history the forecast error decreases on average and the stability of the results increases.

Interpretation

The results shown in the line plots here conform to intuition that, on average, with increasing data quantity in terms of available time history, the quality of the forecasting model increases.

It can also be seen that the availability of the two first full years indicates a strong contribution to model quality. Additional months still provide quality improvement, which is, however, relatively lower. Also, this finding is intuitive: the marginal effect on model quality is higher when only a few months are available for analysis.

Results in numbers

The results that are presented visually in Figures 21.2 and 21.3 are shown in numbers in Table 21.1. Table 21.1 shows the MAPE and the averaged MAPE values for selected time history lengths. The MAPE column shows the MAPE of the respective history month. The averaged MAPE shows the MAPE averaged over the last 6 history months up to this month's (see the range also in the Averaging Period column).

Table 21.1: **MAPE and averaged MAPE results for selected history lengths**

History	MAPE	Averaged MAPE	Averaging Period
6	14.8%	15.15%	1-6
12	14.3%	14.53%	7-12
18	13.6%	13.79%	13-18
24	13.2%	13.35%	19-24
30	12.7%	12.81%	25-30
36	12.8%	12.71%	31-36
42	12.4%	12.46%	37-42
48	12.6%	12.54%	43-48

From the numbers, you can see that the average MAPE decreases by 2.61 percentage points from history 6 to history 48, representing a relative decrease of 17.2%.

Performing linear regression for the MAPE over the history results in a 0.105 decrease in MAPE per additionally available time history month for months 1 through 24 and in a 0.017 decrease for months 25 through 48.

Business case calculation

Relating these results to the reference company introduced in section 20.3 shows that for each additional available month in time history, an additional monthly profit of $6,350 can be gained.

Table 21.2 again shows the averaged MAPE values plus the financial impact for the fictional reference company Quality DataCom. The values of the business case are shown in absolute values relative to the period of 1-6 months. Thus, having a time period of 7–12 months on average identifies the additional achieved MAPE of $61,561 in the business case. The last column shows the marginal gain when moving from one availability range to the other.

Table 21.2: Average MAPE values plus the financial impact for the business case

Averaging Period	Average MAPE	US $ profit relative to period 1-6	Marginal Difference
1-6	15.15%	0	.
7-12	14.53%	61,561	61,561
13-18	13.79%	136,162	74,601
19-24	13.35%	179,766	43,603
25-30	12.81%	234,088	54,323
31-36	12.71%	244,062	9,974
37-42	12.46%	269,131	25,069
43-48	12.54%	260,708	-8,423

21.3 Optimal Length of the Available Time History

General

After the simulations have been performed to deliver the results shown in section 21.2, other results can be generated from these data. For each time series, forecasts on different lengths of time history have been trained. Based on these results, for each time series the optimal history length that best forecasts for future periods can be identified.

This results in a distribution of time histories that deliver the model with the smallest MAPE for the particular time history and shift.

Results

The results are represented in a bar chart in Figure 21.4. The detailed data are shown in Table 21.3.

Figure 21.4: Bar chart for the optimal length of data history

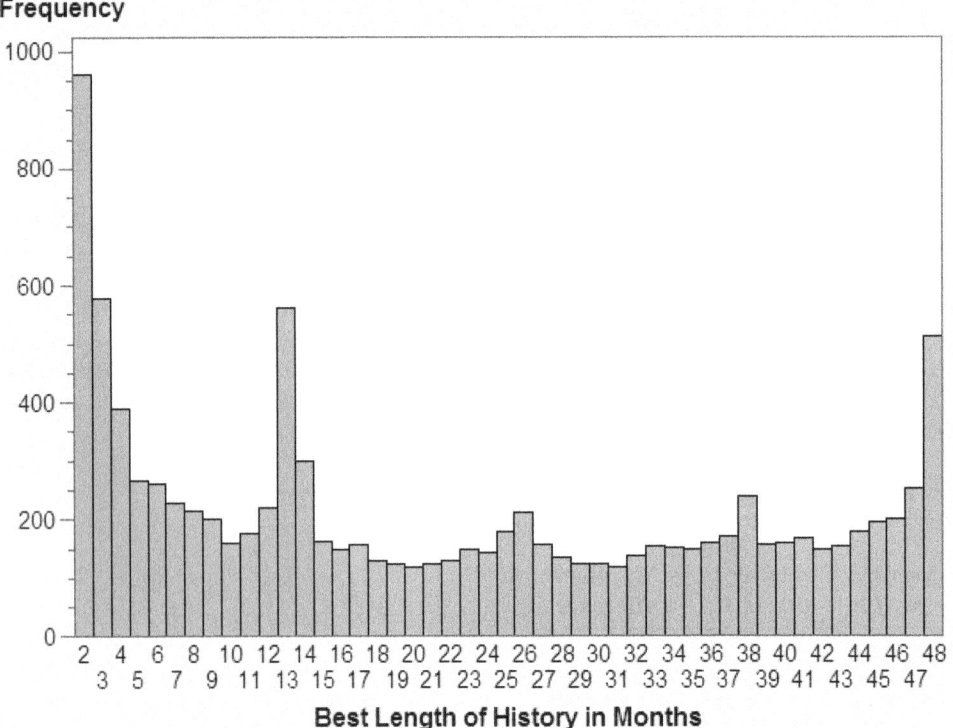

From the bar chart, you can see that

- The optimal history values range from 2 to 48, where each individual history length is represented with a reasonable frequency.
- There is no clear indication that the best forecasts are always performed with the longest available history.
- There is a (surprisingly) high accumulation of values in the range from 2 to 12. This means that there is a large proportion of about a third of the simulation cases (35.68%) where a time history of 12 months or less resulted in the best forecast.
- From the data in Table 21.3, you can see that in half (49.91%) of the simulation cases, the optimal length of the time history is less than 1.5 years. Around one-third of the simulation cases achieve their best results with a time history longer than 2.5 years.
- The spikes at the start of the second, third, and fourth years of history indicate that in some of the simulations the availability of the information for the next seasonal cycle is important for the model.

Table 21.3: Best length of available history grouped by half-years

Best History	Frequency	Percent	Cumulative Percent
Up to 0.5 year	2,455	23.97	23.97
0.5 to 1 year	1,200	11.71	35.68
1 to 1.5 years	1,458	14.23	49.91
1.5 to 2 years	790	7.71	57.62
2 to 2.5 years	929	9.07	66.69
2.5 to 3 years	872	8.51	75.20
3 to 3.5 years	1,044	10.19	85.40
3.5 to 4 years	1,496	14.60	100.00

Interpretation

The results closely mirror the background of the simulation data. For simulation cases where the optimal length is relatively short (up to 12 or 18 months), additional information from older time periods does not improve the forecast quality. Instead, it downgrades it.

From a business point of view, this is often the case with time series that are taken from retail sales, where the demand pattern shows a fluctuating behavior. Here, only the most recent months describe the most recent behavior. In this case, the more recent picture of the data in months 1 through 12 outperforms the additional benefit that would eventually be gained by repeating the seasonal cycle in the data.

21.4 Conclusion

General

This chapter has shown results based on simulation studies on how different available lengths of time series data affect the forecast quality. The results shown in Figure 21.3 correspond to intuition that, on average, a longer time history improves forecast quality. The graph additionally shows the decline in the marginal benefit of additional months if a longer time history is already available. Steps at the completion of another seasonal cycle can also be seen. They correspond to intuition as well.

In addition to these results, the optimal length of the time history was also analyzed. You can see that unlike the average case, where a downward trend in MAPE can be seen with increasing lengths of time history, the **individual optimal lengths** of the time history can also be very short (from 1 to 12 months). This is especially true for time series where the underlying patterns change frequently and quickly and, therefore, the more recent history is more important than the older history.

Data relevancy

This leads back to chapter 9, which discussed data relevancy. Having the relevant data available for the analysis is also a feature of data quality. Data relevancy, however, is in most cases understood as the need to have the relevant data available like having access to additional data sources. Often the data quality problem is "not having the relevant data."

The problem of data relevancy encountered here is somewhat different. With the length of the time history and the information from older periods, it might be that too much data and information are available, which is not as relevant for forecasting the future as more recent data.

It is, therefore, a task for the analyst to make sure that only the relevant segment is being used from the available data. While this is a typical task in predictive modeling where variable selection is performed, it is only rarely the case when the optimal length of the time history for time series forecasting is studied.

This chapter emphasized the need to carefully analyze the optimal length of the time history in time series forecasting. In order to facilitate this for individual analysis data, a part of the program that has been used to run these simulations is provided in the appendix D and can be downloaded.

Self-assessment of time series data

Appendix D shows a macro that can be used to perform the simulations that are shown in this chapter for time series. This macro can be used to analyze the optimal length of the history as shown in section 21.3. It also generates the line chart shown in section 21.2.

Refer to http://www.sascommunity.org/wiki/Data_Quality_for_Analytics for a download of this macro.

Chapter 22: Consequences of Random Disturbances in Time Series Data

22.1 Introduction	**273**
General	273
Simulation procedure	274
Types of random disturbances	274
22.2 Consequences of Random Missing Values	**274**
General	274
Insertion and replacement of missing values	274
Results for missing value imputation with PROC EXPAND	275
Results for missing value imputation with PROC TIMESERIES	276
22.3 Consequences of Random Zero Values	**276**
General	276
Results	277
22.4 Consequences of Random Biases	**277**
General	277
Standard deviation as basis for random bias	277
Code to insert a random bias	278
Results	278
22.5 Conclusion	**279**

22.1 Introduction

General

This chapter analyses the consequences of random disturbances in time series data. The values of the dependent variable in the time series itself have been altered in order to simulate a random disturbance.

The idea is to study the effect of the introduction of a random disturbance on the forecast quality. Results are generated that show the consequences on forecast quality if values that are randomly changed are used.

The results are presented in the form of line plots over different available history lengths, similar to previous chapters. This allows the analyst to see whether the effect has the same shape over the entire range of months. Thus, for example, you can see whether the effect increases or decreases with an increasing number of history months.

Simulation procedure

The simulation procedure is identical to that described in chapter 20. Compared to the simulations that are presented in chapter 21, the procedure here has been extended by inserting random disturbances into the data before they are forecast with PROC HPFENGINE.

Types of random disturbances

The following disturbances are introduced into the data randomly:

- Random missing values: 10%, 20%, 30%, 40%, and 50% of missing values are randomly inserted into the data and imputed with PROC EXPAND or PROC TIMESERIES before they are analyzed with PROC HPFENGINE.
- Random zero values: 10% of the values are randomly set to zero. These data are then directly entered into the analysis without any prior data correction measures.
- Random biases: for each time series, the amount of 0.5, 1, and 2 standard deviations of the respective time series is randomly added or subtracted from the data. The data then enter the forecasting process without any correction.

22.2 Consequences of Random Missing Values

General

This section presents simulations where values in the time series have been randomly set to missing. The results in terms of the influence on the mean absolute percentage error (MAPE) are presented here and a business case calculation is described.

Insertion and replacement of missing values

The following code has been used to randomly insert missing values into the data:

```
retain seed 0;
call ranuni(seed, rnd2);
if rnd2 < &PropMissing then qty = .;
```

The macro variable &PROPMISSING contains the proportion of observations that is set to a missing value. Note that because this random selection is unrestricted, the exact proportion of missing values for a particular time series may deviate slightly from the &PROPMISSING proportion.

Next PROC EXPAND or PROC TIMESERIES replaces the missing values. For more details on using both procedures to replace missing values in time series data, see chapter 11, sections 11.4 and 11.5.

The following code has been used to replace missing values with an imputation value using PROC EXPAND:

```
proc expand data = ts_data
            out  = ts_data from = month;
 convert qty / observed = total;
 by tsid;
run;
```

The following code has been used to replace missing values with the mean value of the time series using PROC TIMESERIES:

```
proc timeseries data = ts_data    out= ts_data;
   id monyear interval=month setmiss=mean;
     var qty;
     by tsid;
run;
```

Results for missing value imputation with PROC EXPAND

The results of the five scenarios with random missing values that were imputed with PROC EXPAND plus the baseline scenario with no missing values are shown in Figure 22.1 in the form of a line plot.

Figure 22.1: Line plot for different proportions of random missing values

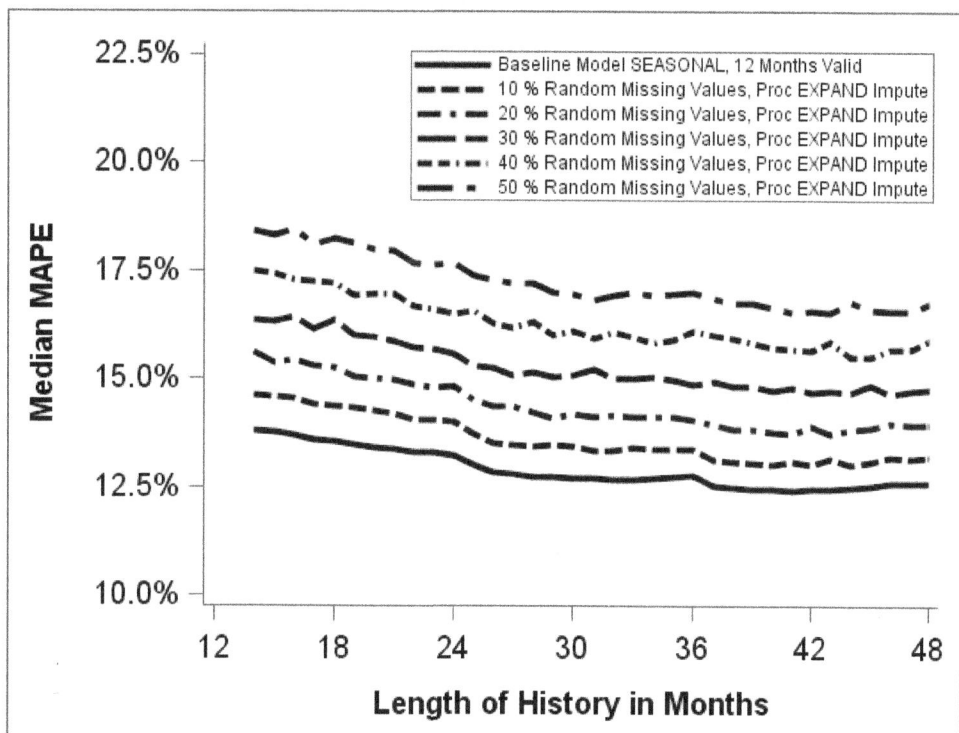

The shape and the course of the MAPE over the available history length are stable across the different scenarios. The result corresponds to intuition that with increasing proportions of random missing values, the MAPE increases.

The MAPE has been averaged over the lengths of available history for 14 to 48 months and is displayed in Table 22.1. The MAPE increases between 0.7% and 1% with each additional 10% of random missing values. Performing linear regression of the MAPE over missing percentages shows an increase of 0.087% per additional missing value percent. For the reference company, presented in section 20.3, this means an expected average loss of $8,700 per missing value percent.

Table 22.1: MAPE averaged over history lengths 14—48 for different proportions of random missing values

Scenario	MAPE	Loss in US $ compared to no missing values
No Random Missing Vaule	12.9%	0
10% Random Missing Values	13.6%	-70,184
20% Random Missing Values	14.4%	-148,149
30% Random Missing Values	15.3%	-237,834
40% Random Missing Values	16.3%	-337,203
50% Random Missing Values	17.2%	-433,340

The last column in this table shows the calculation of the business case in US dollars, comparing each random missing value percentage simulation with the no-missing value simulation. It can, for example, be seen that 10% of missing values for the reference company Quality DataCom means a loss of around $70,000 due to a reduced MAPE.

Results for missing value imputation with PROC TIMESERIES

Figure 22.2 shows the results of random missing values as defined previously for 10% of missing values. Here, however, the missing values are imputed with both methods: PROC EXPAND and PROC TIMESERIES.

You can see that using the mean as the imputation value with PROC TIMESERIES performs better than the imputation with PROC EXPAND, where a spline interpolation is used. This could be because the means represents a more stable estimate of the missing value in the case of strong variability in the data. Similar results can be obtained for 20% of random missing values.

Figure 22.2: Line plot comparing the MAPE for 10% random missing values imputed with PROC EXPAND and PROC TIMESERIES

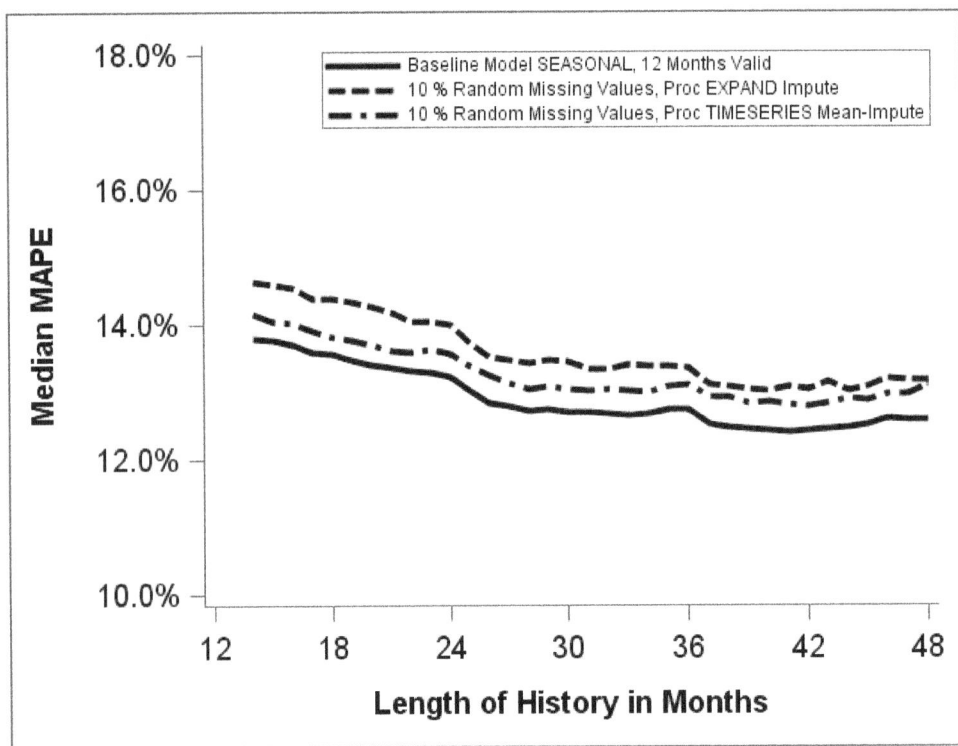

22.3 Consequences of Random Zero Values

General

This scenario shows the influence of random zero values in the data. Note that here random zero values are defined as observations whose values have been set to zero. This is different from the typical case of zero values in time series, where a zero value occurs when no observations have been recorded (for example, not a single unit has been sold) in a certain time period.

The zero values in the simulation case are not treated by any data cleaning steps but enter the forecasting procedure as they are. This creates a strong bias in the data because the true value is replaced by a zero value. Consequently, this has a strong effect on the forecast quality.

Results

The results of the scenario with 10% random zero values are shown in Figure 22.3. The MAPE ranges in the area of 30% compared to the situation of random missing values that are place by PROC EXPAND, which are significantly lower.

Figure 22.3: Line plot comparing the MAPE for 10% random missing values and 10% random zero values

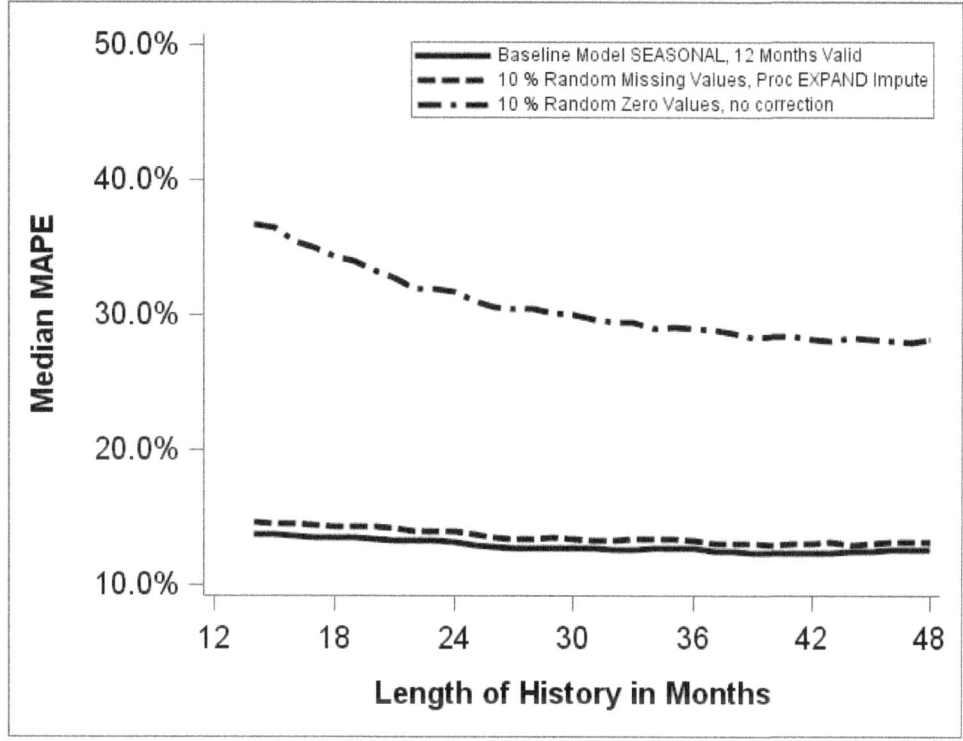

This section shows that zero values that do not represent zero quantities—but are errors in the data and are not handled in a data cleaning step—introduce a large bias into the data and significantly increase the forecast error. These zero values should be treated as missing values and imputed.

Thus, a detailed analysis of the reason for existing zero values is very important. If they truly represent zero quantities, they may be treated as the value zero; otherwise, they should be replaced using business logic.

22.4 Consequences of Random Biases

General

This section shows the effect of random biases on the quality of the forecast.

Standard deviation as basis for random bias

In order to simulate a random bias for the time series data, the standard deviation of the values of the time history has been calculated and randomly added or subtracted from the data. Depending on the scenario, the standard deviation has been multiplied by 0.5, 1, or 2 to represent different sizes of biased data.

Note that the calculated standard deviation in the time series may show a high value if the time series has a strong trend or strong seasonal component. These effects, however, are modeled by the time series model itself anyhow. It is, therefore, difficult to decide whether to use the raw standard deviation or the standard deviation that is corrected for trend and season. For the simulations that are shown here, the raw standard deviation has been used, which is, for simulation data, approximately twice as high as the corrected standard deviation.

Code to insert a random bias

The following code inserts a random bias into the data. The &NSTD macro contains the multiplicative factor, and STDQTY is the variable that contains the standard deviation of the values of the respective time series.

```
retain seed 0;
call ranuni(seed, rnd1);
qty = qty + (1-2*(rnd1>0.5))*&nstd*StdQty
```

The expression `(1-2*(rnd1>0.5))` controls the sign of the bias. This expression checks whether the bias is added or subtracted from the data. A random number RND1 is used to generate an expression `(rnd1>0.5)` that is 0 or 1 in 50% of the cases. The calculation 1-2*expression transforms this value into the set -1 and 1.

Results

The results are shown in Figure 22.4 and Table 22.2. You can see that the results for the different scenarios deviate quite remarkably.

Figure 22.4: Line plot comparing the MAPE for different amounts of random biases

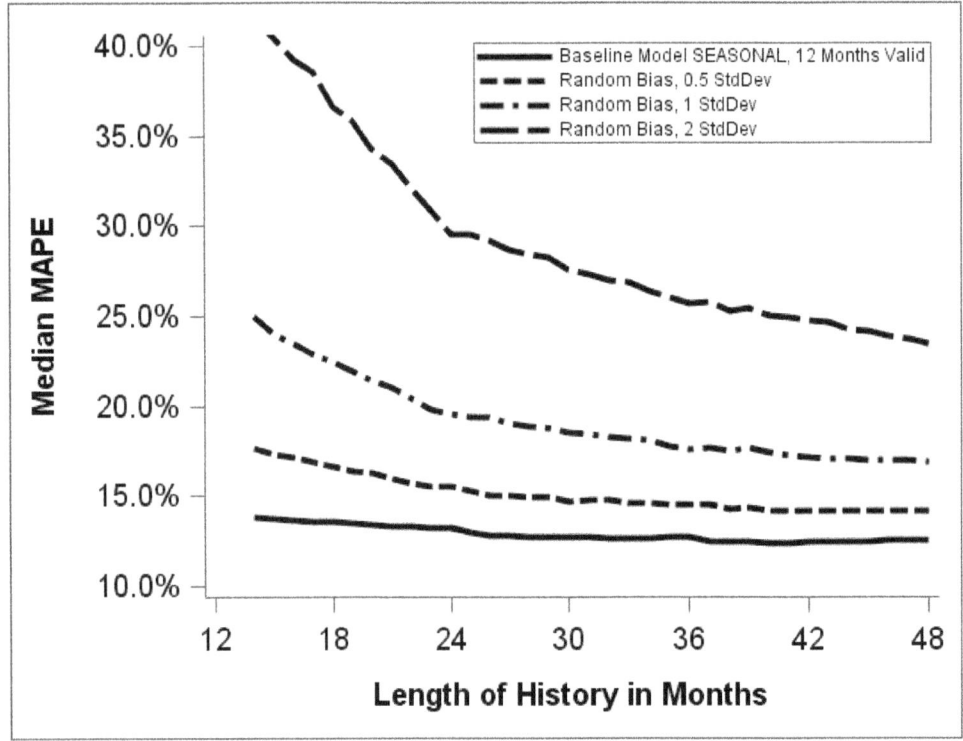

For example, randomly adding or subtracting half a standard deviation causes a relative increase of the MAPE between 12% (from 13.24% to 15.54%) and 18% (12.6% to 14.17%) for the scenarios with a history length of 24, 36, and 48 months.

In the case of a random bias of 1 standard deviation, the relative MAPE increase is between 34.3% and 47.7%. When introducing a random error of 2 standard deviations, the MAPE doubles on average.

Table 22.2: Table comparing the MAPE for different amounts of random biases for history lengths 24, 36, and 48 months

	Length of History in Months		
	24	36	48
No Bias	13.24%	12.76%	12.60%
Random Bias, 0.5 StdDev	15.54%	14.53%	14.17%
Random Bias, 1 StdDev	19.55%	17.62%	16.92%
Random Bias, 2 StdDev	29.56%	25.65%	23.48%

Interpretation

Random biases, represented by the addition or subtraction of standard deviations, significantly increase the MAPE for the time series. When discussing standard deviations in time series, you must decide whether the uncorrected or the corrected standard deviation for seasons and trends is used.

It has also been shown that in the case of a random bias, the available length of the time history gains more importance. For time histories with less than 24 months, the effect of the random bias is much stronger.

The effects of random errors in the time series forecasting scenarios are much stronger than in predictive modeling. This is because in predictive modeling only the input variables have been biased. Here, the dependent variable is biased.

22.5 Conclusion

This chapter has shown the influence of random disturbances in the time series data.

The influence of random missing values that are imputed with different methods has been discussed. Note that the scenarios discussed here assume that the missing values are replaced. Missing values that are not replaced in time series forecasting are a problem for many methods because no forecasting model can be fitted to the data. Simply removing the time periods with missing values from the data also causes a problem because the time series loses the property of contiguity (see also chapter 11).

Different scenarios for biasing the data have also been discussed. Results have been shown for the scenario of replacing values with zero values, which causes a significant decrease in model quality. This chapter has also discussed the importance of differentiating whether a zero value is simply a wrong value or whether it should be considered as a period with (true) zero quantities.

Scenarios for biasing the data with the addition or subtraction of factors of standard deviations have also been discussed. Here you need to define which variations will be included in the standard deviation.

Chapter 23: Consequences of Systematic Disturbances in Time Series Data

23.1 Introduction	**281**
General	281
Simulation procedure	282
Systematically selecting observations from the time series	282
Types of systematic disturbances	282
23.2 Coding Systematic Disturbances in Time Series Data	**283**
Precalculation of values	283
Systematic selection based on the decile group	283
Systematic selection based on the calendar month	283
23.3 Results for the Effect of Systematic Disturbances	**284**
General	284
Systematic disturbances inserted for the top 10% of time series values	284
Systematic disturbances inserted for three consecutive calendar months	286
23.4 Interpretation	**287**
23.5 General Conclusions of the Simulations for Time Series Forecasting Shown in Chapters 21–23	**288**
Increasing length of data history decreases forecast error	288
The marginal effect of additional forecast months decreases	288
For many time series, a short time history causes better forecast accuracy	288
Long time histories can solve data quality problems to some extent	288

23.1 Introduction

General

This chapter shows results of simulations for time series data where the values of the dependent variable have been altered in a systematic way. It studies the effect of systematic disturbances in the data on the forecast quality.

The results are presented in the form of line charts over the different available lengths of historic data, as shown in the previous two chapters. The results show the effect of the systematic disturbance not only at a single point in the available time history; they also illustrate the shape of the line of the forecast quality over time. This, for example, allows you to see how the effect or the difference to the un-disturbed baseline scenario changes with an increasing number of history months.

Simulation procedure

The simulation procedure is identical to that described in chapter 20. Compared to the simulations that are presented in chapter 21, the procedure used here has been extended by inserting random disturbances into the data before the data are forecast with PROC HPFENGINE.

Systematically selecting observations from the time series

To insert systematic disturbances for simulation purposes into the data, a systematic has to be defined that determines the selection of observations in the time series. For the simulations that are presented here, this "systematic" has been defined in two different ways:

- **Systematically selecting observations (time periods) based on the percentile of their value**. For each time series, the observations have been organized into ordered groups of 10%, each according to its value of the time series variable. To insert systematic biases, those observations that are in the top 10%, top 20%, and so on of the time series have been selected.
- **Systematically selecting observations based on the calendar month**. The calendar month has been used to select observations that are consecutive to each other and that are in the same quarter of the year. This procedure has been iterated four times to select each quarter of the year in one of the runs.

Types of systematic disturbances

The following disturbances have been introduced systematically into the data:

- Systematic missing values
  - Business example: People with a salary in a certain salary group do not report their salary.
  - Some 10% of systematic missing values have been inserted for the time series values that are in the top decile.
  - Three consecutive calendar months have been selected and the missing values have been inserted for those periods.
  - The systematic missing values have been imputed using PROC EXPAND before using the data for forecasting in PROC HPFENGINE.
- Systematically set to the mean value
  - Business example: People with a salary in a certain salary group report an average salary instead of their true salary.
  - For the observations in the top decile, the values have been set to the mean value of the time series.
  - Three consecutive calendar months have been selected and the value has been set to the mean value.
- Systematically inverting the value
  - Business example: People with a salary in a certain salary group report the opposite salary (large salary earners report low salaries, low salary earners report high salaries).
  - Here inverting means that the value is mirrored at the mean value. A value much larger than the mean becomes a value much smaller than the mean. A value close to the mean will, after inverting, also be close the mean, only at the other side of the mean. In the case of standardized data where the mean equals 0, this indicates that the sign of the value is changed.
  - For the observations in the top decile, the values have been set to the inverted value.
  - Three consecutive calendar months have been selected and the value has been set to the inverted value.

23.2 Coding Systematic Disturbances in Time Series Data

Precalculation of values

The following code groups the observations of a time series into deciles according to the time series values:

```
PROC RANK DATA=TS_DATA_groups=10
          OUT=TS_DATA;
 by tsid;
 var qty;
 ranks qty_rnk;
RUN;
```

The following statement precalculates the CalendarMonthID into a separate variable:

```
CalMonth=MONTH(monyear);
```

Systematic selection based on the decile group

The following statements change the time series values to missing, the mean, or the inverted value based on a systematic selection on the decile group of the time series value:

```
if (1-&PropSystMissing) *10 <= qty_rnk then qty=.;
if (1-&PropSystMean)    *10 <= qty_rnk then qty=meanqty;
if (1-&PropSystInvert)  *10 <= qty_rnk then qty=qty-2*(qty-meanqty);
```

Note that the macro variables &PROPSYSTMISSING, &PROPSYSTMEAN, and &PROPSYSTINVERT contain the percentage (for example, 0.1) of the observations of the time series that will be disturbed. The QTY_RANK variable contains the decile grouping of the time series values. Here, the values range from 0 for the smallest group to 9 for the largest group.

Depending on the desired proportion of missing values, the periods of one or more clusters are set to missing. Note that the calculation (&PropSystMissing*10)-1 is used to relate the cluster numbers that range from 0 to 9 to the appropriate value between 0 and 1.

Note that the expression -2*(qty-meanqty) adds or subtracts the amount of the difference between the value and the mean.

Systematic selection based on the calendar month

The following statements change the time series values to missing, the mean, or the inverted value based on a systematic selection on the calendar month:

```
if CalMonth in &NCalMonSystMissing then qty=.;
if CalMonth in &NCalMonSystMean    then qty=meanqty;
if CalMonth in &NCalMonSystInvert  then qty=qty-2*(qty-meanqty);
```

Note that the macro variables &NCALMONSYSTMISSING, &NCALMONSYSTMEAN, and &NCALMONSYSTINVERT contain a list of calendar months, for example (1, 2, 3).

The following code replaces missing values with an imputation value using PROC EXPAND:

```
PROC EXPAND DATA = TS_data
            OUT  = TS_data from = month;
 CONVERT qty / OBSERVED = total;
 BY tsid;
RUN;
```

23.3 Results for the Effect of Systematic Disturbances

General

This section presents the simulations and the simulation results, where the data have been systematically disturbed in one of the ways described earlier. The results in terms of the influence on the mean absolute percentage error (MAPE) are presented here.

Systematic disturbances inserted for the top 10% of time series values

Figure 23.1 shows results from simulations where the top 10% of the time series values have been systematically set to missing, to the mean, or to the inverted value.

Figure 23.1: Line plot for different systematic disturbances for the top 10% of time series values

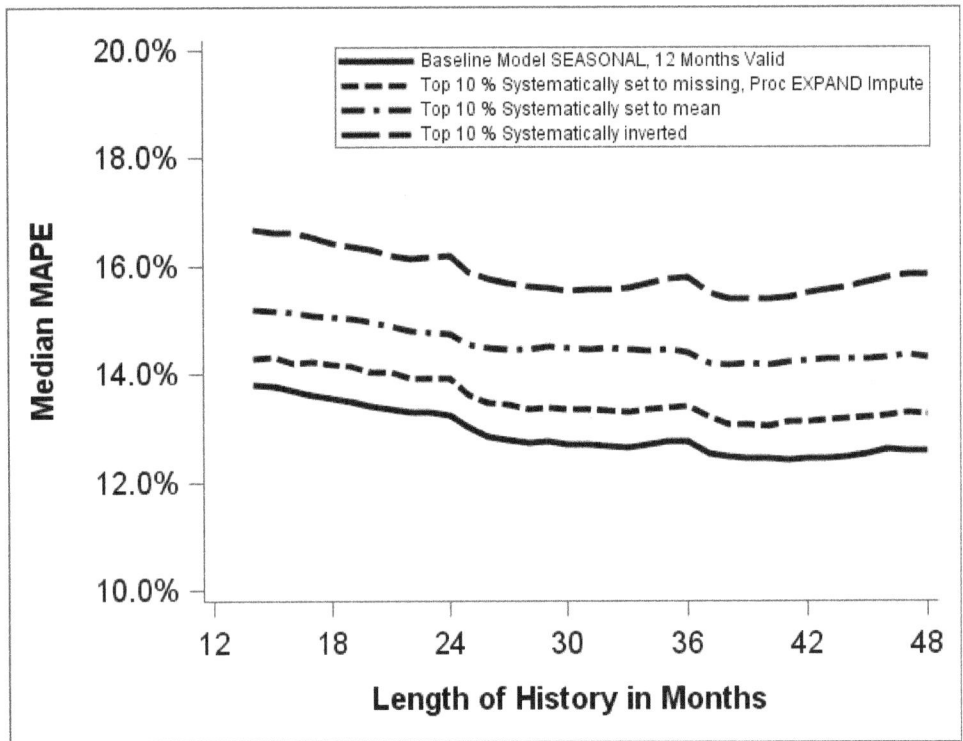

From the chart here and from Table 23.1, you can see that the order of the scenarios in terms of an increase in the MAPE are "Set to Missing" with 13.9%, "Set to Mean" with 14.8% and "Set to inverted value" with 16.2% (for the 24-month history length).

Table 23.1: Comparing the MAPE for different scenarios of systematic disturbances for history lengths 24, 36, and 48 months

	Length of History in Months		
	24	36	48
Baseline Model SEASONAL, 12 Months Valid	13.24%	12.76%	12.60%
Top 10% Systematically set to missing	13.93%	13.40%	13.27%
Top 10% Systematically set to mean	14.76%	14.42%	14.33%
Top 10% Systematically inverted	16.19%	15.79%	15.85%
10 % Random Missing Values	14.00%	13.37%	13.19%
Top 20% Systematically set to missing	15.24%	14.67%	14.74%
Top 20% Systematically set to mean	15.27%	14.97%	14.95%
Top 20% Systematically inverted	18.59%	18.17%	18.34%

The fact that the inversion of the values causes a larger increase in MAPE than setting it to the mean value is also an intuitive finding.

It can also be seen from the table and from the line plot in Figure 23.2 that there is almost no difference in MAPE between the systematic and the random missing scenario. This can be explained by the fact that even as the values are systematically selected and set to missing, this selection sets the largest values of the time series to missing and imputes afterward.

This can be seen as the insertion of a systematic bias on the one hand but also as a data cleaning process on the other hand because it removes all potential large outliers from the data and replaces them with the mean.

Figure 23.2: Line plot for different systematic disturbances for the top 10% of time series values, including the scenario with 10% of random missing values

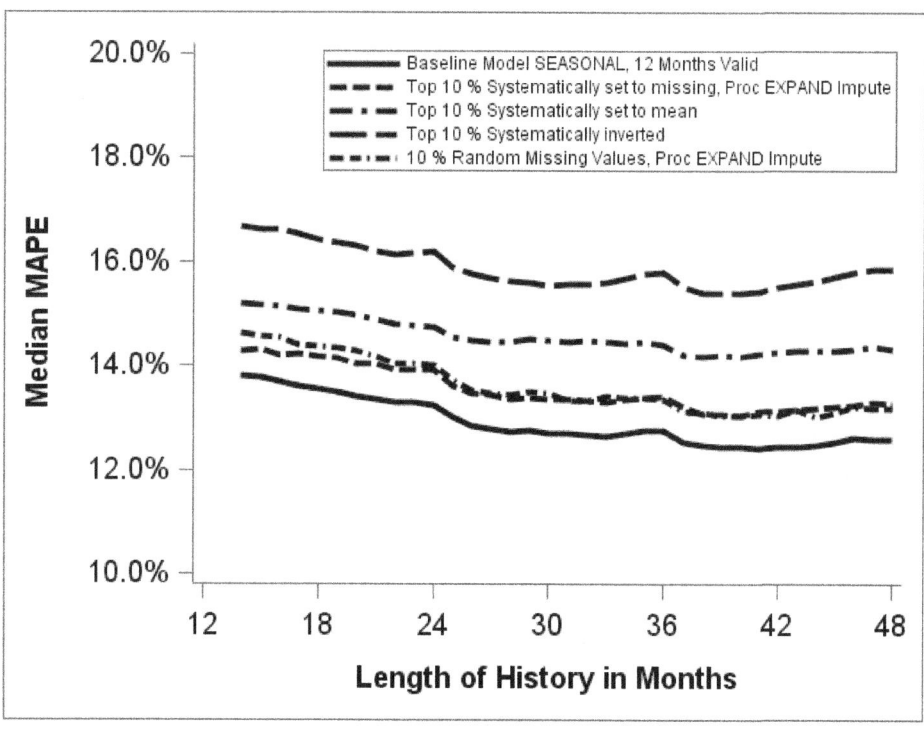

Systematic disturbances inserted for three consecutive calendar months

Figure 23.3 shows results from simulations where the time series values of three consecutive calendar months have been systematically set to missing, to the mean, or to the inverted value.

Figure 23.3: Line plot for different systematic disturbances for three consecutive calendar months

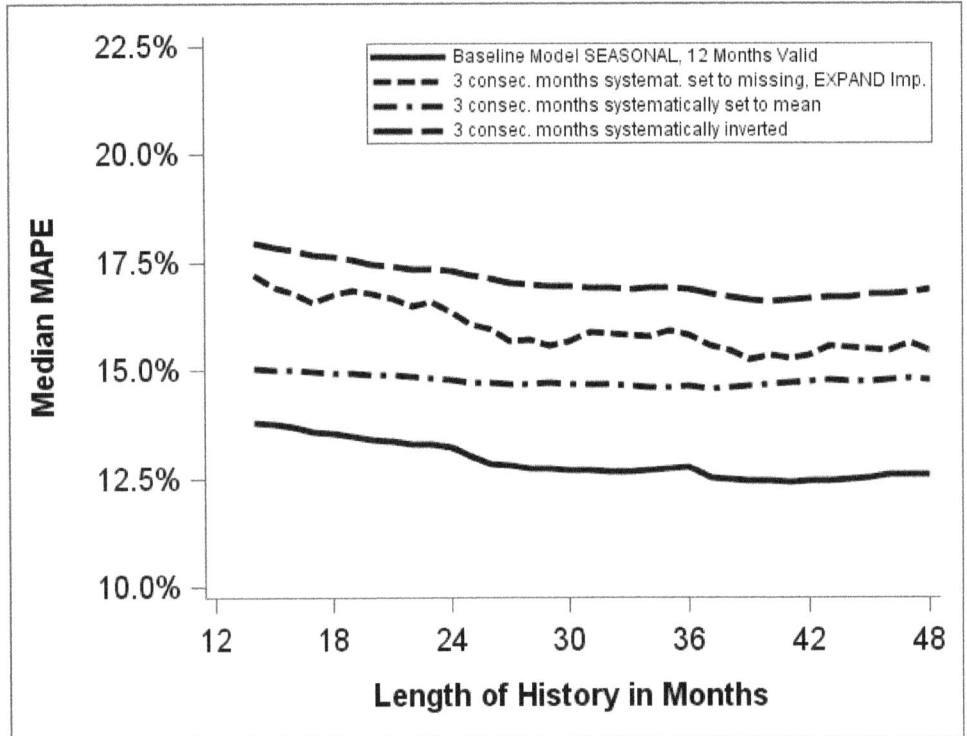

From Figure 23.3 and Table 23.2, you can see that the order of the scenarios in terms of an increase in the MAPE are "Set to Mean" with 14.8%, "Set to Missing" with 16.4%, and "Set to inverted value" with 17.3% (for the 24-month history length).

Table 23.2: Comparing the MAPE for different scenarios of systematic disturbances for history lengths 24, 36, and 48 months

	Length of History in Months		
	24	36	48
Baseline Model SEASONAL, 12 Months Valid	13.24%	12.76%	12.60%
3 consec. months systemat. set to missing	16.37%	15.83%	15.48%
3 consec. months systematically set to mean	14.79%	14.64%	14.81%
3 consec. months systematically inverted	17.33%	16.91%	16.88%
20% Random Missing Values	14.82%	14.05%	13.96%

The fact that the inversion of the values causes a larger increase in the MAPE than the setting to the mean value it intuitively understandable.

However, unlike the scenarios where the top 10% of the values have been biased, the results for biasing three consecutive months show a different behavior.

- The set to missing scenario has a higher MAPE than the set to mean scenario.
  - This can be due to the fact that PROC EXPAND uses a spline interpolation between the existing values. If three consecutive values are missing, such an imputation heavily depends on the first and last observations before and after this period. This may, however, be a less effective imputation that just using the average value.
- As you can see from the line plot in Figure 23.4, there is now a difference in the MAPE between the systematic and the random missing scenario.
  - This can be explained by the fact that the systematic selection here does not systematically filter the largest values of the time series and, therefore, is not some kind of data cleaning procedure for outliers.

Figure 23.4: Line plot for different systematic disturbances for three consecutive calendar months, including the scenario with 20% of random missing values

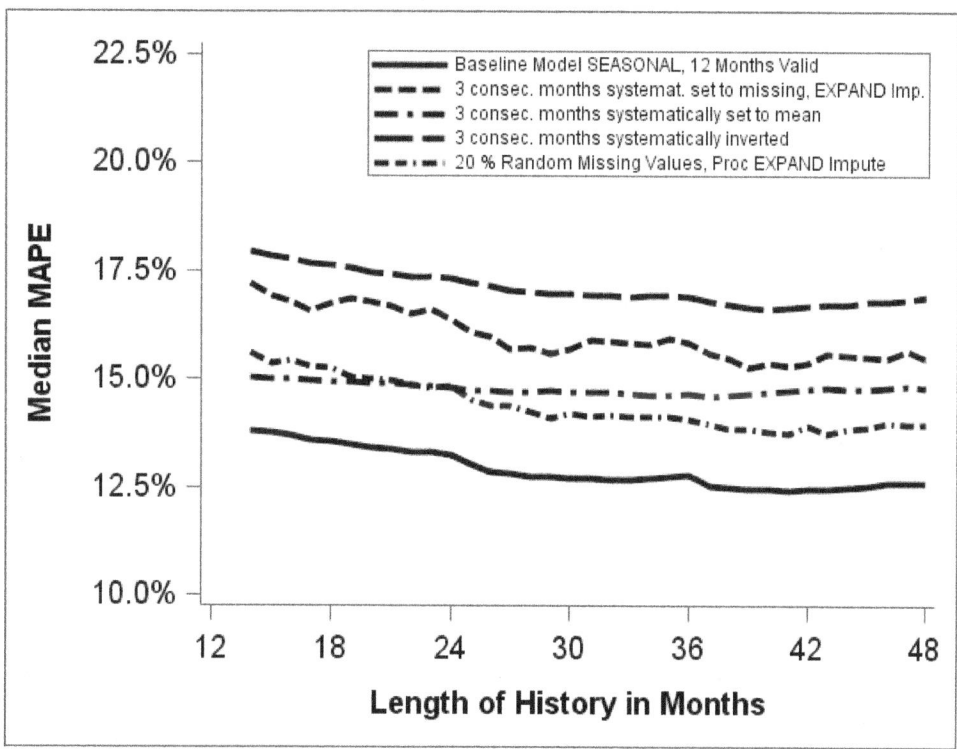

23.4 Interpretation

This chapter has shown the influence of systematic disturbances in time series data.

Two examples for the systematic selection of observations for insertion of data disturbances have been shown: the selection of the top percentage of the time series values and the selection of consecutive calendar months.

Data disturbances like the insertion of mean values and missing values and the inversion of a value have been performed on these systematically selected observations. The respective results have been shown and discussed.

23.5 General Conclusions of the Simulations for Time Series Forecasting Shown in Chapters 21–23

From a data quality point of view, the following conclusions can be drawn from the results presented in chapters 21 through 23.

Increasing length of data history decreases forecast error

The forecast error decreases with an increasing number of months that are available for the data history. This is a result that would have been expected by intuition. The decrease of the MAPE of history months together with the calculations of the business case allows you to quantify the impact of the increased data quality by additional history months on the forecast quality and the financial benefit.

The macro &TS_HISTORY_CHECK that is presented in the appendix allows you to calculate these numbers for other data.

The marginal effect of additional forecast months decreases

While additional data periods decrease forecast error, the effect of additional periods is stronger in short histories than in already long histories.

The results have shown that, on average, the effect of an additional month in histories of up to 24 months is much higher (up to 10 times) than the effect of an additional month in time histories that already have 24 or more months.

In general, it is advisable to collect long data histories because that reduces forecast error. However, if you are confronted with the decision to invest data collection and data quality improvement effort in a longer time history or in a more trustworthy but shorter history, it has been shown that the second option offers better forecast accuracy. Thus, it is advisable to collect and quality check only the last 36 months of data history, but make sure that the data quality level is very high (instead of providing 48 months or more of data with inferior data quality).

For many time series, a short time history causes better forecast accuracy

By analyzing for each time series the optimal length of time history that gives the lowest forecast error for future months, it has been shown that around 35% of the time series achieve their best forecasts with up to 12 months of available time history. Checking the course of these time series on an individual basis, it turns out that the time series with a strong variation in historic values and unstable seasonal patterns are those that have better predictions with short data histories.

Long time histories can solve data quality problems to some extent

In the scenarios with random and systematic data disturbances, it has been shown that there is a downward trend in the average MAPE with increasing time history. Here, from a practical point of view, it becomes visible that on average longer histories allow the analyst to compensate for the increased MAPE values due to data disturbances.

Appendix A: Macro Code

A.1 Introduction **289**
A.2 Code for Macro %COUNT_MV **289**
    General 289
    Macro code 290
A.3 Code for Macro %MV_PROFILING **290**
    General 290
    Parameters 290
    Macro code 291
    Comments 294
A.4 Code for Macro %PROFILE_TS_MV **294**
    General 294
    Parameters 295
    Macro code 295
A.5 Code for Macro %CHECK_TIMEID **298**
    General 298
    Parameters 298
    Macro code 299
    Comments 299

A.1 Introduction

This chapter shows the code of the macros %COUNT_MV, %MV_PROFILING, %PROFILE_TS_MV, and %CHECK_TIMEID, which were presented and discussed in chapters 10 and 11.

A.2 Code for Macro %COUNT_MV

General

The code of macro %COUNT_MV, which was introduced in section 10.2, is shown here.

The following parameters can be specified with the macro:

DATA
    Name of the input data set.

VARS
    List of variables in the data set to be analyzed. Variables should be separated by a blank.

Macro code

```
%MACRO COUNT_MV(data=,vars=);
*** LOAD THE NUMBER OF ITENMS IN &VARS INTO MACRO VARIABLE NVARS;
%LET C=1;
%DO %WHILE(%SCAN(&vars,&c) NE);
    %LET C=%EVAL(&c+1);
%END;
%LET NVARS=%EVAL(&C-1);
*** CALCULATE THE NUMBER OF OBSERVATIONS IN THE DATASET;
DATA _NULL_;
  CALL SYMPUT('N0',STRIP(PUT(nobs,8.)));
  STOP;
  SET &data NOBS=NOBS;
RUN;
PROC DELETE DATA = work._CharMissing_;RUN;
%DO I = 1 %TO &NVARS;
 PROC FREQ DATA = &data(KEEP =%SCAN(&VARS,&I))  NOPRINT;
    TABLE %SCAN(&vars,&I) / MISSING OUT = DATA_%SCAN(&vars,&I)(WHERE
=(%SCAN(&vars,&I) IS MISSING));
 RUN;
 DATA DATA_%SCAN(&vars,&i);
 FORMAT VAR $32.;
   SET data_%SCAN(&vars,&i);
   VAR = "%SCAN(&vars,&i)";
   DROP %SCAN(&vars,&i) PERCENT;
 RUN;
 PROC APPEND BASE = work._CharMissing_ DATA = DATA_%SCAN(&vars,&i) FORCE;
 RUN;
 PROC DELETE DATA=DATA_%SCAN(&vars,&i);RUN;
%END;
PROC PRINT DATA = _CharMissing_;
RUN;
TITLE;TITLE2;

%MEND;
```

A.3 Code for Macro %MV_PROFILING

General

The code of macro %MV_PROFILING, which was introduced in section 10.3, is shown here.

Parameters

The following parameters can be specified with the macro:

DATA
    Name of the input dataset.

VARS
    List of variables in the dataset to be analyzed. Variables should be separated by a blank. If _ALL_ is specified, all variables of the dataset are used. Default value is _ALL. A "wild card" notation with a colon ":" can also be used. For instance, DEMO: causes all variables starting with "DEMO" to be used.

ORDER
    Only relevant when VARS=_ALL_ is specified. Possible values are POS or ALPHA. Default value is ALPHA. ALPHA means that the variables of the dataset are used in alphabetical order for the MV_VALUE_CHAIN. POS means that the variables are used in the same order like in the dataset for MV_VALUE_CHAIN.

ODS
: Only when ODS = YES the Tile-chart, the Varclus-tree-plot, and the principal component analysis are created. Default = YES.

VARCLUS
: Only when VARCLUS = YES, the Varclus-tree-plot is created. Default = YES.

PRINCOMP
: Only when PRINCOMP=YES, the principal component analysis is created. Default = YES.

NCOMP
: Defines the number of principal components that are used for the plots of components. NCOMP = 2 creates one plot for the first and second component. COMP = 3 creates 3 plots; one for the 1st and 2nd, one for the 1st and 3rd, and one for the 2nd and 3rd component, and so forth.

SAMPLE
: Specifies a sample proportion to allow running the missing value profiling on a sample rather than the entire data. Default = 1 (100 %), no sampling. Values from 0 to 1 are valid.

SEED
: Seed value that shall be used for sampling. Default = 18419.

Macro code

The code for macro %MV_PROFILING is the following:

```
%macro MV_Profiling (data=,vars=_ALL_,ODS=YES,varclus=YES,
                     princomp=YES,ncomp=2,sample=1,
                     seed=18419,order=ALPHA);
%* Analyses Frequencies and the structure of missing values in a dataset;
%* Dr. Gerhard Svolba - 2009-01-10;

title "Missing Value Profiling for data = &data";

%*** 1. Prepare a list of variables, if necessary;
%if %upcase(&vars) = _ALL_ or %INDEX(&vars,:) ne 0 %then %do;

%* Retrieve Variables from PROC Contents;
proc contents data = &data
              out = work._Vars_content_(keep = name npos)
                    noprint;
run;

%if %INDEX(&vars,:) ne 0 %then %do;
 data work._Vars_content_;
   set work._Vars_content_;
   if upcase(name)=:"%upcase(%scan(&vars,1,:))" then output;
 run;

%end;

%if %upcase(&order) = POS %then %do;
   proc sort data = work._Vars_content_;
     by npos;
   run;
%end;

%*Initialize vars;
%let vars = ;
```

```
proc sql noprint;
 select name
 into :vars separated by " "
 from _vars_content_ ;
quit;

%put vars = &vars;

%end; %* END: Prepare a list of variables, if necessary;



%*** 2. Calculate number of variables in the macro list;
%LET c=1;
%DO %WHILE(%SCAN(&vars,&c) NE );
       %LET c=%EVAL(&c+1);
%END;
%LET nvars=%EVAL(&c-1);
%put nvars = &nvars;

%put nvars = &nvars;
%if &nvars ne 0 %then %do;

%*** 3. Create Flags for Missing Values;

data _profile_mv_tmp_;
 set &data(keep = &vars);
 format MV_PROFILE_CHAIN $%eval(&nvars.+8).;   %* Format length = Number of
Variables;
 %if &sample ne 1 %then %do;
      if uniform(&seed) gt &sample then delete;
 %end;
%* Initialize List;
MV_PROFILE_CHAIN='';
%let vars_mv=;

%* Create derived variables and list;
%do i = 1 %to &nvars;
  %SCAN(&vars,&i)_MV = missing(%SCAN(&vars,&i));
  %let vars_mv = &vars_mv %SCAN(&vars,&i)_MV;
  MV_PROFILE_CHAIN=cats(MV_PROFILE_CHAIN,%SCAN(&vars,&i)_MV);
%end;
N_MV = sum(of &vars_mv);
MV_PROFILE_CHAIN=catx('_',MV_PROFILE_CHAIN,put(n_mv,best.));
run;

proc means data = _profile_mv_tmp_  noprint;
 var &vars_mv;
 output out = _profile_mv_sum_ sum=;
run;

proc transpose data = _profile_mv_sum_(drop = _type_) out = _profile_mv_sum_tp;
run;

data _profile_mv_sum_tp;
 set _profile_mv_sum_tp;
 retain nobs;
 if _n_ = 1 then nobs = col1;
 else do;
       Variable = substr(_name_,1,length(_name_)-3);
     format Missing_Rel percent8.2;
        Missing_Rel = col1/nobs;
   end;
```

```
  rename col1 = Missing_Abs;
  if _n_=1 then delete;
run;

*** Create Lookup for MV_Profiles;
proc sql;
  title2 Lookup for variable names and MV_PROFILE patterns;
  select upcase(Variable) as Variable,
           %do i = 1 %to &nvars;put((variable="%scan(&vars,&i)"),1.)||%end;''
as MV_PROFILE
  from _profile_mv_sum_tp
  order by 2 desc;
quit;

%*** 4. Create Frequencies for Missing Values per Variable and Missing Value
Profile;
%* Missing Value Profile;

proc freq data = _profile_mv_tmp_ order = freq;
  title2 Distribution of MV_PROFILE_CHAIN;
  table MV_PROFILE_CHAIN/out = _profile_mv_freq_;
run;

%* Display patterns of missing values graphically with a TILE Chart;
%if %upcase(&ods) = YES %THEN %DO;
proc gtile data=_profile_mv_freq_;
    title2 Tile-Chart for the distribution of MV_PROFILE_CHAIN;
    tile count tileby = (MV_PROFILE_CHAIN);
run;
quit;
%end; %* ODS TILE CHART YES/NO;

%put vars = &vars;
%put vars_mv = &vars_mv;

data _profile_mv_tmp_vc;
set _profile_mv_tmp_ (drop = &vars);
 where scan(MV_PROFILE_CHAIN,2,'_') ne '0';
 %do i = 1 %to &nvars;
     rename %SCAN(&vars_mv,&i) = %SCAN(&vars,&i);
        %put i = &i;
 %end;
run;

proc freq data = _profile_mv_tmp_;
  title2 Distribution of Number of Missing Values per Observation;
  table n_mv;
run;

%* Frequencies for Missing Values per Variable;

proc sql;
  title2 Variable list ordered by number of missing values;
  select upcase(Variable) as Variable,
         Missing_Abs,
             Missing_Rel,
             %do i = 1 %to &nvars;put((variable="%scan(&vars,&i)"),1.)||%end;''
as MV_PROFILE
  from _profile_mv_sum_tp
  order by 2 desc;
quit;

%if %upcase(&varclus) = YES and %upcase(&ods) = YES %then %do;
proc varclus data=_profile_mv_tmp_vc centroid outtree = _mv_profile_tree_
noprint;
```

```
    var &vars;
run;

axis1 order=(0.5 to 1 by 0.1);
axis2 label=none;
proc tree data = _mv_profile_tree_ horizontal haxis=axis1 vaxis=axis2;
 title2 Missing Value based on Clustering of Variables;
      height _propor_;
      id _label_;
run;

%end; %* Varclas YES/NO;

%if %upcase(&princomp) = YES and %upcase(&ods)=YES %then %do;
ods select  eigenvalues patternplot;
ods graphics on;
PROC PRINCOMP DATA = _profile_mv_tmp_vc  out=pc_out outstat = pc_out_stat
        PREFIX='PRIN'n
        SINGULAR=1E-08
        VARDEF=DF
        plots(ncomp=&ncomp) = pattern
   ;
 title2 Principal component analysis based on missing values;
 var &vars;
 where scan(MV_PROFILE_CHAIN,2,'_') ne '0';
RUN;
ods select all;
ods graphics off;
%end; %* Princompo YES/NO;

%end; %* IF elements in VARS;

title2;title;
%mend;
```

Comments

Note the following from the code:

- A macro variable &VARS_MV is created to hold the variable names of the indicator variables for each variable in &VARS. For each variable in &VARS a new variable <variable name>_MV is created in dataset PROFILE_MV_TMP. For instance, AGE_MV for variable AGE. The length of the variable name must not exceed 29 characters.
- Variable MV_PROFILE_CHAIN is created by concatenating the _MV variables.
- Note the efficient programming of the creation of the MV_PROFILE LOOKUP table where the pattern for each variable is created by concatenating the results of 0s, and 1s at the ith digit if the ith variable in the &VARS list matches the actual variable.

```
select %do i = 1 %to &nvars;put(((variable="%scan(&vars,&i)"),1.)||%end;'' as
MV_PROFILE
```

- PROC VARCLUS and PROC PRINCOMP use the indicator values stored in the _MV variables. For better readability in the graphs, however, the variables are renamed to the original name.

A.4 Code for Macro %PROFILE_TS_MV

General

The code of macro %PROFILE_TS_MV, which was introduced in section 11.2, is shown here.

Parameters

The following parameters can be specified with the macro:

DATA
    Name of the input dataset (mandatory).

ID
    Name of the time series ID variable (mandatory).

CROSS
    List of variables for the cross-sectional dimensions (optional). The values of these variables are concatenated to a string and then concatenated to the ID variable in order to allow the analysis of the same time series ID for different subgroups.

DATE
    Name of the time-id variable (mandatory).

VALUE
    Name of the variable that holds the values of the time series (mandatory).

MV
    List of values that shall be considered as missing values. Values need to be specified in brackets, separated by a comma. Examples are mv=(.), mv=(.,9). The default value is (.). Note that this list is used with an IN-operator in the DATA step.

ZV
    List of values that shall be considered as zero values. Values need to be specified in brackets, separated by a comma. Examples are mv=(0), mv=(0,-1). The default value is (0). Note that this list is used with an IN-operator in the DATA step.

PLOT
    A profile plot is only produced if PLOT is set to YES. Default = YES. This option should be turned to NO if your data contains more than 500 time series in order to avoid excessive runtimes.

NMAX_TS
    Depending on the number of time series in the data, this option controls whether a profile plot shall be produced or not. The default number is 100. Thus, if your data contains more than 100 time series a profile plot is not produced and you need to set higher values. In this case, the following message in the log file is produced:

```
Number of time series =     1026, is higher then nmax_ts value(= 100).
No plot has been created. Reset parameter NMAX_TS to a value of at least 1026
```

W
    Define the thickness of the lines used in the profile plot to represent one time series.

Macro code

The code for macro %PROFILE_TS_MV is the following:

```
%macro Profile_TS_MV (data=,id=,cross=,date=,value=,mv=(.),
plot=YES,w=1,nmax_ts=100);

proc sql noprint;
 select count(distinct &date) into :maxlength from &data;
 select count(distinct &id) into :nseries from &data;
quit;
```

```sas
proc sort data = &data(rename = (&id = _id_)) out = sorted_input_data;
 by _id_;
run;

** if cross sectional dimension are specified, create a concatenated _ID_
variable ;
%if &cross ne %then %do;

%LET c=1;
%DO %WHILE(%SCAN(&cross,&c) NE);
        %LET c=%EVAL(&c+1);
%END;
%LET ncrossvars=%EVAL(&c-1);

data sorted_input_data;
   set sorted_input_data;
   %do i= 1 %to &ncrossvars;
      _id_ = catx('_',_id_,%scan(&cross,&i));
    %end;
   drop &cross;
   run;
%end;

** Resort may be necessary as '_' is concatenated to a string of variable
length;
proc sort data =  sorted_input_data;
 by _id_;
run;

data MV_PROFILE_TS(keep = _id_ TS_Profile_Chain TS_Profile_Chain_Unique n
nmiss)
      MV_PROFILE_TS_PLOT(keep = _id_ &date idx2);
 set sorted_input_data;
 by _id_;
 format TS_Profile_Chain TS_Profile_Chain_Unique $%eval(&maxlength.+9).;
 retain TS_Profile_Chain TS_Profile_Chain_unique;

   *** Init;
   if first._id_ then do;
               TS_Profile_Chain = '';
               TS_Profile_Chain_unique='';
                     N=1;NMiss=0;
                          end;

   *** Missing;
   n+1;
   _ActualMV_ = put(1-(&value in &mv.),1.);
   if _ActualMV_ = 0 then nmiss+1;
   TS_Profile_Chain = cats(TS_Profile_Chain,_ActualMV_);
   lag_actual_MV = lag(_ActualMV_);
   if first._id_ or _ActualMV_ ne lag_actual_MV then
 TS_Profile_Chain_Unique=cats(TS_Profile_Chain_Unique,_ActualMV_);
   if _ActualMV_ = 0  then idx2 = .; else idx2 = 1;

   if last._id_ then do;
            TS_Profile_Chain=catx('_',TS_Profile_Chain,n,nmiss);
            output MV_PROFILE_TS;
   end;
   output MV_PROFILE_TS_PLOT;
 run;

 title Time Series Profiling for Data = &data;
```

```sas
proc freq data = MV_PROFILE_TS order = freq;
 title2 Frequencies of TS_PROFILE_CHAIN;
 table TS_Profile_Chain;
run;

proc freq data = MV_PROFILE_TS order = freq;
 title2 Frequencies of TS_PROFILE_CHAIN_UNIQUE;
 table TS_Profile_Chain_Unique;
run;

proc freq data = MV_PROFILE_TS;
 title2 Distribution of time series length and number of missing values;
 table  N NMiss;
run;

%if %upcase(&plot) EQ YES %then %do;

proc sql;
 select count(*)
 into :n_ts
 from  MV_PROFILE_TS
 ;
quit;
%put n_ts = &n_ts;

proc sql;
 create table MV_PROFILE_TS_PLOT
 as select catx('_',b.TS_Profile_Chain,a._id_) as _ID_,
           a.&date,
               a.idx2,
               a.*
    from MV_PROFILE_TS_PLOT as a
        left join MV_PROFILE_TS as b
        on a._id_ = b._id_
        order by calculated _ID_, a.&date
;
quit;

data MV_PROFILE_TS_PLOT;
 set MV_PROFILE_TS_PLOT;
 by _ID_;
 retain idx_tmp 0;
 if first._ID_ then idx_tmp = idx_tmp + 1;
 if idx2 = . then idx = .; else  idx = -idx_tmp;
run;

%* Only plot if number of time series is lower eqal than threshold;
%if &n_ts <= &nmax_ts %then %do;

PROC GPLOT DATA = MV_PROFILE_TS_PLOT;
 title2 Profile plot for the time series structure;
 symbol i = join c=black v=none w = &w r=&nseries;
 PLOT idx * &date    =_ID_ / nolegend SKIPMISS;
run;
quit;

%end; %* check max ts OK;

%else %do;
 %put -----------------------------------------------------------------------
---------------;
 %put ---  Number of time series = &n_ts, is higher then nmax_ts value (=
&nmax_ts).          ;
 %put ---  No plot has been created. Reset parameter NMAX_TS to a value of at
least &n_ts.    ;
```

```
    %put ------------------------------------------------------------------
    ---------------;
%end;  %* check max ts non OK;

%end; %*plot=yes;

title; title2;
%mend Profile_TS_MV;
%mend Profile_TS_MV;
```

A.5 Code for Macro %CHECK_TIMEID

General

The code of macro %CHECK_TIMEID, which has been introduced in section 11.3, is shown here.

Parameters

The following variables can be specified with the macro:

DATA
    Name of the input data set that holds the time series data (mandatory).

OUT
    Name of the output data set that holds the original data plus the inserted records. Default = TIMEID_INSERTED.

OUT_CHECK
    Name of the output data set that holds those records that are missing. Default = TIMEID_MISSING.

MODE
    Defines whether the macro shall run in CHECK or in INSERT mode. Valid values are CHECK or INSERT. Default = INSERT.

TIMEID
    Name of the time id variable (mandatory).

INTERVAL
    Time interval of the time series data. This interval is used to check for contiguity.

VALUE
    Name(s) of the value variable(s) that receives the inserted value (mandatory). Note that a list of variables or the global variable _NUMERIC_ can be specified here that denotes all numeric variables in the dataset.

INSERTVALUE
    Value that shall be inserted into the value variable for missing records. Default = 0.

BY
    Optional variable that holds the names of the variables that define the cross-sectional dimensions.

CHECKDUMMYVALUE
    Arbitrary value that is inserted into the data in CHECK mode to identify the missing records. Default = 123456789.123456789.

Macro code

```
%macro CHECK_TIMEID (data=,out=TIMEID_INSERTED,
                    out_check=TIMEID_MISSING,
                    timeid=, Interval=MONTH,
                    value=,by=,mode=INSERT,
                    Insertvalue=0,
                    CheckDummyValue=-123456789.123456789);

%IF &by ne %THEN %DO;
  %*** If a BY-Statement is used, data needs to be sorted for
       PROC Timeseries
       In this case, the data macro variable is set
       to the sorted data;

  proc sort data = &data out = TIMEID_DATA_SORT;
   by &by;
  run;
  %let data = TIMEID_DATA_SORT;
  %put data = &data;
%END;

%IF %upcase(&mode) = CHECK %THEN %DO;
  %*** Insert the CheckDummyValue for those
       TIMEIDs that are missing
       output only those observations that hold the
       CheckDummyValue to see
       those observations that were inserted;

  proc timeseries data = &data out = &out_check
                  (where=(&value=&CheckDummyValue));
    id &timeid interval =&interval setmiss=&CheckDummyValue;
    var &value;
    by &by;
  run;

%END;
%ELSE %IF %upcase(&mode) = INSERT %THEN %DO;
  %*** Insert the InsertValue for those observations
       that are missing
       Output the a table that holds the existing
       and the inserted observations;

  proc timeseries data = &data out = &out;
    id &timeid interval =&interval setmiss=&InsertValue;
    var &value;
    by &by;
  run;
%END;

%mend;
```

Comments

Note the following from the syntax:

- PROC TIMESERIES is used to check for missing records and to insert the missing records.
- Inserting the missing records is done using typical PROC TIMESERIES syntax. An insert value different from 0 can be specified with the INSERTVALUE macro variable.
- Checking for missing records is also done with PROC TIMESERIES.
  - Here records with an arbitrary value of -123456789.123456789 are inserted if a record is missing. The inserted value is then used to identify the inserted records.

- If the default value of -123456789.123456789 conflicts with real data a different value using the CHECKDUMMYVALUE variable can be specified.
- Cross-sectional dimensions are treated using the BY statement in PROC TIMESERIES.
- If BY variables are specified with the macro, the input data are sorted for these variables, which may take some time for large datasets.
- Note that PROC TIMESERIES is part of SAS/ETS.

Appendix B: General SAS Content and Programs

B.1 Calculating the Number of Records with at Least One Missing Value 301

B.2 The SAS Power and Sample Size Application 302

B.1 Calculating the Number of Records with at Least One Missing Value

The following code has been used to create the effective number of non-missing observations for various missing percentages and number of variables in Table 4.1.

```
*** 1. Define Parameters;
%let prob_min = 0;      *** Minimum Probability;
%let prob_max = 0.3;    *** Maximum Probability;
%let prob_by  = 0.01;   *** By value for probabilities;
%let nvars = 100;       *** Maximum number of variables;

*** 2. Loop in a dataset over probabilities and number of
        variables. Use the PROBBNML function to calculate
        probabilities from the binomial distribution.;
data binomial;
  do Prob = &prob_min to &prob_max by &prob_by;
    do nvars = 1 to &nvars;;
        PercentNonMissing=probbnml(prob,nvars,0);
        output;
    end;
  end;
run;

*** 3. Prepare the output table;
proc transpose data = binomial out = binomial_tp;
 by  prob;
 var PercentNonMissing;
 id  nvars;
run;
data binomial_tp;
  set binomial_tp(drop = _name_);
  format prob percent8.
         _: percent8.1;
  if prob=0 then delete;
run;

*** 4. Create output table with prob print and ODS HTML;
ods html;
proc print data = binomial_tp noobs;
  var Prob _1 _2 _3 _4 _5 _10 _15 _20 _25 _30 _40 _50 _60 _70 _80 _90 _100;
  where round(Prob*100) in (1, 2, 3, 4, 5, 6, 7, 8, 9, 10, 11, 12, 13, 14, 15, 20, 25, 30);
run;
ods html close;
```

Note the following from the code:

- DO loops and the OUTPUT statement have been used to create the desired simulation data.
- The PROBBNML function has been used to calculate the probabilities from the binomial distribution.
- The value k (= number of success) is set to zero in the PROBBNML function as the probability that a record containing no missing value will be calculated.
- The PRINT procedure and ODS HTML have been used to create the output.

B.2 The SAS Power and Sample Size Application

The SAS Power and Sample Size Application is a graphical user interface to calculate necessary sample sizes and the resulting power for configuration in various statistical methods. The calculation kernel for the application is PROC POWER of SAS/STAT.

Note that JMP also provides methods for Power and Sample Size calculations. Compare also section 14.11.

The following methods are supported by the SAS Power and Sample Size Application:

- Comparing means with a one-sample, two-sample, or paired t-test
- Calculating confidence intervals for one proportion, a one-sample mean, a two-sample mean, or paired means
- Comparing proportions for one-proportion, two correlated proportions, or two independent proportions
- Equivalence testing for one-proportion, one-sample mean, paired means, and two-sample means
- Correlation and regression analysis for one Pearson correlations coefficient, for the logistic regression with a binary response, and for multiple regression
- Linear models and analysis of variance for the general linear univariate model and the one-way ANOVA
- Survival analysis for a two-sample survival rank test
- Distribution tests for the Wilcoxon Mann-Whitney test for two distributions

The SAS Power and Sample Size Application can either be used with the graphical user interface, as shown in Figure B.1, or can be programmed with SAS procedure statements:

```
proc power;
.logistic
 alpha = 0.05
 vardist('Duration') = normal(4, 1.5)
 testpredictor = 'Duration'
 testoddsratio = 1.7
 responseprob = 0.65
 ntotal = 50 60 70
 power = . ;
run;
```

In this example the resulting power values for logistic regression are calculated for sample sizes 50, 60, and 70 based on settings for odds ratios and covariates.

Figure B.1: SAS Power and Sample Size Application for logistic regression

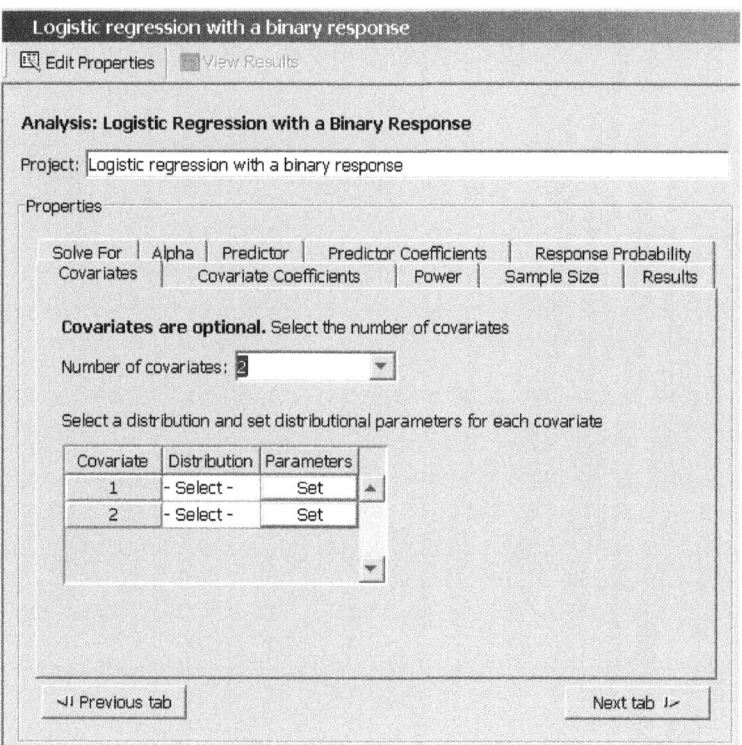

More details on the SAS Power and Sample Size Application can be found in the SAS online documentation or in [21].

Appendix C: Using SAS Enterprise Miner for Simulation Studies

C.1 Introduction 305
C.2 Preparation of SAS Enterprise Miner for SEED=0 Random Numbers..................305
    General 305
    Changing the settings..306
C.3 Simulation Environment 306
    General 306
    Features...307
    Deriving the parameter setting from the node name 307
    Programming details...307
C.4 Discussion of the Suitability of SAS Enterprise Miner for a Simulation Environment...308
    General 308
    Advantages 308
C.5 Selected Macros and Macro Variables Available in a SAS Enterprise Miner Code Node..309

C.1 Introduction

The simulations that have been performed for predictive modeling in part III of this book are technically based on SAS Enterprise Miner. Against the background of the simulation of the consequences of poor data quality in predictive modeling as shown in chapters 16–19, this chapter shows specifics of using SAS Enterprise Miner for simulation studies.

The technical basis of the simulation studies is SAS Enterprise Miner 6.2 running on a SAS 9.2 platform. Please check the website for this book, http://www.sascommunity.org/wiki/Data_Quality_for_Analytics, for updates and changes to new versions of SAS and SAS Enterprise Miner.

C.2 Preparation of SAS Enterprise Miner for SEED=0 Random Numbers

General

Negative seed values are important in simulations, as they cause a different split into training and validation data each time the node is run.

SAS Enterprise Miner does not allow, in these standard settings, a non-positive seed in the SAMPLE or the PARTITION node. In order to be able to set a non-positive seed, the following changes to the XML file of the PARTITION or SAMPLE node are necessary.

Note that in general it is not advisable to edit the standard XML files that define the properties of the standard SAS Enterprise Miner nodes. Thus, the following steps should only be taken with care.

306 *Data Quality for Analytics Using SAS*

Changing the settings

- Navigate to the "components" directory on your analytics platform. This is in the directory <sasinstall>\Lev1\AnalyticsPlatform\apps\EnterpriseMiner\conf\components where <sasinstall> is the directory that has been specified during the SAS installation.
- Make a backup copy of the PARTITION.XML file.
- Edit PARTITION.XML.
- Navigate to the Property RANDOMSEED.
- Make sure that RANGE MIN is set to 0 and the EXCLUDEMIN is set to "N."

```
<Property type="int" name="RandomSeed"
        displayName="properties.common.randomseed.txt"
        description="randomseed.desc.txt"
        initial="12345">
  <Control>
     <Range min="0" excludeMin="N" />
  </Control>
</Property>
```

- Close and save the PARTITION.XML file.

An example result is shown in Figure C.1.

Figure C.1: Screenshot of the property sheet of the partition node

Train	
Variables	
Output Type	Data
Partitioning Method	Default
Random Seed	0

C.3 Simulation Environment

General

In section 16.5 the simulation environment has already been described. Here only the most important features are described.

Figure C.2: Example process flow chart for the simulation in SAS Enterprise Miner

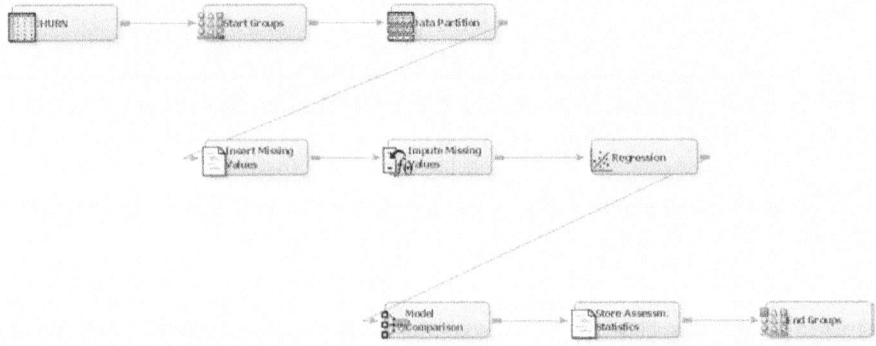

Features

- The simulations are being looped with START/END Groups Nodes. Note the MODE in the START GROUPS Nodes has to be set to INDEX. The number of iterations must be specified with the "Index Count" property.
- The Model Comparison node is used to calculate the assessment statistics. These results are then prepared and output to a separate dataset by a consecutive SAS Code Node.
    - Note that after each iteration such a dataset is being output.
    - The name of this dataset needs to be unique for each iteration run in order to avoid overwriting.
    - The advantage of having results immediately available after the first run allows for an early look into the results to check whether the simulation produces plausible results or whether the simulation will be stopped because some parameters have possibly been set in a wrong way.
- After the last iteration of the simulations the results are appended to a central result repository. This is important as it provides a central storage of simulation results that can be prepared and analyzed in a standardized way.

Deriving the parameter setting from the node name

The parameter setting that differentiates the scenarios from each other has been used to name the modeling node in the scenario. See Figure C.3.

Figure C.3: Naming of the modeling node

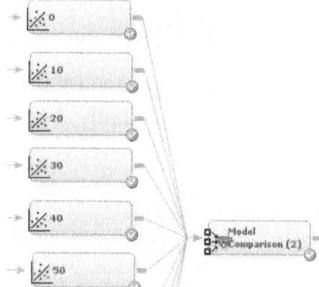

As a result, for each scenario the setting 0, 10, 20, ... is written into the MODELDESCRIPTION variable in the output data of the "Model Comparison Node" and can therefore be used for further processing.

Programming details

- All macros that control the preparation of the data for the simulations—for example, insertion of missing values, the deletion of observations, and the insertion of biases into the data—have been written and maintained in a central program. This program is included in the START UP code of the SAS Enterprise Miner project and, thus, makes the macros available to each code node. This allows for central maintenance of the whole simulation project.
- Missing values with flexible start values have been generated in the SAS code node with the CALL MISSING statement as described in section 15.4.
- Macro variables and macros in the code nodes of SAS Enterprise Miner have been used to keep the program as flexible as possible. For example:

```
select count(*)
into :ne_trn
from &em_import_data
where %em_target = 1;

%do i = 1 %to &em_num_input;
  call ranuni(seed, rnd);
  if rnd le &pct_missing then call missing(%scan(%em_input,&i));
%end;
```

C.4 Discussion of the Suitability of SAS Enterprise Miner for a Simulation Environment

General

There is a single disadvantage of using SAS Enterprise Miner for the simulations that have been described in chapters 16–19; the definition of the different scenarios is a little bit more bulky than it would be if all settings were to be maintained in a single code file.

If a simulation with 10 different parallel arms—for example, percentages of missing values with 0%, 10%, 20%, ... 90%—is defined, the nodes for each arm need to be copied and named accordingly and the settings need to be made in the code node.

Advantages

The advantages of using SAS Enterprise Miner for these simulations weigh, however, much higher:

- SAS Enterprise Miner provides tools that perform many elements of the tasks that are needed for the respective simulation runs. Otherwise all these tasks would need to be programmed manually. For example:
  - Imputation of missing values
  - Standardization of missing values
  - Partition of the data in training, validation, and test partitions
  - Creation of the forecast model on the training data and application of the model to the other data partitions
  - Calculation of the assessment statistics with the Model Comparison Node
- The process flow in SAS Enterprise Miner visualizes the simulation flow and breaks it down to a modular structure that helps to document the content better.
- SAS Enterprise Miner automatically runs parallel streams in the flow on different processors. This saves a lot of time for the simulation runs. Outside SAS Enterprise Miner this would need to be programmed manually.
- SAS Enterprise Miner allows STOPPING a process in a save and clean way. This is important when a process needs to be stopped because the preliminary analysis of the first results shows that the settings need to be changed.

C.5 Selected Macros and Macro Variables Available in a SAS Enterprise Miner Code Node

The following macros and macro variables that are available in the code node in SAS Enterprise Miner have been used for the simulation environment:

Table C.4: Selected macros and macro variables that have been used in a code node in SAS Enterprise Miner

Macro or macro variable	Explanation
%EM_TARGET	resolves to the variables that have a model role of target. The target variable is the dependent or the response variable.
&EM_IMPORT_DATA	resolves to the name of the training data set.
&EM_EXPORT_TRAIN	resolves to the name of the export training data set.
&EM_IMPORT_VALIDATE	resolves to the name of the validation data set.
&EM_EXPORT_VALIDATE	resolves to the name of the export validation data set.
&EM_IMPORT_TEST	resolves to the name of the test data set.
&EM_EXPORT_TEST	resolves to the name of the export test data set.
%EM_INPUT	resolves to the variables that have a model role of input. The input variables are the independent or predictor variables.
&EM_NUM_INPUT	resolves to the number of input variables.
&EM_NUM_INTERVAL_INPUT	resolves to the number of interval input variables.
%EM_INTERVAL_INPUT	resolves to the interval variables that have a model role of input.

Appendix D: Macro to Determine the Optimal Length of the Available Data History

D.1: Introduction 311
D.2: Example Call and Results..311
    Preparation of the data...311
    Example call...312
    Results..312
D.3: Macro Parameters..314
    Parameters for macro %TS_HISTORY_CHECK..314
    Parameters for macro %TS_HISTORY_CHECK_ESM..315
D.4: Macro Code...315
    Macro code for %TS_HISTORY_CHECK...315
    Macro code for %TS_HISTORY_CHECK_ESM...318
    Comments...318

D.1: Introduction

This appendix shows the macro that can be used to analyze the optimal length of the history of time series data. The background for this procedure has been shown in chapter 21.

Two macros are presented here.

- Macro %TS_HISTORY_CHECK that uses PROC HPFENGINE and requires SAS High-Performance Forecasting
- Macro %TS_HISTORY_CHECK_ESM that used PROC ESM and requires SAS/ETS

Please check the website for this book, http://www.sascommunity.org/wiki/Data_Quality_for_Analytics, for downloads of these macros.

D.2: Example Call and Results

Preparation of the data

An example call of the macro is shown based on the SASHELP.AIR data.

As the macro requires finding time series ID for each time series, this variable has to be provided, even if the data only contain one time series.

```
data air;
 set sashelp.air;
 tsid = 1;
run;
```

Example call

Example call for the data is shown here. Note that some of the parameters are defaults anyhow, but are presented here for illustrative purposes.

```
%ts_history_check(data=air,tsid=tsid,y=air,
                  timeid=date,interval=month,
                  minhist=6,maxhist=36,
                  shiftfrom=0,shiftto=2,shiftby=2,
                  periodvalid=12,
                  mrep=sashelp.hpfdflt,sellist=tsfsselect,
                  stat=mape,aggrstat=median);
```

Results

The macro outputs a table with the median RMSE and MAPE for different lengths of the time series. See Table D.1.

Table D.1: RMSE and MAPE for the different lengths (up to 12 months) of the time series

History Length	RMSE	MAPE
1	.	.
2	82875.19275410	0.41496676
3	34418.26640664	0.21654528
4	32535.02943198	0.19699401
5	35225.76060197	0.21115719
6	36309.29466752	0.21458678
7	37378.73082962	0.22657719
8	38049.30660346	0.22595523
9	36978.26828051	0.21775359
10	35331.10132530	0.20633470
11	25831.88747121	0.13985875
12	13088.69167134	0.07010186

Additionally, a line chart is generated that shows the median MAPE over the available history months.

Figure D.1: Line plot for MAPE over the different history months

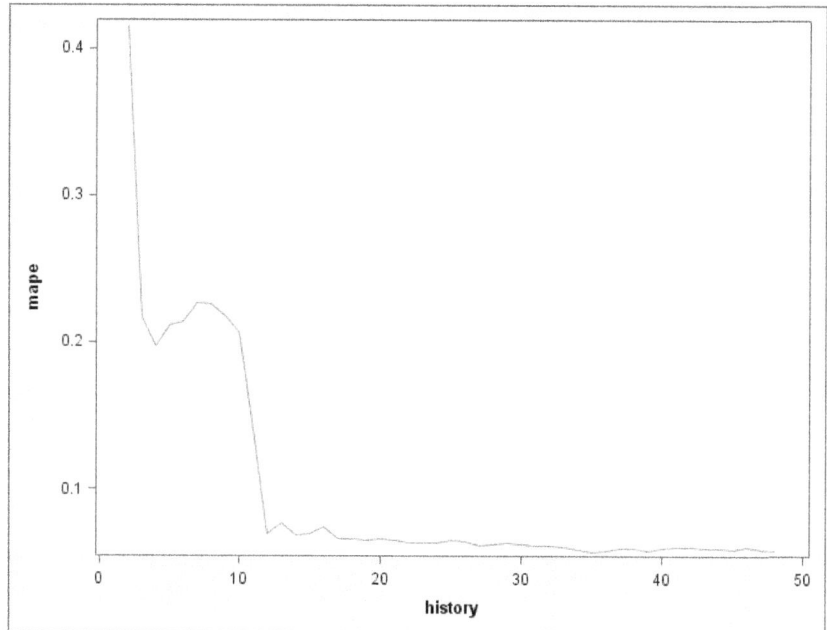

For each time series and SHIFT the optimal length of the time history to achieve the minimum MAPE is calculated and displayed in a bar chart. See Figure D.2.

Figure D.2: Best length of the time history for a minimum MAPE

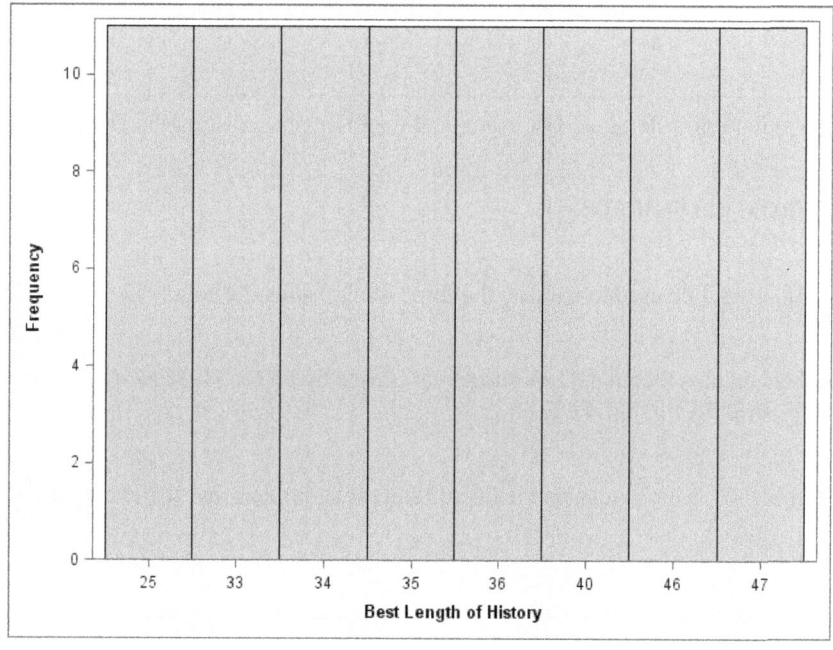

D.3: Macro Parameters

Parameters for macro %TS_HISTORY_CHECK

The following parameters can be specified with the macro %TS_HISTORY_CHECK:

DATA
  Name of the data set that contains the time-series data. Note that the data need to be in a longitudinal format and that there needs to be an ID variable that identifies the time series.

TSID
  Name of the variable that contains the time-series ID variable. Default = TSID.

Y
  Name of the variable that contains the dependent variable of the time series. Default = Y.

TIMEID
  Name of the variable that contains the time values of the variable.

INTERVAL
  Time interval definition of the TIMEID variable; for example, MONTH, WEEK, ... Default = MONTH.

MINHIST
  Minimum history length that will be tested.

MAXHIST
  Maximum history length that willl be tested.

SHIFTFROM
  Shift start value that will be used to run the scenario with different subsets of the data. Default = 0.

SHIFTTO
  Shift end value that will be used to run the scenario with different subsets of the data. Set this value to the SHIFTFROM value to only run on shift iteration. For example 0 for a non-shifted scenario. Default = 12.

SHIFTBY
  Shift-BY value for SHIFTFROM and SHIFTTO.

PERIODVALID
  Number of (future) periods that shall be used to validate the forecasting results. Default = 12.

MREP
  Name of the SAS catalog that contains the model repository that is used by PROC HPFENGINE. Specify with library name. Default = SASHELP.HPFDFLT.

SELLIST
  Name of the selection list in MPRE, which contains the list of forecasting models that shall be used by PROC HPFENGINE.

STAT
  Statistic that shall be used for model selection in HPFENGINE and calculation of the validation statistics. Valid values are MAPE and RMSE. Default = MAPE.

AGGRSTAT
  Aggregation statistic that is used to aggregate the results over shifts and time series. Refer to the documentation of the means procedure for the possible values. Default = MEDIAN.

Parameters for macro %TS_HISTORY_CHECK_ESM

The following parameters are additionally available with the %TS_HISTORY_CHECK_ESM macro:

SEASONALITY
   Specifies the length of the seasonal cycle. For example, 12 for a season with 12 months. Default = 12.

MODEL
   Specifies the forecasting model to be used to forecast the time series. Possible values include WINTERS, ADDWINTERS, SIMPLE, SEASONAL. Refer to the documentation of the ESM procedure for a complete list of possible values. Default = SEASONAL.

The SELLIST and the MREP parameter are not available with the %TS_HISTORY_CHECK_ESM macro.

D.4: Macro Code

Macro code for %TS_HISTORY_CHECK

The macro code of the %TS_HISTORY_CHECK macro is as follows:

```
%macro TS_History_Check(data=,tsid=tsid,y=qty,timeid=monyear,
                 interval=month, minhist=1,maxhist=48,
                 hiftfrom=0,shiftto=12,shiftby=1,periodvalid=12,
                 mrep=sashelp.hpfdflt,sellist=tsfsselect,
                 stat=mape,aggrstat=median);

/*** Part I - Prepare Data for the Analysis ***/
*** Calculate the number of observations per time series;
proc means data=&data noprint nway;
 class &tsid;
 var &y;
 output out=ts_count(drop = _type_ _freq_) n=_count_obs_;
run;

*** Join the number of observations to the base table;
proc sql noprint;
 create table _Hist_check_tmp_
 as select a.*, b._count_obs_
    from &data as a
        left join ts_count as b
        on a.tsid = b.tsid
    order by a.&tsid., a.&timeid.;
quit;

*** Generate _LEAD_ variable;
data _Hist_check_tmp_;
 set _Hist_check_tmp_;
 by &tsid;
 if first.tsid then _idx_=1;
 else _idx_+1;
 _lead_ = -(_count_obs_ - _idx_)-1;
 drop _idx_;
run;

*** Generate Empty Template table for appending the MAPE values;
data mape_base;
 set &data(keep=&tsid);
 format mape 16.8 rmse 16.8 shift 8. history 8.;
 delete;
run;
```

```sas
/*** Iterate Scenarios ***/

%do shift = &shiftfrom. %to &shiftto. %by &shiftby.;
  %do history = &minhist. %to &maxhist.;
            *** only leave the observations in the table that are needed for
the analysis;
             data _Hist_check_tmp_input_;
             set _Hist_check_tmp_ (where = (_count_obs_ >= &minHist.));
             _y_valid_ = &y.;
             if _lead_*(-1) <= &periodvalid then &y.      = .;
             if _lead_*(-1) > (&periodvalid + &history) then delete;
             run;

          /*** Forecast the Szenario ***/
          proc hpfengine   data = _Hist_check_tmp_input_
                           out=_out_
               outfor = _Hist_check_tmp_fc_
            (drop = lower upper error std)
                                     modelrepository = &mrep
                            globalselection = &sellist
                                _lead_ = &periodvalid back=0
                            task = select(criterion = &stat
                            minobs=(season=1)
                            seasontest = none
                            ) ;
          by &tsid;
          id &timeid interval = &interval;
          forecast &y;
          run;

       proc delete data=_out_;run;

          *** Join Forecast Values with original
             values for MAPE calculation;
              proc sql;
               create table _Hist_check_tmp_fc_xt_
                as select a.*,b._lead_,b._y_valid_
                  from _Hist_check_tmp_fc_ (drop = _name_) as a
                  left join _Hist_check_tmp_input_ as b
                    on a.&tsid. = b.&tsid.
                    and a.&timeid. = b.&timeid.
              order by &tsid., &timeid.;
                quit;

             /**** Validate the Scenario ****/
             ** Calculate the mean;
             data _ape_;
               set _Hist_check_tmp_fc_xt_;
               _FC_Period_ = (_lead_*(-1) <= &periodvalid);
               _APE_ = abs(predict-_y_valid_)/_y_valid_;
               _MS_  = (predict-_y_valid_)**2;
             run;

             proc means data = _ape_ (rename =
                            (_ape_ = mape _ms_ = _mse_))
                          noprint nway;
              class &tsid _fc_period_;
              var mape _mse_;
              output out = _mape_ (drop=_type_ _freq_
                     where=(_fc_period_=1)) mean=;
             run;
             data _mape_;
```

```
                        set _mape_;
                        rmse = _mse_ ** 0.5;
                        drop _mse_;
                        shift=&shift;
                        History=&history;
                        drop _fc_period_;
                   run;

       *** Append the results of this run;
     proc append base=mape_base  data=_mape_  force nowarn;
        run;

  %end; ** Hist Loop;
%end; ** SHIFT LOOP;

*** Aggregate data per time history;
proc means data = mape_base noprint nway;
 class history;
 var rmse mape;
 output out = _mape_aggr_(drop=_type_ _freq_) &aggrstat=;
run;

*** Output the results;
proc print data=_mape_aggr_ noobs;
run;

*** Lineplot;
proc sgplot data = _mape_aggr_;
 series x=history y=&stat;
run;

*** Calculate the number of optimal history months;
proc sort data = mape_base;
 by &tsid shift history;
run;

data BestHistory;
 set mape_base(where=(mape ne .));
 by &tsid shift;
 retain BestHistory MinMAPE;
 if first.shift then do; BestHistory = History;
                         MinMAPE = MAPE;
                                end;
 if MAPE < MinMAPE then do;
                              BestHistory = History;
                              MinMAPE = MAPE;
                                 end;
 if last.shift then output;
run;

proc sgplot data=BestHistory;
 vbar besthistory/ barwidth=1;
 xaxis label="Best Length of History";
 yaxis label="Frequency";
run;

%mend;
```

Macro code for %TS_HISTORY_CHECK_ESM

The code for the %TS_HISTORY_CHECK_ESM macro differs by the following parameters from the %TS_HISTORY_CHECK macro.

Macro Header

```
%macro TS_History_Check_ESM(data=,tsid=tsid,y=qty,timeid=monyear,
interval=month, minhist=1,maxhist=48,shiftfrom=0,shiftto=12, shiftby=1,
periodvalid=12,
stat=mape,aggrstat=median,seasonality=12,model=seasonal);
```

Call of the forecasting procedure

```
/*** Forecast the Scenario ***/
   proc esm data=_Hist_check_tmp_input_ out=_out_
      outfor = _Hist_check_tmp_fc_ (drop = lower upper error std)
      seasonality=&seasonality lead=&periodvalid;
        by &tsid;
        id &timeid interval = &interval;
        forecast &y/model=&model;
   run;
```

Comments

Note the following regarding the code:

- PROC HPFENGINE and PROC ESM are used to generate the forecasts in the two macros. They can also be replaced by any other forecasting procedure, such as PROC ARIMA, as long as an output is produced that corresponds to the output table structure of PROC HPFENGINE.
- A macro loop is used to iterate over SHIFTS and TIMEHISTORIES.
- The results are presented using PROC SGPLOT.

Appendix E: A Short Overview on Data Structures and Analytic Data Preparation

E.2 Wording: Analysis Table and Analytic Data Mart	319
E.3 Normalization and De-normalization	320
General	320
Normalization	320
De-normalization	320
Example	320
E.4 Analysis Subjects	321
Definition	321
Representation in the data set	322
E.5 Multiple Observations	322
General	322
Examples	322
Repeated measurements over time	323
Multiple observations because of hierarchical relationships	323
E.6 One-Row-per-Subject Data Mart	323
E.7 The Multiple-Rows-per-Subject Data Mart	324
E.8 The Technical Point of View	325
Transposing	325
Aggregating	326

E.1 Introduction

Because some aspects of data preparation and data structures for analytics are also relevant for data quality for analytics and some terms are used throughout this book, this chapter gives a short overview on data preparation for analytics. More details can be found in my previous book, *Data Preparation for Analytics* [1]. The text in the following sections is partly taken from this book.

E.2 Wording: Analysis Table and Analytic Data Mart

In data mining it has become common to call the analysis table a *data mart*. This name is not necessarily correct, because a data mart is not always only one table; it can also be a set of tables that hold data for a certain business domain. For example, tables of a data mart might need to be joined together in order to have the appropriate format. However, *data mart* has become a synonym for the table that holds the data for data mining.

In non-data mining areas, the term *data mart* is almost unknown. Names such as analysis table, data table, or data matrix are common here and mean the same thing as data mart in the preceding example, which is a table with data for analysis.

In this book the term *analysis table* is used in most cases. In some chapters, the term *data mart* is used as well. If not separately specified, this term also defines a table with data for analysis.

E.3 Normalization and De-normalization

General

I will illustrate the concept of normalization and de-normalization on the basis of a CUSTOMER table and an ACCOUNT table.

- The CUSTOMER table holds a unique customer identifier and all relevant information directly related to the customer.
- The ACCOUNT table holds a unique account identifier, all the information about accounts, and the customer key.
- The unique account identifier in the account table and the unique customer identifier in the customer table are called *primary key*s. The customer identifier in the account table denotes which customer the account belongs to and is called the *foreign key*.

The process of combining information from several tables based on the relationships expressed by primary and foreign keys is called *joining* or *merging*.

Normalization

In a *normalized relational model* no variables, aside from primary and foreign keys, are duplicated among tables. Each piece of information is stored only once in a dedicated table. In data modeling theory this is called the *second normal form*.

Normalization is important for transactional systems. The rationale is that certain information is stored in a single table only, so that updates on data are done in only one table. These data are stored without redundancy.

De-normalization

The opposite of normalization is de-normalization. *De-normalization* means that information is redundantly stored in the tables. The same column appears in more than one table. In the case of a one-to-many relationship, this leads to repeated values.

There are two reasons for de-normalization of data:

- De-normalization is necessary for analytics. All data must be merged together into one single table.
- De-normalization can be useful for performance and simple handling of data. In reporting, for example, it is more convenient for the business user if data are already merged together. For performance reasons, in an operational system, a column might be stored in de-normalized form in another table in order to reduce the number of table merges.

Example

We return to our preceding example and see the content of the CUSTOMER and ACCOUNT tables.

Table E.1: Content of CUSTOMER table

CustID	Birthdate	Gender
1	16.05.1970	Male
2	19.04.1964	Female

Table E.2: Content of ACCOUNT table

AccountID	CustID	Type	OpenDate
1	1	Checking	05.12.1999
2	1	Savings	12.02.2001
3	2	Savings	01.01.2002
4	2	Checking	20.10.2003
5	2	Savings	30.9.2004

Tables E.1 and E.2 represent the normalized version. Besides column CustID, which serves as a foreign key in the ACCOUNT table, no column is repeated.

Merging these tables creates the de-normalized CUSTOMER_ACCOUNT table.

Table E.3: Content of the de-normalized CUSTOMER_ACCOUNT table

CustID	Birthdate	Gender	AccountID	Type	OpenDate
1	16.05.1970	Male	1	Checking	05.12.1999
1	16.05.1970	Male	2	Savings	12.02.2001
2	19.04.1964	Female	3	Savings	01.01.2002
2	19.04.1964	Female	4	Checking	20.10.2003
2	19.04.1964	Female	5	Savings	30.09.2004

In the de-normalized version, the variables BIRTHDATE and GENDER appear multiple times per customer. This version of the data can be used directly for analysis of customers and accounts because all information is stored in one table.

E.4 Analysis Subjects

Definition

Analysis subjects are entities that are being analyzed, and the analysis results are interpreted in their context. Analysis subjects are therefore the basis for the structure of our analysis tables.

The following are examples of analysis subjects:

- Persons: Depending on the domain of the analysis, the analysis subjects have more specific names such as "patients" in medical statistics, "customers" in marketing analytics, or "applicants" in credit scoring.
- Animals: Piglets, for example, are analyzed in feeding experiments; rats are analyzed in pharmaceutical experiments.
- Parts of the body: In medical research analysis the subject might be body parts such as arms (the left arm compared to the right arm), shoulders, or hips. Note that from a statistical point of view, the validity of the assumptions of the respective analysis methods has to be checked if dependent observations per person are used in the analysis.
- Things: Objects such as cash machines in cash demand prediction, cars in quality control in the automotive industry, or products in product analysis can be the subjects.
- Legal entities: Companies, contracts, accounts, and applications can be analysis subjects.
- Geographical features and regions: In agricultural studies, regions or plots can be subjects, or reservoirs in the maturity prediction of fields in the oil and gas industry.

Analysis subjects are the heart of each analysis because their attributes are measured, processed, and analyzed. In deductive (inferential) statistics the features of the analysis subjects in the sample are used to infer the properties of the analysis subjects of the population. Note that we use feature and attribute interchangeably here.

Representation in the data set

When we look at the analysis table that we want to create for our analysis, the analysis subjects are represented by rows, and the features that are measured per analysis subject are represented by columns. See Table E.4 for an illustration.

Table E.4: Results of ergonometric examinations for 21 runners

	PersonNr	Age in years	Weight in kg	Oxygen consumption	Min. to run 1.5 miles	Heart rate while resting	Heart rate while running	Maximum heart rate	Experimental group
1	1	44	89.47	44.609	11.37	62	178	182	2
2	2	40	75.07	45.313	10.07	62	185	185	2
3	3	44	85.84	54.297	8.65	45	156	168	2
4	4	42	68.15	59.571	8.17	40	166	172	2
5	5	38	89.02	49.874	9.22	55	178	180	2
6	6	47	77.45	44.811	11.63	58	176	176	2
7	7	40	75.98	45.681	11.95	70	176	180	2
8	8	43	81.19	49.091	10.85	64	162	170	2
9	9	44	81.42	39.442	13.08	63	174	176	2
10	10	38	81.87	60.055	8.63	48	170	186	2
11	11	44	73.03	50.541	10.13	45	168	168	2
12	12	45	87.66	37.388	14.03	56	186	192	1
13	13	45	66.45	44.754	11.12	51	176	176	1
14	14	47	79.15	47.273	10.6	47	162	164	1
15	15	54	83.12	51.855	10.33	50	166	170	1
16	16	49	81.42	49.156	8.95	44	180	185	1
17	17	51	69.63	40.836	10.95	57	168	172	1
18	18	51	77.91	46.672	10	48	162	168	1
19	19	48	91.63	46.774	10.25	48	162	164	1
20	20	49	73.37	50.388	10.08	67	168	168	1
21	21	57	73.37	39.407	12.63	58	174	176	1

In this table 21 runners have been examined, and each one is represented by one row in the analysis table. Features such as age, weight, and time spent running have been measured for each runner, and each feature is represented by a single column. Analyses, such as calculating the mean age of our population or comparing the runtime between experimental groups 1 and 2, can start directly from this table.

E.5 Multiple Observations

General

The analysis Table E.4 is simple in that we have only one observation per analysis subject. It is therefore straightforward to structure the analysis table in this way.

There are, however, many cases where the situation becomes more complex, namely, when we have multiple observations per analysis subject.

Examples

- In the preceding example we would have multiple observations when each runner does more than one run, such as a second run after taking an isotonic drink.
- A dermatological study in medical research where different creams are applied to different areas of the skin.
- Evaluation of clinical parameters before and after surgery.
- An insurance customer with insurance contracts for auto, home, and life insurance.
- A mobile phone customer with his monthly aggregated usage data for the last 24 months.
- A daily time series of overnight stays for each hotel.

In general there are two reasons that multiple observations per analysis subject can exist:

- repeated measurements over time
- multiple observations because of hierarchical relationships

We will now investigate the properties of these two types in more detail.

Repeated measurements over time

Repeated measurements over time are obviously characterized by the fact that for the same analysis subject the observation is repeated over time. From a data model point of view, this means that we have a one-to-many relationship between the analysis subject entity and a time-related entity.

Note that we are using the term *repeated measurement* where observations are recorded repeatedly. We are not necessarily talking about measurements in the sense of numeric variables per observation of the same analysis subject; only the presence or absence of an attribute (yes or no) would be noted on each of X occasions.

The simplest form of repeated measurements is the two-observations-per-subject case. This case happens most often when we compare observations before and after a certain event and are interested in the difference or change in certain criteria (pre-test and post-test). Examples of such an event include the following:

- giving a certain treatment or medication to patients
- execution of a marketing campaign to promote a certain product

If we have two or more repetitions of the measurement, we will get a measurement history or a time series of measurements:

- Patients in a clinical trial make quarterly visits to the medical center where laboratory values and vital signs are collected. A series of measurement data such as the systolic and diastolic blood pressure can be analyzed over time.
- The number and duration of phone calls of telecommunications customers are available on a weekly aggregated basis.
- The monthly aggregated purchase history for retail customers.
- The weekly total amount of purchases using a credit card.
- The monthly list of bank branches visited by a customer.

Multiple observations because of hierarchical relationships

If we have multiple observations for an analysis subject because the subject has logically related child hierarchies, we call this *multiple observations because of hierarchical relationships*. The following are examples:

- One insurance customer can have several types of insurance contracts (auto insurance, home insurance, life insurance). He can also have several contracts of the same type, for example, if he has more than one car.
- A telecommunications customer can have several contracts; he or she could subscribe to one or more lines. (In this case, we have a one-to-many relationship between the customer and contract and another one-to-many relationship between the contract and the line.)
- In one household, one or more persons can each have several credit cards.
- Both eyes of a patient are investigated in an ophthalmological study.
- A patient can undergo several different examinations (laboratory, x-ray, vital signs) during one visit.

E.6 One-Row-per-Subject Data Mart

In the one-row-per-subject data mart, all information per analysis subject is represented by one row. Features per analysis subject are represented by a column. When we have no multiple observations per analysis subject, the creation of this type of data mart is straightforward—the value of each variable that is measured per analysis subject is represented in the corresponding column. The one-row-per-subject data mart usually has only one ID variable, namely that of identifying the subjects.

Table E.5: Content of CUSTOMER table

CustID	Birthdate	Gender
1	16.05.1970	Male
2	19.04.1964	Female

Table E.6: Content of ACCOUNT table

AccountID	CustID	Type	Open Date
1	1	Checking	05.12.1999
2	1	Savings	12.02.2001
3	2	Savings	01.01.2002
4	2	Checking	20.10.2003
5	2	Savings	30.09.2004

In the case of the presence of multiple observations per analysis subject, we have to represent them in additional columns. Because we are creating a one-row-per-subject data mart, we cannot create additional rows per analysis subject. See the following example.

Table E.7: One-row-per-subject data mart for multiple observations

CustID	Birthdate	Gender	Number of Accounts	Proportion of Checking Accounts	Open date of oldest account
1	16.05.1970	Male	2	50 %	05.12.1999
2	19.04.1964	Female	3	33 %	01.01.2002

Table E.7 is the one-row-per-subject representation of Tables E.5 and E.6. We see that we have only two rows because we have only two customers. The variables from the CUSTOMER table have simply been copied to the table. When aggregating data from the ACCOUNT table, however, we experience a loss of information. Information from the underlying hierarchy of the ACCOUNT table has been aggregated to the customer level by completing the following tasks:

- counting the number of accounts per customer
- calculating the proportion of checking accounts
- identifying the open date of the oldest account

We have used simple statistics on the variables of ACCOUNT in order to aggregate the data per subject.

E.7 The Multiple-Rows-per-Subject Data Mart

In contrast to the one-row-per-subject data mart, one subject can have multiple rows. Therefore, we need one ID variable that identifies the analysis subject and a second ID variable that identifies multiple observations for each subject. In terms of data modeling we have the child table of a one-to-many relationship with the foreign key of its master entity. If we also have information about the analysis subject itself, we have to repeat this with every observation for the analysis subject. This is also called de-normalizing.

- In the case of multiple observations because of hierarchical relationships, ID variables are needed for the analysis subject and the entities of the underlying hierarchy. See the following example of a multiple-rows-per-subject data mart. We have an ID variable CUSTID for CUSTOMER and an ID variable for the underlying hierarchy of the ACCOUNT table. Variables of the analysis subject such as birth date and gender are repeated with each account.

Table E.8: Multiple-rows-per-subject data mart as a join of the CUSTOMER and ACCOUNT tables

CustID	Birthdate	Gender	AccountID	Type	OpenDate
1	16.05.1970	Male	1	Checking	05.12.1999
1	16.05.1970	Male	2	Savings	12.02.2001
2	19.04.1964	Female	3	Savings	01.01.2002
2	19.04.1964	Female	4	Checking	20.10.2003
2	19.04.1964	Female	5	Savings	30.09.2004

- In the case of repeated observations over time the repetitions can be enumerated by a measurement variable such as a time variable or, if we measure the repetitions only on an ordinal scale, by a sequence number. See the following example with PATNR as the ID variable for the analysis subject PATIENT. The values of CENTER and TREATMENT are repeated per patient because of the repeated measurements of CHOLESTEROL and TRIGLYCERIDE at each VISITDATE.

E.8 The Technical Point of View

Central to this type of data mart is that we have to put all information per analysis subject into one row. Multiple observations per analysis subject must not appear in additional rows; they have to be converted into additional columns of the single row.

From a technical point of view we can solve the task of putting all information into one row by using two main techniques:

- Transposing: Here we transpose the multiple rows per subject into columns. This technique can be considered the "pure" way because we take all data from the rows and represent them in columns.

- Aggregating: Here we aggregate the information from the columns into an aggregated value per analysis subject. We perform information reduction by trying to express the content of the original data in descriptive measures that are derived from the original data.

Transposing

We now look at a very simple example with three multiple observations per subject.

Table E.9: Base table with static information per patient (= analysis subject)

PatNr	Gender
1	Male
2	Female
3	Male

And we have a table with repeated measurements per patient.

Table E.10: Table with multiple observations per patient (= analysis subject)

PatNr	Measurement	Cholesterol
1	1	212
1	2	220
1	3	240
2	1	150
2	2	145
2	3	148
3	1	301
3	2	280
3	3	275

We have to bring data from the table with multiple observations per patient into a form so that the data can be joined to the base table on a one-to-one basis. We therefore transpose Table E.10 by patient number and bring all repeated elements into columns.

Table E.11: Multiple observations in a table structure with one row per patient

PatNr	Cholesterol 1	Cholesterol 2	Cholesterol 3
1	212	220	240
2	150	145	148
3	301	280	275

This information is then joined with the base table, and we retrieve our analysis data mart in the requested data structure.

Table E.12: The final one-row-per-subject data mart

PatNr	Gender	Cholesterol 1	Cholesterol 2	Cholesterol 3
1	Male	212	220	240
2	Female	150	145	148
3	Male	301	280	275

Note that we have brought all information on a one-to-one basis from the multiple-rows-per-subject data set to the final data set by transposing the data. This data structure is suitable for analyses such as repeated measurements analysis of variance.

Aggregating

If we aggregate the information from the multiple-rows-per-subject data set, for example, by using descriptive statistics such as the median, the minimum and maximum, or the interquartile range, we do not bring the original data on a one-to-one basis to the one-row-per-subject data set. Instead we analyze a condensed version of the data. The data might then look like Table E.13.

Table E.13: Aggregated data from Tables E.9 and E.10

PatNr	Gender	Cholesterol_Mean	Cholesterol_Std
1	Male	224	14.4
2	Female	148	2.5
3	Male	285	13.8

We see that aggregations do not produce as many columns as transpositions because we condense the information. The forms of aggregations, however, have to be carefully selected because the omission of an important and, from a business point of view, relevant aggregation means that information is lost. There is no ultimate truth for the best selection aggregation measures over all business questions. Domain-specific knowledge is key when selecting them. In predictive analysis, for example, we try to create candidate predictors that reflect the properties of the underlying subject and its behavior as accurately as possible.

References

[1] Svolba, G. 2006. *Data Preparation for Analytics Using SAS®*. Cary, NC: SAS Publishing.

[2] Richtig E., H. P. Soyer, M. Posch, et al. 2005. Prospective, randomized, multicenter, double-blind placebo-controlled trial comparing adjuvant interferon alfa and isotretinoin with interferon alfa alone in stage IIA and IIB melanoma: European Cooperative Adjuvant Melanoma Treatment Study Group. *Journal of Clinical Oncology* 23(34): 8655–63.

[3] Schubert, S. 2010. Fraud detection with SAS Data Mining. SAS Global Forum. Paper 345-2010. www.support.sas.com/resources/papers/proceedings10/345-2010.pdf (accessed 12 March 2012).

[4] Gloskin, G. 2009. Data quality for analytics. *ISO Review*, December. Available at http://www.iso.com/Research-and-Analyses/ISO-Review/Data-Quality-for-Analytics.html (accessed 22 February 2012).

[5] Barkaway, D. 2010. dATa qWaliti 4 Analytics. SAS Global Forum. Paper 328-2010. http://support.sas.com/resources/papers/proceedings10/328-2010.pdf (accessed 19 March 2011).

[6] Burzykowski, T., J. Carpenter, C. Coens, et al. 2010. Missing data: Discussion points from the PSI missing data expert group. *Pharmaceutical Statistics* 9(4): 288–97.

[7] Orli, Richard J. Data Quality Methods, Kismet. http://www.kismeta.com/cleand1.html (accessed 19 March 2012).

[8] Berka, C., S. Humer, M. Lenk, et al. 2010. A quality framework for statistics based on administrative data sources using the example of the Austrian census 2011. *Austrian Journal of Statistics* 39: 299–308.

[9] Cody, Ron. 2010. *Cody's Data Cleaning Techniques Using SAS, Second Edition*. Cary, NC: SAS Publishing.

[10] ICH Topic E 6 (R1). Guideline for Good Clinical Practice (GCP). http://www.ema.europa.eu/docs/en_GB/document_library/Scientific_guideline/2009/09/WC500002874.pdf (accessed 21 February 2012).

[11] Siddiqi, N. 2005. *Credit Risk Scorecards: Developing and Implementing Intelligent Credit Scoring*. Hoboken, NJ: Wiley.

[12] JMP User Manual. 2010. *Using JMP*. http://www.jmp.com/support/downloads/pdf/jmp9/using_jmp.pdf (accessed 1 March 2012).

[13] Rubin, D. B. 1996. Multiple imputation after 18+ years. *Journal of the American Statistical Association* 91: 473–89.

[14] SAS/STAT 9.22 User's Guide. http://support.sas.com/documentation/onlinedoc/stat/ (accessed 1 March 2012).

[15] SAS/ETS 9.22 User's Guide. http://support.sas.com/documentation/cdl/en/etsug/63348/PDF/default/etsug.pdf (accessed 1 March 2012).

[16] Cohen, R. A. 2006. Introducing the GLMSELECT PROCEDURE for model selection. SUGI San Francisco. Paper 207-31. http://www2.sas.com/proceedings/sugi31/207-31.pdf (accessed 12 March 2012).

[17] Svolba, G. 1999. Statistical quality control in clinical trials. *Facultas*.

[18] Benford, F. 1938. The law of anomalous numbers. *Proceedings of the American Philosophical Society* 78(4): 551–72.

[19] Gilliland, M. 2010. *The Business Forecasting Deal*. Hoboken, NJ: Wiley.

[20] Brocklebank, J., and D. Dickey. 2003. *SAS for Forecasting Time Series, Second Edition*. Cary, NC: SAS Publishing.

[21] Watson, W. 2008. Updates to SAS Power and Sample Size Software in SAS/STAT 9.2. SAS Global Forum. Paper 368-2008. Available at http://www2.sas.com/proceedings/forum2008/368-2008.pdf (accessed 20 March 2012).

Index

A

actuality   40
aggregated data   104
alignment of data   34
analytical systems and tools
    Base SAS   191
    DataFlux Data Management Platform   195–198
    defined   23
    historic data and   35–36
    JMP   191–195
    SAS Enterprise Miner   135–137, 171, 176, 182–186
    SAS/ETS software   190–191
    SAS Forecast Server   188–190
    SAS Forecast Studio   155, 174–175, 188–190
    SAS Model Manager   187
    SAS Rapid Predictive Modeler   169, 185
    SAS/STAT software   187–188
    SAS Text Miner   185
Andrew's wave imputation method   136–137
ARIMA procedure   149, 191
ARIMA(X) models   174–175, 189
ASSESSMENT node (SAS Enterprise Miner)   185
association
    See correlation
ASSOCIATION node (SAS Enterprise Miner)   184
availability of data
    See data availability

B

base data   34–35
Base SAS   191
Benford, Frank   178
Benford's law for checking data   178–179
bias
    optimistic   212
    random   274, 277–279
    simulation studies and   209–213, 243–254, 257
    treating   253
box-and-whisker plots
    about   217
    data availability simulation study   222, 225
    data completeness simulation study   235–238
    data correctness simulation study   248–249, 251–252
    data quantity simulation study   222, 225
business data quality   108
BY statement
    EXPAND procedure   156
    STANDARD procedure   134

C

case-record forms (CRFs)   11
case studies
    data management and analysis in clinical trials   4, 10–14
    data quality features summarized   17–19
    demand forecasting   4, 14–17
    race boat performance in sailing regattas   4–10, 191–193
categorical data
    association between   96
    available imputation methods   137
    defined   41
    distribution of values   51–52
    grouping sparse categories   102–103
    plausibility checks and   74
    rare events and   102
    reference value and   171
changes, tracing for data management in clinical trials case study   13
%CHECK_TIMEID macro   150–152
cleaning data
    race boats in sailing regattas case study   9
    typical steps   42
clinical trials
    data management and analysis case study   4, 10–14
    outlier detection with predictive modeling example   171–174
clustering observations
    about   103
    outlier detection with cluster analysis   168, 176–177
Cody, Ron   161
collecting data for data management in clinical trials case study   12
COMPARE procedure   191
completeness of data
    See data completeness
component plots   131
content of variables   40–42
CONTENTS procedure   191
contiguity of time series data   147–152
correctness of data
    See data correctness
correlation
    about   96
    derived variables for customer behavior   99
    derived variables from transactional data   98–99
    imputing missing values based on variables   97
    independent variables and   97
    multicollinearity and   97
    problem and benefit of   96–99
    sign inversion   98
    substituting effect of variables   97
correlation coefficients   96

count imputation method   137
$COUNT_MV macro   124–125
CRFs (case-record forms)   11
customer age missing values examples   63–64
customer demand analysis example   79–80
customer event histories   35

D

data alignment   34
data availability
    actuality   40
    completeness versus   60–62
    data with different meanings   44–45
    defined   32
    demand forecasting case study   16
    format and content of variables   40–42
    format and structure   42–44
    general considerations   32–34
    granularity of   40
    historic snapshots of data   34, 36–39
    of historic data   34–36
    periodic   39–40
    predictive modeling and   86–88, 219–230
    race boats in sailing regattas case study   9
    simulation studies and   209, 219–230
    usability and   32–33
databases, linking   96, 105–106
    See also data marts and warehouses
data bias
    optimistic   212
    random   274, 277–279
    simulation studies and   209–213, 243–254, 257
    treating   253
data cleaning
    race boats in sailing regattas case study   9
    typical steps   42
data collection for data management in clinical trials case study   12
data completeness
    across tables   64–67, 160–161
    availability versus   60–62
    customer age example   63–64
    defined   59–60
    demand forecasting case study   16
    duplicate records and   67–69
    in parent-child relationships   163–164
    in relational models   160–161
    predictive modeling and   231–242
    race boats in sailing regattas case study   10
    random missing values   62–63
    simulation studies and   231–242, 265–272
data correctness
    business rules and   75
    criteria for good data quality   29
    data entry and   78–79
    data management in clinical trials case study   14
    defined   72

    demand forecasting case study   16
    domain-specific   81
    in data transfer and retrieval   72–73
    interviewer effects   80–81
    plausibility checks   73–76
    psychological and business effects on data entry   79–80
    race boats in sailing regattas case study   9
    random errors and   76–77
    selecting same value in data entry   78–79
    simulation studies and   243–254
    systematic errors and   77–78
    time-dependent   81–82
data entry
    data management in clinical trials case study   12
    data transfer and retrieval and   72
    default values in   78–79
    psychological and business effects on   79–80
    selecting same value in   78–79
data fields   22
DataFlux Data Management Platform   195–198
data formats
    See formats
data linkage   96
data management and analysis in clinical trials case study   4, 10–14
data marts and warehouses
    characteristics of   35–36
    data quality responsibilities   114
    data structures for analytics   104
    defined   23
    demand forecasting case study   4, 14–17
    examples of data quality in   24
    linking databases   96, 105–106
    missing values in date of birth variable example   26
    profiling structure of missing values   126–133
    structure considerations   50
    time granularity and   36
data matrix
    See data tables
data mining
    sample size calculation   54–55
    SAS Enterprise Miner   135–137, 171, 176, 182–186
    SEMMA methodology   135, 182
DATA PARTITION node (SAS Enterprise Miner)   214
data process considerations
    See process considerations for data quality
data profiling
    See profiling
data quality
    See also analytical systems and tools
    about   22–25, 96
    across related tables   159–165
    benefits of analytics in general   168–169

classical outlier detection 169–170
completeness and plausibility checks 160–162
correlation 96–99
criteria for 28–30
data profiling examples 178–179
defined 27–28
distribution 101–103
hashes and 162–165
importance of 27
level of detail 104–105
linking databases 96, 105–106
merging tables 161–162
missing values example 25–27
modeling features for 184
outlier detection in time series analysis 168, 174–175
outlier detection with cluster analysis 168, 176–177
outlier detection with predictive modeling 168, 171–174
predictive modeling and 84
process considerations and 108
quantity versus 47–48
recognition of duplicates 168, 177–178
scoring logic and 90–91
simulation studies and 255–264
sparseness 101–103
variability 96, 100–101
data quantity
　benefits of analytics for 168–169
　considerations about 48–49
　data management in clinical trials case study 14
　demand forecasting case study 17
　dimensions of 50–53
　effect of missing values on 55–56
　handling small amounts 185
　predictive modeling and 48–49, 219–230
　quality versus 47–48
　sample size planning 53–55
　SAS Forecast Studio and 190
　simulation studies and 219–230, 265–272
data records
　completeness check of 160
　defined 22
　duplicate 67–69, 168, 177–178
　plausibility check of 161
data relevancy 108–111, 271–272
Data Source Wizard 186
DATA step
　Base SAS support 191
　CALL functions in 205
　IN operator 147
　merging tables 161–162
　random numbers with changing start values 205
data structure, available data and 42–44

data synchronization, race boats in sailing regattas case study 9
data tables
　data completeness across 64–67, 160–161
　data quality control across 159–165
　defined 23
　merging 161–162
　plausibility checks in 75, 160–161
　replacement values for entire 133–134
data transfer and retrieval
　comparing data between systems 73
　data correctness in 72–73
　data management in clinical trials case study 12
　minimizing steps 73
　race boats in sailing regattas case study 9
data usability
　considerations for 41
　data availability and 32–33
data warehouses
　See data marts and warehouses
date of birth examples 25–27, 64
decision support systems
　See analytical systems
decision trees, reference values and 171
default constant imputation method 136–137
default values in data entry 78–79
deleting duplicate records 69
demand forecasting case study 4, 14–17
derived variables
　for customer behavior 99
　from transaction data 98–99
detailed data 104
DEVICE system option 132, 146
dimensions of analytical data 50–53
distribution
　about 96
　categorical variables and rare events 102
　changes in 89–91
　clustering observations 103
　imputation for categorical variables 137
　imputation for interval variables 136
　missing values in 101
　of interval data 102
　of length of time series 145
　of number of missing values 145
　random number generators 205
　segmenting observations 103
　transforming variables to 169
domain-specific data correctness 81
duplicate records
　consequences of 68
　data quality control and 168, 177–178
　defined 67–68
　deleting 69
　reasons for 68–69
　treating 69

E

electronically available data 42–43, 72–73
END GROUPS node (SAS Enterprise Miner) 185, 214
entering data
    See data entry
Enterprise Miner
    about 182–183
    assessing importance of variables 184
    ASSESSMENT node 185
    ASSOCIATION node 184
    DATA PARTITION node 214
    data quality checks and correction 183, 186
    description of simulation environment 213–215
    END GROUPS node 185, 214
    FILTER node 183
    gaining insight into data relationships 184
    handling small data quantities 185
    IMPUTE MISSING VALUES node 214
    IMPUTE node 135–137, 183
    INSERT MISSING VALUES node 214
    LARS node 49, 184
    MODEL COMPARISON node 215–216
    modeling features for data quality 184
    MODIFY tab 135, 183
    outlier detection with cluster analysis 176
    outlier detection with predictive modeling 171
    PLS node 184
    REGRESSION node 184–185, 214
    REPLACEMENT node 183
    RULE INDUCTION node 185
    SAMPLE tab 183
    simulation studies and 204
    START GROUPS node 185, 214
    STAT EXPLORE node 184
    text mining 185
    TRANSFORM VARIABLES node 183
    TREE node 184–185
    VARIABLE SELECTION node 184
    WARN variable 186
    what-if analysis 185
ESM procedure 149
ETS software 190–191
events
    as dimension of analytical data 51
    categorical variables and 102
    customer event histories 35
    influence of data availability on 223–227
    influence of data quality on 209
    influence of data quantity on 223–227
EXPAND procedure
    about 190
    BY statement 156
    interpolating missing values in time series data 155–156, 260
    METHOD=JOIN statement 156
    METHOD=SPLINE statement 155
    METHOD=STEP statement 156
    random missing values and 274–276
    systematic missing values and 282–283, 287

F

filtering data 102
FILTER node (SAS Enterprise Miner) 183
FORECAST procedure 149
Forecast Server 188–190, 204
Forecast Studio 155, 174–175, 188–190
FORMAT procedure 191
formats
    available data 42–44
    lookup tables and 162
    variables 40–42
FREQ procedure 125
future periods, stable data definition for 88–91
fuzzy matching methods 177

G

geographic effects on data correctness 80
GLM procedure 171–172, 188
GLMSELECT procedure 49, 187
GPS devices in race boats case study 4–10, 191–193
granularity of data
    about 40
    data warehouses and 36
    predictive modeling and 96

H

hashes
    Base SAS support 191
    data quality control and 162–165
    lookup tables and 162
heterogeneity
    consequences of 45
    international bank example 44–45
High-Performance Forecasting 190, 204, 259
historic data
    availability of 34–36
    optimal length of available time history 269–271
    predictive modeling and 88, 92
    simulation studies and 257
historic snapshots
    about 34, 36–39
    predictive modeling and 86
HPFDIAGNOSE procedure 260
HPFENGINE procedure
    about 259, 266
    random missing values and 274
    systematic missing values and 282
HPFIDMSPEC procedure 190
Huber's imputation method 136–137

I

ID statement, TIMESERIES procedure
    SETMISS= option 150, 152–153
    ZEROMISS= option 154–155
IMPUTE MISSING VALUES node (SAS Enterprise Miner) 214
IMPUTE node (SAS Enterprise Miner) 135–137, 183
incomplete data 61
independent variables 97
IN operator 147
INSERT MISSING VALUES node (SAS Enterprise Miner) 214
interquartile ranges
    interval values and 74
    of MAPE statistics 268
interval data
    available imputation methods 136–137
    data correctness and 74
    defined 41
    distribution of 102
interviewer effects on data correctness 80–81

J

JMP 49, 191–195

K

key performance indicators (KPIs) 117–118
KPIs (key performance indicators) 117–118

L

LARS node (SAS Enterprise Miner) 49, 184
LASSO method 187
legal considerations
    criteria for good data quality 30
    for poor data quality 111
line plots
    data completeness simulation study 240
    data correctness simulation study 250
    outlier detection in time series data 175
    random disturbances simulation study 276–277
linking databases 96, 105–106
LOGISTIC procedure 138–139, 187

M

macros 125–126
    See also specific macros
MAPE (mean absolute percentage error)
    about 263
    effect of length of available time history 266–269
    random disturbances example 275–279
    systematic disturbances example 284–287
MDS procedure 187
mean
    effect of outliers on 102
    interval values and 74
    outlier detection and 170
mean absolute percentage error (MAPE)
    about 263
    effect of length of available time history 266–269
    random disturbances example 275–279
    systematic disturbances example 284–287
mean imputation method 136
MEANS procedure
    about 191
    counting missing values with 125
    NMISS option 125
    VAR statement 125
measurement gathering
    criteria for good data quality 28
    poor data quality and 113
    variability and 100
median imputation method 136
merging tables 161–162
METHOD=JOIN statement, EXPAND procedure 156
METHOD=SPLINE statement, EXPAND procedure 155
METHOD=STEP statement, EXPAND procedure 156
MIANALYZE procedure 138, 140, 188
mid-minimum imputation method 136
midrange imputation method 136
MI procedure 137–140, 188
missing data pattern 193–195
missing values
    benefits of analytics for 168
    categories of data 60–61
    changing zero values to 154–155
    contiguity of time series data 147–152
    counting with MEANS procedure 125
    criteria for good data quality 29
    customer age example 63–64
    date of birth example 25–27, 64
    defined 124
    differentiating between types of 240–241
    effect on data quantity 55–56
    general profiling with macros 125–126
    imputing based on other variables 96–97
    incomplete data and 61
    in date of birth variable example 25–27
    in distributions 101
    in scoring data partition 232
    interpolating in time series data with EXPAND procedure 155–156
    in time series data 141–143
    missing data pattern for 193–195
    multiple imputation with MI procedure 137–140
    profiling structure for time series data 142–147
    profiling structure of 126–133
    race boats in sailing regattas case study 10
    random 62–63, 232–237, 274–276
    replacement values for 133–134, 168
    replacing for entire table 133–134
    replacing for subgroups 134

missing values (*continued*)
    replacing in time series data with TIMESERIES procedure   143, 152–155
    replacing with IMPUTE node   135–137
    scoring logic   89
    simple profiling of   124–126
    simulation studies and   209, 231–242, 257
    systematic   62–64, 232–233, 237–239, 281–288
    univariate imputation of   133–135
%MISS_OBS_TS macro   142
MODEL COMPARISON node (SAS Enterprise Miner)   215–216
Model Manager   187
monitoring data quality   117–119
multicollinearity   97
multivariate data
    outlier detection and   176–177
    plausibility checks   75–76, 105
    quantifying effects in simulation studies   241
%MV_PROFILE_CHAIN macro   127–131
%MV_PROFILING macro   124, 132–133, 193
%MV_TS_PROFILING macro   142

N

Newcomb, Simon   178
NMAX_TS macro option   146
NMISS option, MEANS procedure   125
non-electronic format of data   42
NUMERIC global variable   152

O

observations
    clustering   103
    effective number of   91–93, 168
    influence of data availability on   220–223, 226–227
    influence of data quality on   209
    influence of data quantity on   220–223, 226–227
    in time series data   141–142
    required number of   50–51
    segmenting   103
    systematically selecting from time series   282
ODS HTML statement   132
ODS OUTPUT statement   139
operational systems
    defined   23
    historic data and   35–36
    missing values in date of birth variable example   25
optimistic bias   212
outlier detection
    benefits of analytics for   168
    classical   169–170
    defining validation limits   169
    in time series analysis   168, 174–175
    purpose of   169–170
    SAS Forecast Studio and   189
    statistical   170
    with cluster analysis   168, 176–177
    with predictive modeling   168, 171–174
outliers
    moment-based parameters and   102
    plausibility checks and   75–76
    trends and   76

P

panel data sets   53
parent-child relationships   65–67, 106
    completeness control in   163–164
    plausibility checks in   164–165
perfect multicollinearity   97
periodic availability   34, 39–40
PERL regular expressions   185, 191
plausibility checks
    about   73
    business rules and   75
    categorical variables and   74
    criteria for good data quality   28–29
    in parent-child relationships   164–165
    in relationships between tables   75, 160–161
    interval values and   74
    multivariate   75–76, 105
    process steps and   75
PLOT macro option   146
PLS node (SAS Enterprise Miner)   184
PLS procedure   187
Power and Sample Size application   188, 195
POWER procedure   188
predictive modeling
    about   83, 96
    assessment of model quality   169
    business case calculations   209–210
    correlation in   96–99
    data availability in   86–88, 219–230
    data completeness in   231–242
    data correctness in   243–254
    data linkage in   105–106
    data quality in   84
    data quantity in   48–49, 219–230
    distribution and   101–103
    effective number of observations in   91–93, 168
    granularity in   96
    historic data and   88, 92
    historic snapshots and   37
    importance of   208
    level of detail   104–105
    multiple target windows   87–88
    outlier detection with   168, 171–174
    process of   84–85
    simulation studies   207–217, 219–230
    sparseness and   101–103
    stable data definition for future periods   88–91
    variability in   96, 100–101

what-if analysis 169
process considerations for data quality
    about 108
    consequences of poor data quality 111–114
    data availability and 33
    DataFlux Data Management Platform 197
    data quality monitoring 117–119
    data quality responsibilities 114–116
    data relevancy and 108–111
    demand forecasting case study 16
    improving quality 117
    maintaining status 116–117
    plausibility checks 75
    SAS Enterprise Miner and 185
    treating duplicates 69, 168, 177–178
%PROFILE_TS_MV macro 146–147, 259
profiling
    See also analytical systems and tools
    DataFlux support 196
    examples of 178–179
    missing values 124–126
    missing values with macros 125–126
    reports for a time series 143–146
    structure of missing values 126–133
    structure of missing values for time series data 142–147
    univariate data 168
p-values 100

Q

quality, data
    See data quality
quantiles
    interval values and 74
    statistical outlier detection and 170

R

race boat performance in sailing regattas case study 4–10, 191–193
random bias
    about 244, 274
    consequences of 277–279
    inserting into input variables 245
    inserting into target variables 246
    simulation study results 247–253
random errors 77
random missing values
    about 62–63, 232
    consequences of 274–276
    inserting 233
    simulation scenario results 234–237
random number generators 204–205
random numbers
    code example 205–206
    creating in SAS 205
    generating 204–205
    in simulation environment 204
    with changing start values 205
random sampling 49
random zero values 274, 276–277
RANUNI function 205
Rapid Predictive Modeler 169, 185
records
    See data records
reference models for simulations 210–213
reference value for variables 171
REG procedure 138, 171, 187–188
REGRESSION node (SAS Enterprise Miner) 184–185, 214
regular expressions 185, 191
regulatory considerations
    data linkage and 106
    for poor data quality 111
relational models
    about 159–160
    data completeness in 160–161
    plausibility checks in 160–161
repeated measurement data sets 53
REPLACEMENT node (SAS Enterprise Miner) 183
REPLACE option, STANDARD procedure 133–134
reports
    based on %MV_PROFILE_CHAIN macro 130–131
    for profiling a time series 143–146
RULE INDUCTION node (SAS Enterprise Miner) 185

S

sailing regattas, race boat performance case study 4–10, 191–193
samples, drawing from transactional data 104–105
sample size planning 53–55, 188, 195
SAS Enterprise Miner
    See Enterprise Miner
SAS/ETS software 190–191
SAS Forecast Server 188–190, 204
SAS Forecast Studio 155, 174–175, 188–190
SAS High-Performance Forecasting 190, 204, 259
SAS Model Manager 187
SAS Power and Sample Size application 188, 195
SAS Rapid Predictive Modeler 169, 185
SAS/STAT software 187–188
SAS Text Miner 185
scatterplots 173
scoring
    biased values and 244–245
    categories of 89
    change in distributions 89–91
    defined 88
    missing values and 89, 232
    output data quality 90–91
    requirements for 89
    SAS Enterprise Miner and 186
    SAS Model Manager and 187

segmenting observations   103
SEMMA methodology   135, 182
SETMISS= option, ID statement (TIMESERIES)   150, 152–153
SGPLOT procedure   174
SHEWART procedure   170
sign inversion, correlation and   98
simulation studies
    availability of variables and   227–229
    consequences of random disturbances in time series data   273–279
    consequences of systematic disturbances in time series data   281–288
    data availability and   219–230
    data completeness and   231–242, 265–272
    data correctness and   243–254
    data quality and   255–264
    data quantity and   219–230, 265–272
    downloading   206
    effects of missing values   209, 231–242
    generalizability of   203–204
    interpretability of   203–204
    predictive modeling   207–217, 219–230
    random numbers in   204–206
    rationale for   201–203
    reference models for   210–213
    results based on   203
    simulation environment   213–215
    time series forecasting   204, 255–272, 288
    validation method   215–216
sparseness
    grouping categories   102–103
    time series forecasting and   103
standard deviations
    as basis for random bias   277
    effect of outliers on   102
    interval values and   74
    outlier detection and   170
STANDARD procedure
    BY statement   134
    REPLACE option   133–134
    univariate imputation of missing values   133–135
START GROUPS node (SAS Enterprise Miner)   185, 214
STAT EXPLORE node (SAS Enterprise Miner)   184
statistical outlier detection   170
STAT software   187–188
STORE ASSESSM. STATISTICS node (SAS Enterprise Miner)   215
stratified sampling   49
structure of available data   42–44
SURVEYSELECT procedure   188
synchronizing data, race boats in sailing regattas case study   9
systematic bias
    about   244
    inserting into input variables   245–246
    inserting into target variables   246–247
    simulation study results   247–253
systematic errors   77–78
systematic missing values
    about   62–64, 232, 282
    inserting   233
    in time series data   281–288
    simulation scenario results   237–239

T

tables
    See data tables
TABLES statement, FREQ procedure   125
technical data quality   108
text data   41
Text Miner   185
text mining   169, 185
time-dependent data correctness   81–82
timeliness as criteria for good data quality   29
time series data and forecasting
    consequences of poor data quality in   255–264
    consequences of random disturbances in   273–279
    consequences of systematic disturbances in   281–288
    contiguity of   147–152
    data completeness in   66–67, 265–272
    data quantity in   265–272
    decomposition and smoothing of   175
    defined   147, 174
    historic snapshots in   38–39
    interpolating missing values with EXPAND procedure   155–156
    length of time history   52–53
    level of detail in   53
    methods supporting   256–257
    missing values in   141–143
    optimal length of available time history   269–271
    outlier detection in   168, 174–175
    profiling structure of missing values for   142–147
    purpose and application of   255–257
    replacing missing values with TIMESERIES procedure   143, 152–155
    reports profiling   143–146
    SAS Forecast Server and   188
    self-assessment of   272
    simulation conclusions for   288
    sparse values in   103
    transaction data and   147–148
TIMESERIES procedure
    about   188, 191
    changing zero values to missing values   154–155
    %CHECK_TIMEID macro   150–152
    contiguity of time series data   149–150
    functionality of   152–154
    ID statement   150, 152–155
    random missing values and   274, 276

replacing missing values in time series data   143, 152–155, 260
tracing changes for data management in clinical trials case study   13
transaction data
    defined   34, 147
    derived variables from   98–99
    samples from   104–105
    time series data and   147–148
transferring data
    See data transfer and retrieval
TRANSFORM VARIABLES node (SAS Enterprise Miner)   183
tree imputation method   136–137
tree map diagrams   193–194
TREE node (SAS Enterprise Miner)   184–185
tree plots   130–131, 133
tree surrogate imputation method   136–137
TREE SURROGATE method   233
TS_PROFILE_CHAIN   143–144
Tukey's biweight imputation method   136–137

U

UCM procedure   191
univariate data
    gaining insight into relationships   184
    imputation of missing values   133–135
    profiling   168
usability of data
    considerations for   41
    data availability and   32–33

V

validation considerations
    assessing forecast accuracy   263
    data correctness and validation rules   75
    defining validation limits   169
    simulation studies   215–216
VARCLUS procedure   187
variability
    about   96, 100
    in business processes   100–101
    statistical measures and   100
    undescribed   101
variables
    See also specific types of variables
    assessing importance of   184
    data quantity and   49, 52
    effect of availability of   227–229
    format and content of   40–42
    imputing missing values based on   96–97
    inserting random bias into   245–246
    inserting systematic bias into   245–246
    reference value for   171
    selection considerations   169
    substituting effect of   97
    transforming to distributions   169

VARIABLE SELECTION node (SAS Enterprise Miner)   184
VAR statement, MEANS procedure   125
Velocitek software   6–7

W

WARN variable   186
what-if analysis   169, 185

Z

ZEROMISS= option, ID statement (TIMESERIES)   154–155
zero values
    changing to missing values   154–155
    random   274, 276–277

CPSIA information can be obtained at www.ICGtesting.com
Printed in the USA
LVOW02s2004050614

388837LV00002B/7/P